VOLUME NINETY SEVEN

Current Topics in DEVELOPMENTAL BIOLOGY

Growth Factors in Development

VOLUME NINETY SEVEN

CURRENT TOPICS IN DEVELOPMENTAL BIOLOGY

Growth Factors in Development

Edited by

CARMEN BIRCHMEIER
Department of Neuroscience
Developmental Biology/Signal Transduction
Max-Delbrück-Center for Molecular Medicine
Berlin-Buch
Germany

ELSEVIER

AMSTERDAM • BOSTON • HEIDELBERG • LONDON
NEW YORK • OXFORD • PARIS • SAN DIEGO
SAN FRANCISCO • SINGAPORE • SYDNEY • TOKYO
Academic Press is an imprint of Elsevier

Academic Press is an imprint of Elsevier
525 B Street, Suite 1900, San Diego, CA 92101-4495, USA
225 Wyman Street, Waltham, MA 02451, USA
32, Jamestown Road, London NW1 7BY, UK
Linacre House, Jordan Hill, Oxford OX2 8DP, UK

First edition 2011

ISBN: 978-0-12-385975-4
ISSN: 0070-2153

For information on all Academic Press publications
visit our website at elsevierdirect.com

Printed and bound by CPI Group (UK) Ltd, Croydon, CR0 4YY

Transferred to digital print 2013

Contents

CONTRIBUTORS

Suat Özbek
Department of Molecular Evolution and Genomics, Centre for Organismal Studies, Heidelberg University, Heidelberg, Germany

Michael Boutros
German Cancer Research Center (DKFZ), Division of Signaling and Functional Genomics, and Faculty of Medicine Mannheim, Department of Cell and Molecular Biology, University of Heidelberg, Heidelberg, Germany

Tina Buechling
German Cancer Research Center (DKFZ), Division of Signaling and Functional Genomics, and Faculty of Medicine Mannheim, Department of Cell and Molecular Biology, University of Heidelberg, Heidelberg, Germany

Thomas W. Holstein
Department of Molecular Evolution and Genomics, Centre for Organismal Studies, Heidelberg University, Heidelberg, Germany

Jane E. Johnson
Department of Neuroscience, UT Southwestern Medical Center, Dallas, Texas, USA

Raymond J. MacDonald
Department of Molecular Biology, UT Southwestern Medical Center, Dallas, Texas, USA

Michael P. Matise
UMDNJ/Robert Wood Johnson Medical School, Piscataway, New Jersey, USA

Stefan Mundlos
Institute for Medical Genetics, Charité University Medicine Berlin, Berlin, Germany

Alya R. Raphael[1]
Department of Developmental Biology, Stanford University School of Medicine, Stanford, California, USA

[1]Current address: Department of Cell and Developmental Biology, University of Pennsylvania School of Medicine, Philadelphia, Pennsylvania, USA.

Macarena Sahores
Department of Cell and Developmental Biology, University College London, London, United Kingdom

Patricia C. Salinas
Department of Cell and Developmental Biology, University College London, London, United Kingdom

Sigmar Stricker
Development and Disease Group, Max Planck-Institute for Molecular Genetics, Berlin, Germany

William S. Talbot
Department of Developmental Biology, Stanford University School of Medicine, Stanford, California, USA

Hui Wang
UMDNJ/Robert Wood Johnson Medical School, Piscataway, New Jersey, USA

Hiroshi Watanabe
Department of Molecular Evolution and Genomics, Centre for Organismal Studies, Heidelberg University, Heidelberg, Germany

Preface

Signaling between cells is the prerequisite for the development of multicellular organisms and controls patterning, proliferation, and morphogenesis, as well as differentiation. In this issue of *Current Topics in Developmental Biology*, experts discuss the role of signaling mechanisms in a wide range of developmental events that encompasses the initial determination of the body axis, patterning of the nervous system, formation of the skeleton, or differentiation of Schwann cells. The authors do not only discuss the function of these signaling systems in developmental events but also review the current state of knowledge on the molecular mechanisms employed by various developmental signaling pathways.

In Chapter 1, Talbot and Raphael discuss myelinization and neuronal signals that control myelination. Myelin is a vertebrate-specific structure that allows for the rapid propagation of action potentials along the nerve, and its formation is controlled by axonal signals. A long line of cell culture experiments point toward a pivotal role of cAMP in the initiation of myelination, but the G-protein-coupled receptor that activates cAMP production was only recently identified using zebrafish genetics. This breakthrough provides a beautiful example of the power of forward genetics in a vertebrate model organism and has led to the identification of a receptor whose function is conserved in mammals.

The Wnt pathway is an ancient signaling pathway that is well conserved throughout evolution, and its functions are discussed in several chapters. In Chapter 2, Buechling and Boutros lay out the basic regulation of Wnt signaling and highlight recent findings in posttranslational modification, trafficking, and secretion of Wnts. Holstein and Özbek, in Chapter 6, present evidence that, already in lower metazoans, Wnt signaling controls the establishment of the anterior–posterior body axis. They discuss the evolution of the signaling pathway, a field that has greatly profited by the availability of many metazoan genome sequences. Finally, in Chapter 5, Sahores and Salinas review Wnts and their role during synapse formation, a function of the signaling pathway that was only recently fully appreciated.

Sonic Hedgehog (Shh) takes over important roles in patterning of the nervous system. Matise and Wang discuss the role of Shh in patterning and provide a concise review not only of the biology but also of the molecular mechanism of Shh signaling that is still incompletely understood (Chapter 4).

Notch provides an example of a developmental signaling system whose molecular mediators have recently revealed unexpected twists in their function. In Chapter 3, Johnson and MacDonald discuss the transcriptional mediators of Notch signaling, the transcription factors of the CSL family. Unexpectedly, CSL factors also function in Notch-independent pathways and co-operate with the basic helix-loop-helix transcription factors Ptf1a to control neuronal specification and differentiation of the exocrine pancreas.

Approaching developmental biology from a viewpoint of human pathology, Stricker and Mundlos, in Chapter 7, review recent advances in the field of skeletal malformations that are frequently caused by mutations in components of the FGF receptor family. They further discuss the role of ROR2, a tyrosine kinase receptor that participates in Wnt signaling.

Taken together, the reviews in this issue highlight the complex, and at times dazzling, signaling that shapes development. The fact that the handful of signaling cascades is used repeatedly not only during development but also in the maintenance and regeneration of organs might make the reading interesting to researchers outside the immediate field. I hope that these reviews can convey some of the intellectual satisfaction that was provided to those that unravel the functions and mechanisms of signaling pathways.

Carmen Birchmeier

CHAPTER ONE

New Insights into Signaling During Myelination in Zebrafish

Alya R. Raphael*,[1] *and* William S. Talbot*

Contents

Abstract

Myelin is a vertebrate adaptation that allows for the rapid propagation of action potentials along axons. Specialized glial cells—oligodendrocytes in the central nervous system (CNS) and Schwann cells in the peripheral nervous system (PNS)—form myelin by repeatedly wrapping axon segments. Debilitating diseases result from the disruption of myelin, including multiple sclerosis and Charcot-Marie-Tooth peripheral neuropathies. The process of myelination involves extensive communication between glial cells and the associated neurons. The past few years have seen important progress in understanding the molecular basis of the signals that coordinate the development of these fascinating cells. This review highlights recent advances in myelination deriving from studies in the zebrafish model system, with a primary focus on the PNS. While Neuregulin1-ErbB signaling has long been known to play important roles in peripheral myelin development, work in zebrafish has elucidated its roles in Schwann cell migration and radial sorting of axons *in vivo*. Forward genetic screens in zebrafish have also uncovered new genes required for development of myelinated axons, including *gpr126*, which encodes a G-protein coupled receptor required for Schwann cells to progress beyond the promyelinating

* Department of Developmental Biology, Stanford University School of Medicine, Stanford, California, USA
[1] Current address: Department of Cell and Developmental Biology, University of Pennsylvania School of Medicine, Philadelphia, Pennsylvania, USA.

Current Topics in Developmental Biology, Volume 97
ISSN 0070-2153, DOI: 10.1016/B978-0-12-385975-4.00007-3

stage. In addition, work in zebrafish uncovered new roles for Schwann cells themselves, including in regulating the boundary between the PNS and CNS and positioning a nerve after its initial outgrowth.

1. Introduction

The myelin sheath increases axonal conduction velocity by reducing capacitance of the axonal membrane and allowing saltatory conduction (Hodgkin, 1964; Stampfli, 1954). Thus, myelinated axons of small diameter can transmit information as rapidly as much larger unmyelinated axons. Myelin therefore is an evolutionary innovation that allows the nervous system to increase in speed and complexity without a corresponding increase in size and energy requirements. Although some invertebrate species have myelinated axons, myelin is ubiquitous among the gnathostomes (jawed vertebrates), and this adaptation has surely been essential for the formation of the large, complex nervous systems that distinguish the vertebrates from other groups (Bunge, 1968; Hartline and Colman, 2007).

Disruption of the myelin sheath underlies many debilitating diseases including Multiple Sclerosis, Charcot Marie Tooth disease, and others (Berger *et al.*, 2006; McQualter and Bernard, 2007). Specialized glial cells, oligodendrocytes in the central nervous system (CNS) and Schwann cells in the peripheral nervous system (PNS), wrap their membranes many times around a segment of an axon to form the myelin sheath (Bunge, 1968; Geren and Raskind, 1953; Peters, 1964). Along its length, each axon is ensheathed by multiple myelin segments, which are separated by unmyelinated gaps called nodes of Ranvier (Bunge, 1968; Stampfli, 1954; Tasaki, 1959). Oligodendrocytes interact with and elaborate myelin sheaths around many different axons; in contrast, Schwann cells myelinate only one segment of one axon (Bunge, 1968).

2. Schwann Cell Development

As extensively documented in mammals, Schwann cells originate from the neural crest and undergo a series of developmental transitions that culminates with myelination (Fig. 1.1; Jessen and Mirsky, 2005). Schwann cell precursors form from neural crest progenitors that have delaminated from the dorsal neural tube and associated with axons in peripheral nerves (Jessen *et al.*, 1994; Sauka-Spengler and Bronner, 2010). Schwann cell precursors migrate with axons as they grow toward their targets (Dong *et al.*, 1999). Once migration is complete, Schwann cell precursors differentiate into immature Schwann cells, which associate with many axons organized as a bundle (Jessen and Mirsky, 2005). Through a process termed

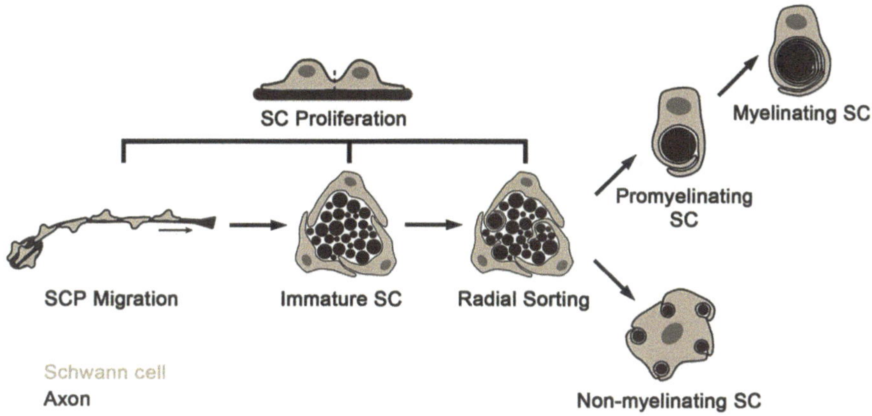

Figure 1.1 Multiple roles for Nrg1/ErbB signaling during Schwann cell development. Schwann cells (yellow) migrate along axons (blue) as Schwann cell precursors (SCP; Jessen and Mirsky, 2005). Once migration is complete, SCPs differentiate into immature Schwann cells (SC) and begin the process of radial sorting by inserting their processes into the axon bundle (Jessen and Mirsky, 2005; Webster *et al.*, 1973). Immature Schwann cells can become either promyelinating and then myelinating Schwann cells, if they are associated with one axon, or non-myelinating Schwann cells, which associate with multiple small axons (Nave and Salzer, 2006). Schwann cell proliferation occurs in SCPs and in immature Schwann cells and during radial sorting; however, once immature Schwann cells differentiate, they exit the cell cycle (Jessen and Mirsky, 2005; Martin and Webster, 1973; Webster *et al.*, 1973). Schwann cells divide parallel to the axons (Martin and Webster, 1973). Steps that require Nrg1/ErbB signaling are underlined and include Schwann cell migration, Schwann cell proliferation, radial sorting, myelination, and the formation of Remak bundles—non-myelinating Schwann cells ensheathing small caliber axons (Dong *et al.*, 1995; Garratt *et al.*, 2000a,b; Lyons *et al.*, 2005; Michailov *et al.*, 2004; Morrissey *et al.*, 1995; Perlin *et al.*, 2011; Raphael *et al.*, 2011; Riethmacher *et al.*, 1997; Taveggia *et al.*, 2005; Woldeyesus *et al.*, 1999). (See Color Insert.)

radial sorting, Schwann cells progressively sort axons from the main bundle, such that a Schwann cell associates with a single axonal segment that is to be myelinated (Webster *et al.*, 1973). At this point, the immature Schwann cells associated with a single large caliber axon will become promyelinating Schwann cells, initiate the myelination transcription pathway, and begin to wrap their cytoplasm around the axon to create a myelin sheath (Jessen and Mirsky, 2005; Webster *et al.*, 1973). In contrast, immature Schwann cells associated with several small caliber axons will become nonmyelinating Schwann cells, ensheathing each axon in a pocket of its cytoplasm, forming a Remak bundle (Aguayo and Bray, 1975; Aguayo *et al.*, 1976; Hahn *et al.*, 1987; Jessen and Mirsky, 2005; Peters and Muir, 1959).

Much of our current understanding of myelination comes from mammalian studies, and a number of genes essential for Schwann cell development are known. *sox10*, *erbb2*, *erbb3*, and *nrg1* are required at multiple steps of Schwann cell development, regulating neural crest migration and specification, as well as Schwann cell proliferation, survival, specification, and myelination (Fig. 1.1; Britsch *et al.*, 2001; Chen *et al.*, 2006; Dong *et al.*, 1995; Finzsch *et al.*, 2010; Garratt *et al.*, 2000b; Jessen and Mirsky, 2005; Kuhlbrodt *et al.*, 1998; Michailov *et al.*, 2004; Morrissey *et al.*, 1995; Newbern and Birchmeier, 2010; Riethmacher *et al.*, 1997; Schreiner *et al.*, 2007; Woldeyesus *et al.*, 1999). Other genes are necessary at discrete stages—*oct6, brn2,* and *krox20* are required for Schwann cells to progress beyond the promyelinating stage and make myelin (Svaren and Meijer, 2008). Components of the extracellular matrix and cytoskeletal regulators are required for radial sorting, including *rac1, cdc42, FAK, ILK, β1-integrin, laminin-γ1, laminin-2,* and *laminin-8* (Benninger *et al.*, 2007; Chen and Strickland, 2003; Feltri *et al.*, 2002; Grove *et al.*, 2007; Nodari *et al.*, 2007; Pereira *et al.*, 2009; Yu *et al.*, 2009, 2005).

Despite important progress in defining key genes that regulate Schwann cell and oligodendrocyte development and myelination, our understanding of the pathways that regulate glial development and myelination is still incomplete (Emery, 2010; Jessen and Mirsky, 2005). In this review, we seek to highlight advances in understanding myelination in the zebrafish model system, which combines powerful genetics with exquisite *in vivo* imaging (Box 1.1). We focus mainly on the PNS, where a more complete pathway of Schwann cell development has emerged, and in particular, on the roles of Neuregulin1/ErbB signaling and Gpr126. In addition, we highlight new roles for Schwann cells that have been discovered in zebrafish.

3. Neuregulin1/ErbB Signaling

Neuregulin1 (Nrg1) is an EGF-related signal that activates ErbB receptor tyrosine kinases on Schwann cells to regulate several aspects of Schwann cell development, including proliferation, survival, myelination, and the formation of Remak bundles (Birchmeier and Nave, 2008; Dong *et al.*, 1995; Garratt *et al.*, 2000b; Lyons *et al.*, 2005; Michailov *et al.*, 2004; Morrissey *et al.*, 1995; Riethmacher *et al.*, 1997; Taveggia *et al.*, 2005; Woldeyesus *et al.*, 1999). ErbB2 and ErbB3 are the main receptors for Nrg1 ligands in Schwann cells (Citri *et al.*, 2003; Newbern and Birchmeier, 2010). There are over 15 different isoforms of Nrg1, but a single isoform, Nrg1 type III, has emerged as a primary regulator of Schwann cell development (Falls, 2003; Michailov *et al.*, 2004; Taveggia *et al.*, 2005). For example, the amount of Nrg1 type III expressed on an axon determines whether the associated Schwann cell will become a myelinating or nonmyelinating Schwann cell, as well as the thickness of the myelin sheath (Michailov *et al.*, 2004; Taveggia *et al.*, 2005). Nrg1 can also

BOX 1.1 Experimental advantages of the zebrafish model system

The zebrafish model system offers several experimental advantages that facilitate the investigation of myelination and other aspects of vertebrate biology.

Forward Genetic Screens

Forward genetic screens in zebrafish are a standard tool in the field, using either chemicals or insertional mutagenesis (Amsterdam and Hopkins, 2006). Two forward genetic screens for myelin mutants have been reported—a screen for defects in the expression of *mbp* mRNA that led to the discovery of the *erbb2*, *erbb3*, and *gpr126* mutants discussed in the main text and a "shelf" screen, where known mutants with neural defects were similarly screened, that found four mutants, *neckless*, *motionless*, *iguana*, and *doc*, required for myelination (Kazakova *et al.*, 2006; Pogoda *et al.*, 2006). Forward genetic screens are unbiased in that they are not looking for a specific gene, but rather are looking for a specific phenotype that defines gene function. Additionally, because current screens have not yet reached saturation, future screening for myelin mutants will yield new genes (Pogoda *et al.*, 2006). Finally, with improvements in the zebrafish genome sequence and rapid advances in sequencing technologies, positional cloning of mutated genes has become much easier and faster in recent years.

Reverse Genetics: Morpholinos, TILLING, and Zinc Fingers

Zebrafish also offers several approaches to study the function of a particular gene of interest, such as a human disease gene, in a developmental context. Overexpression of genes can be achieved by simply injecting RNA into the embryo or by generating transgenic fish with the Tol2 vector system (Kawakami, 2007; Xu *et al.*, 2008). Knockdown of a gene can be performed with antisense morpholino oligonucleotides that block either mRNA splicing or translation (Shestopalov and Chen, 2010). Two methods are available to generate a mutation in a gene of interest. TILLING (targeting induced local lesions in genomes) works by rapidly screening the sequences of many chemically generated mutations to find a mutation in the gene of interest (Stemple, 2004). For example, the mutation in zebrafish *krox20* was generated using TILLING (Monk *et al.*, 2009). By contrast, zinc finger mutagenesis targets a gene of interest using zinc finger nucleases that generate double stranded breaks in the gene; loss-of-function mutations often result from imprecise repair of these breaks (Doyon *et al.*, 2008; Meng *et al.*, 2008).

Small Molecules

Small molecules have been used extensively in zebrafish to activate or inactivate different signaling pathways (e.g., AG1478 and forskolin described in main text; Lyons *et al.*, 2005; Monk *et al.*, 2009; Raphael *et al.*, 2011). Additionally, small molecule screens have been conducted to look for

(*continued*)

BOX 1.1 *(continued)*

compounds with certain activities *in vivo*—these screens have uncovered compounds that rescue a cardiac defect, protect hair cells, and have behavioral effects, among others (Ou *et al.*, 2009; Peterson *et al.*, 2004; Rihel *et al.*, 2010). A recent small molecule screen combined *in vivo* imaging of oligodendrocytes precursor cells with high-throughput screening; several compounds were found that affected the number of *olig2:EGFP* expressing oligodendrocyte precursor cells, or the amount of *mbp* expression in the spinal cord (Buckley *et al.*, 2010). Small molecule screens for compounds that ameliorate myelin-related mutant phenotypes could be an approach toward finding strategies for therapeutic remyelination.

In vivo imaging

Several examples in the main text demonstrate power of *in vivo* imaging for studying glial development in zebrafish. Future studies will take advantage of *in vivo* imaging to generate better probes to watch the localization of molecules during myelination and to follow the behavior of individual cells (Yoo *et al.*, 2010). A recent study used "brainbow" to label different neurons in the zebrafish brain with different combinations of red, green, and blue fluorescent proteins—this technique could be adapted to address questions in myelination and many other areas (Pan *et al.*, 2011). Additionally, new microscopic techniques, such as "Selective Plane Illumination Microscopy" (SPIM), are allowing for even more detailed *in vivo* imaging studies (Arrenberg *et al.*, 2010; Huisken and Stainier, 2009). Finally, advances in the field of optogenetics allow for the control and monitoring of neural activity at a very fine scale (Wyart and Del Bene, 2011). Combined with the power of zebrafish imaging, this has allowed researchers to study the activity of certain neurons or to create an optically controlled pacemaker in the zebrafish heart, among other work (Arrenberg *et al.*, 2010; Wyart *et al.*, 2009). New imaging methods and transgenic reporter systems will continue to exploit the optical clarity of the zebrafish embryo.

induce demyelination (Syed *et al.*, 2010; Zanazzi *et al.*, 2001), however, suggesting that the level or location of the signal may help determine how a Schwann cell will respond to it. Additionally, the signaling pathways downstream of Nrg1/ErbB signaling may modulate the response to the signal at different stages; these pathways include PI3K/Akt, MEK/ERK, Calcineurin/NFAT, Cdc42, Shp2, and others (Benninger *et al.*, 2007; Cotter *et al.*, 2004; Goebbels *et al.*, 2010; Grossmann *et al.*, 2009; Kao *et al.*, 2009; Ogata *et al.*, 2004; for a more extensive review of Nrg1/ErbB signaling in Schwann cells, see Newbern and Birchmeier, 2010).

Analyses in zebrafish have investigated the role of Nrg1/ErbB signaling in Schwann cell migration and subsequent steps in Schwann cell development using mutants, small molecule inhibitors, and timelapse imaging *in vivo*. A genetic screen for mutants with defects in *myelin basic protein* (*mbp*) expression isolated mutations in *erbb2* and *erbb3*, among other genes (Lyons *et al.*, 2005; Pogoda *et al.*, 2006). Similar to previous studies of mouse mutants in *ErbB2* and *ErbB3*, the zebrafish *erbb2* and *erbb3* mutants lack Schwann cells along peripheral axons, although some Schwann cells do associate with neuronal cell bodies at cranial ganglia (Lyons *et al.*, 2005; Pogoda *et al.*, 2006; Woldeyesus *et al.*, 1999). BrdU incorporation studies revealed that Schwann cell proliferation is reduced (but not eliminated) in *erbb* mutants, consistent with many mammalian studies showing that Nrg1 is a Schwann cell mitogen (Garratt *et al.*, 2000b; Lyons *et al.*, 2005; Morrissey *et al.*, 1995; Newbern and Birchmeier, 2010). The use of timelapse imaging combined with chemical inhibitors of ErbB activity also showed that ErbB signaling is required continuously during Schwann cell migration, in addition to its role in Schwann cell proliferation (Lyons *et al.*, 2005). When the ErbB inhibitor AG1478 was applied after the start of migration, some Schwann cells continued to move, but in a misdirected fashion, in some instances even switching nerves. These studies support the possibility that ErbB signaling is important for directed migration of Schwann cells, rather than simply promoting the motility of these cells. A recent paper showed that the ErbB receptor ligand Nrg1 Type III is required for Schwann cell migration and that its overexpression in the spinal cord is sufficient to attract Schwann cells away from peripheral nerves and into the spinal cord, from which they are normally excluded (Perlin *et al.*, 2011).

A recent study has added to the understanding of ErbB signaling in radial sorting, by analyzing the effects of ErbB inhibitors added after Schwann cell migration is complete (Raphael *et al.*, 2011). Migrating Schwann cells associate with bundles of many axons, but shortly after the end of migration, radial sorting begins (Fig. 1.1). Radial sorting occurs similarly in zebrafish as in mammals, with immature Schwann cells surrounding a bundle of axons, extending processes into the axon bundle and "radially" sorting axons to the periphery of the bundle, at which point one Schwann cell is interacting with one axon and the myelination program can begin (Raphael *et al.*, 2011; Webster *et al.*, 1973). During radial sorting, Schwann cells proliferate extensively, so that they are present in numbers corresponding to the many axonal segments that will be myelinated (Jessen and Mirsky, 2005; Webster *et al.*, 1973). Inhibitors of cell division interfere with radial sorting and the onset of myelination (Lyons *et al.*, 2005; Raphael *et al.*, 2011). Inhibition of ErbB signaling also disrupted radial sorting, as expected in light of the role of Nrg1-ErbB signaling in Schwann cell proliferation (Raphael *et al.*, 2011). Supporting a previous study of mammalian Schwann cells in culture (Taveggia *et al.*, 2005), ErbB signaling also has a role that is independent of Schwann cell number: Schwann cell processes do not extend into axon bundles in fish treated with ErbB inhibitors, in contrast to fish treated with inhibitors of cell division or untreated controls (Raphael *et al.*, 2011). This indicates that, in addition to regulating Schwann cell proliferation during radial sorting, ErbB signaling is also required for Schwann cell process extension. It will be interesting to learn how Nrg1 signals from axons are coordinated with signaling from the basal lamina, which is generated by the Schwann cells themselves, to bring about radial sorting and myelination. In addition, time-lapse imaging of the interaction between Schwann cells and their axons in living zebrafish may generate new insights into radial sorting, a dynamic process that is currently understood from electron micrograph time courses.

Combined, the zebrafish and mammalian studies have revealed roles for Nrg1-ErbB signaling in Schwann cell migration, survival, proliferation, Remak bundle formation, radial sorting, and myelination (Birchmeier and Nave, 2008; Dong *et al.*, 1995; Garratt *et al.*, 2000b; Lyons *et al.*, 2005; Michailov *et al.*, 2004; Morrissey *et al.*, 1995; Perlin *et al.*, 2011; Raphael *et al.*, 2011; Riethmacher *et al.*, 1997; Taveggia *et al.*, 2005; Woldeyesus *et al.*, 1999). Future investigation is required to understand how one signal controls so many different aspects of Schwann cell development, but important factors likely include the concentration and source of the ligand, the developmental stage of the Schwann cell receiving the signal, and process specific downstream factors, such as Calcineurin/NFAT, PI3K/Akt, MEK/ERK, Shp2, and Cdc42 (Benninger *et al.*, 2007; Cotter *et al.*, 2010; Goebbels *et al.*, 2010; Grossmann *et al.*, 2009; Kao *et al.*, 2009; Newbern and Birchmeier, 2010; Ogata *et al.*, 2004; Syed *et al.*, 2010).

4. Gpr126

In addition to finding new roles for genes initially characterized in mammals, zebrafish screens have also uncovered novel genes that regulate myelin formation. A screen for mutants with abnormal expression of *mbp* identified two mutations in *gpr126* (Monk *et al.*, 2009; Pogoda *et al.*, 2006). Gpr126 is an orphan member of the adhesion subfamily of GPCRs, which are characterized by long, extracellular segments N-terminal to the 7-pass transmembrane domain (Bjarnadottir *et al.*, 2004; Fredriksson *et al.*, 2003). In *gpr126* mutants, expression of the early Schwann cell marker *sox10* is normal, but markers of later stages are significantly reduced, including the promyelinating genes *oct6* and *krox20* (Monk *et al.*, 2009). These marker studies suggested that *gpr126* is dispensable for early stages of Schwann cell development but essential for the onset of myelination. Transmission electron microscopy analysis revealed that Schwann cells arrest at the promyelinating stage in *gpr126* mutants, with no more than one and a half wraps of Schwann cell cytoplasm around axons in peripheral nerves.

This phenotype is reminiscent of *krox20* mutants in both mammals and zebrafish, where mutant Schwann cells also arrest with only one and a half wraps of cytoplasm around an axon (Monk *et al.*, 2009; Topilko *et al.*, 1994). It had been known for many years that cAMP is an important second messenger during myelination and that the addition of forskolin, which elevates cAMP, could initiate myelination in cultured Schwann cells and mimic the presence of an axon (Jessen *et al.*, 1991; Monuki *et al.*, 1989). The endogenous regulation of cAMP, however, was not understood. Many G-protein coupled receptors signal through cAMP (Jalink and Moolenaar, 2010), raising the possibility that Gpr126 activates myelination by elevating levels of cAMP. Application of forskolin rescues myelination in *gpr126* mutants but not in *krox20* mutants (Monk *et al.*, 2009). These results support the possibility that Gpr126 acts upstream of cAMP, so that adding forskolin to artificially increase cAMP bypasses the requirement for Gpr126 (Monk *et al.*, 2009). In contrast, *krox20* is downstream of cAMP signaling, and these mutants cannot be rescued by elevating the levels of cAMP (Monk *et al.*, 2009). Future studies are required to investigate the pathway downstream of Gpr126, the interaction of Gpr126 and other key regulators such as Nrg1, and the identity of ligands that may activate Gpr126.

Characterization of a *Gpr126* mutant mouse revealed conservation of its function in myelination, and also new roles for Gpr126 in the regulation of other aspects of peripheral nerve development (Monk *et al.*, 2011). *Gpr126* mutant mice have severe congenital hypomyelinating peripheral neuropathy. Similar to the situation in zebrafish, mouse *Gpr126* mutants have decreased expression of *Oct6*, *Krox20*, and *Mbp*, and the mutant Schwann cells arrest at the promyelinating stage (Monk *et al.*, 2009, 2011). The analysis also revealed that there is a delay in radial sorting and a loss of Remak bundles, and thus nonmyelinating Schwann cells (Monk *et al.*, 2011). Additionally, ectopic perineurial fibroblasts were found inappropriately within the nerve and these fibroblasts segregated the axon fibers in small bundles, forming "minifascicles." Several aspects of the *Gpr126* mutant phenotype are similar to mutants for *Adam22* and *Lgi4/claw paw*, which also have Schwann cells arrested at the promyelinating stage and axons organized into minifascicles (Darbas *et al.*, 2004; Henry *et al.*, 1991; Nishino *et al.*, 2010; Ozkaynak *et al.*, 2010). Adam22 binds Lgi4, and it is possible that the functions of these proteins are in some way related to Gpr126.

5. New Roles for Schwann Cells

Mutations in genes that have critical functions in Schwann cells, including *erbb2*, *erbb3*, and *sox10*, have led to important progress in the understanding of the pathways that regulate Schwann cell development. In addition, these mutants have provided useful tools to analyze the roles of Schwann cells in peripheral nerve development. The study of zebrafish mutants lacking Schwann cells has defined several new roles for these cells, including preventing the ectopic accumulation of axonal sodium channels in internodal axonal segments, preventing the premature differentiation of sensory organs, and in the proper fasciculation of the nerve (Gilmour *et al.*, 2002; Grant *et al.*, 2005; Voas *et al.*, 2009). Additionally, more recent studies have uncovered roles for Schwann cells in preventing oligodendrocytes from improperly exiting the spinal cord, and repositioning a peripheral nerve across a basement membrane (Kucenas *et al.*, 2009; Raphael *et al.*, 2010).

Timelapse imaging of oligodendrocytes in the absence of Schwann cells revealed that oligodendrocytes can inappropriately exit the spinal cord in the absence of Schwann cells, suggesting that Schwann cells are required to keep oligodendrocytes from crossing the nerve root transition zones between the PNS and CNS (Kucenas *et al.*, 2009). Normally, only axons cross these zones, either exiting or entering the spinal cord, with oligodendrocytes forming a heminode of myelin on the CNS side and Schwann cells forming one on the PNS side (Fraher, 1999).

Extensive timelapse imaging revealed that, in the absence of Schwann cells, oligodendrocyte processes project out of the spinal cord along the motor axons, and the cell bodies ultimately follow (Kucenas *et al.*, 2009). These data support the idea that the glial cells themselves regulate the transition zones and that, in the absence of Schwann cells, oligodendrocytes are free to exit the spinal cord and myelinate peripheral axons.

Two recent studies in mammals also investigated the role of Schwann cells in restricting the migration of oligodendrocytes into peripheral nerves and suggest that Schwann cells must progress beyond the promyelinating stage to limit ectopic migration of oligodendrocytes. In *Krox20* mutant mice, where Schwann cells are arrested at the promyelinating stage, oligodendrocytes exited the CNS and entered peripheral nerves (Coulpier *et al.*, 2010). Similarly, in the study of the *Gpr126* mutant mouse described above, oligodendrocytes were expanded into peripheral territory at the transition zone of the auditory nerve, despite the presence of Schwann cells arrested at the promyelinating stage (Monk *et al.*, 2011). Interestingly, oligodendrocytes did not extend into the periphery in *TremblerJ/PMP20* mutant mice, which express Krox20 but lack PNS myelination (Coulpier *et al.*, 2010). This expansion of CNS glia into the PNS has also been observed in a human patient with congenital amyelinating neuropathy, which is characterized by deficits in Krox20 protein in Schwann cells (Coulpier *et al.*, 2011). These results suggest that Krox20, downstream of Gpr126, may play a role in mediating the transition zone and that oligodendrocytes may be sensitive to the myelination state of adjoining Schwann cells.

A new role for Schwann cells in the repositioning of a peripheral nerve has recently been described in zebrafish (Raphael *et al.*, 2010). The posterior lateral line nerve innervates sensory organs that detect changes in water currents and, in larvae and adults resides, just below the basement membrane of the epidermis (Ghysen and Dambly-Chaudiere, 2007; Raphael *et al.*, 2010; Winklbauer, 1989). Recent work revealed that the nerve (both axons and Schwann cells) initially grows within the epidermis, and then rapidly transitions across the epidermal basement membrane to its mature location in the subepidermal space (Fig. 1.2; Raphael *et al.*, 2010).

Schwann cells are required for this process, as mutants lacking Schwann cells have the nerve improperly located within the epidermis, and transplantation of wildtype Schwann cells into these mutants is sufficient to restore the nerve to its correct position. Significant defects arise when the posterior lateral line nerve is mislocalized in the epidermis, including defasciculation of the nerve and mispositioning of the nerve along the dorsal–ventral axis. This is apparently the result of the nerve being pulled by the ventrally migrating sensory organs that it innervates. In wildtype animals, the epidermal basement membrane separates the main body of the nerve from the sensory organs; in mutants lacking Schwann cells, however, the nerve remains improperly located within the epidermis in close contact with its target sensory organs. This anatomical organization, with a sensory organ within an epidermal layer and the main nerve fascicle located below a basement membrane, also occurs in many other sensory tissues including the tongue, skin, nasal epithelium, and vestibular organ (Boulais and Misery, 2008; Fernandez *et al.*, 1990; Nedelec *et al.*, 2005; Northcutt, 2004; Oakley and Witt, 2004; Purcell and Perachio, 1997; Si *et al.*, 2003), suggesting that this may be a conserved method of protecting axons from the remodeling or frequent turnover of their targets.

6. Oligodendrocytes

While this review has focused primarily on Schwann cell development, advances have also been made in the understanding of oligodendrocyte development. Forward genetic screens have uncovered new genes required during oligodendrocyte development and myelination, while detailed timelapse imaging studies have elucidated the complex behavior of oligodendrocytes *in vivo* (Almeida *et al.*, 2011; Kirby *et al.*, 2006; Larson *et al.*, 2010; Lyons *et al.*, 2009; Parichy and Turner, 2003; Parichy *et al.*, 2003; Pogoda *et al.*, 2006; Takada *et al.*, 2010). For example, two genes required for oligodendrocyte development that were found in forward genetic screens are *kif1b* and *tuba8l3a*—a kinesin molecular motor and a tubulin gene (Larson *et al.*, 2010; Lyons *et al.*, 2009; Parichy and Turner, 2003; Parichy *et al.*, 2003). Both mutants have defects in the localization of myelin-specific mRNAs in

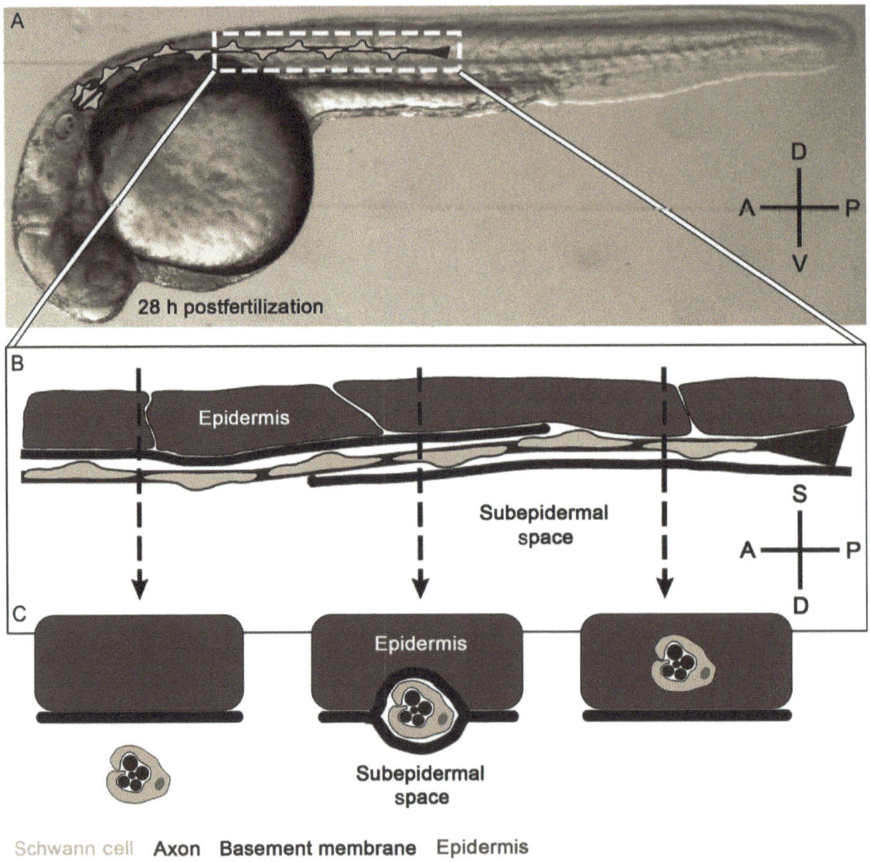

Figure 1.2 Schwann cells are required for repositioning a peripheral nerve. (A) The posterior lateral line nerve in cartoon superimposed on a 28-h postfertilization zebrafish embryo. Schwann cells (yellow) comigrate with axons (blue; Gilmour *et al.*, 2002). Anterior (A) is to the left, posterior (P) to the right, dorsal (D) up, ventral down (v). (B) Zoom of dashed region in (A), showing the transition of the posterior lateral line nerve across the epidermal basement membrane (dark gray) from the epidermis (light gray) into the subepidermal space (unlabeled), with anterior-most portions of the nerve transitioning prior to posterior portions closer to the outgrowing axonal growth cones (Raphael *et al.*, 2010). Dashed lines indicate cross sections shown in (C). Superficial is up (S), deep (D) down, anterior (A) left, posterior (P) right. (C) Cross sections through the posterior lateral line nerve showing the position of the nerve with respect to the basement membrane (Raphael *et al.*, 2010). Anterior (far left), the nerve has transitioned across the basement membrane and is embedded in the subepidermal space. Middle, the nerve is transitioning, with basement membrane both superficial and medial to the nerve. Posterior (far right), the nerve is still within the epidermis, superficial to the basement membrane. (A, B) adapted from Raphael *et al.* (2010). (See Color Insert.)

glial processes. Further, *kif1b* mutant oligodendrocytes have inappropriate, myelin-like membrane compaction in proximal processes and around the cell body, supporting the possibility that myelin mRNA localization within oligodendrocyte distal processes prevents ectopic membrane compaction in other parts of the cell (Lyons *et al.*, 2009).

In vivo timelapse imaging studies have also revealed that oligodendrocyte precursor cells (OPCs) actively repel each other through contact inhibition (Kirby *et al.*, 2006), presumably to ensure the proper spacing of oligodendrocytes throughout the CNS. Many questions about oligodendrocyte development remain unanswered, including the axonal signals that regulate oligodendrocytes, the cell–cell contact signals that OPCs use to repulse each other, and the cues that coordinate the organization of the cytoskeleton to make myelin. Future studies combining genetics with *in vivo* imaging studies in zebrafish will address these questions.

7. Conclusions and Future Directions

The zebrafish is a genetically tractable vertebrate model organism that is now being used to study the development of myelinating glial cells. Many of the genes required for myelination are conserved from zebrafish to mammals. Recent studies in zebrafish have uncovered new roles for previously known myelin genes and have found new genes that regulate Schwann cell and oligodendrocyte development. *In vivo* imaging is being used to study the behavior of developing Schwann cells and oligodendrocytes; advances in imaging techniques may soon allow for the study of glial cells in adult zebrafish (Blackburn *et al.*, 2011). Future studies will combine these powerful techniques for high-throughput screening using *in vivo* imaging (Buckley *et al.*, 2010).

New high-throughput sequencing methods will greatly accelerate the discovery of mutations in zebrafish genes. The zebrafish model system will continue to make important contributions to the understanding of myelination, as screens identify new essential regulators and *in vivo* imaging provides a new view of dynamic cellular interactions.

ACKNOWLEDGMENTS

We thank Julie Perlin, Kelly Monk, and David Lyons for helpful comments on the manuscript. A. R. R. is supported by a Stanford Graduate Fellowship and an NIH training grant. W. S. T. is supported by NIH grant NS050223.

REFERENCES

Aguayo, A. J., and Bray, G. M. (1975). Experimental pathology of unmyelinated fibers. *In* "Peripheral Neuropathy," (P. J. Dyck, P. K. Thomas, and E. H. Lambert, eds.).W. B. Saunders Co., Philadelphia.

Aguayo, A. J., Bray, G. M., Terry, L. C., and Sweezey, E. (1976). Three dimensional analysis of unmyelinated fibers in normal and pathologic autonomic nerves. *J. Neuropathol. Exp. Neurol.* **35,** 136–151.

Almeida, R. G., Czopka, T., Ffrench-Constant, C., and Lyons, D. A. (2011). Individual axons regulate the myelinating potential of single oligodendrocytes in vivo. *Development* **138,** 4443–4450.

Amsterdam, A., and Hopkins, N. (2006). Mutagenesis strategies in zebrafish for identifying genes involved in development and disease. *Trends Genet.* **22,** 473–478.

Arrenberg, A. B., Stainier, D. Y., Baier, H., and Huisken, J. (2010). Optogenetic control of cardiac function. *Science* **330,** 971–974.

Benninger, Y., Thurnherr, T., Pereira, J. A., Krause, S., Wu, X., Chrostek-Grashoff, A., Herzog, D., Nave, K.-A., Franklin, R. J. M., Meijer, D., Brakebusch, C., Suter, U., *et al.* (2007). Essential and distinct roles for cdc42 and rac1 in the regulation of Schwann cell biology during peripheral nervous system development. *J. Cell Biol.* **177,** 1051–1061.

Berger, P., Niemann, A., and Suter, U. (2006). Schwann cells and the pathogenesis of inherited motor and sensory neuropathies (Charcot-Marie-Tooth disease). *Glia* **54,** 243–257.

Birchmeier, C., and Nave, K. A. (2008). Neuregulin-1, a key axonal signal that drives Schwann cell growth and differentiation. *Glia* **56,** 1491–1497.

Bjarnadottir, T. K., Fredriksson, R., Hoglund, P. J., Gloriam, D. E., Lagerstrom, M. C., and Schioth, H. B. (2004). The human and mouse repertoire of the adhesion family of G-protein-coupled receptors. *Genomics* **84,** 23–33.

Blackburn, J. S., Liu, S., Raimondi, A. R., Ignatius, M. S., Salthouse, C. D., and Langenau, D. M. (2011). High-throughput imaging of adult fluorescent zebrafish with an LED fluorescence macroscope. *Nat. Protoc.* **6,** 229–241.

Boulais, N., and Misery, L. (2008). The epidermis: A sensory tissue. *Eur. J. Dermatol.* **18,** 119–127.

Britsch, S., Goerich, D. E., Riethmacher, D., Peirano, R. I., Rossner, M., Nave, K. A., Birchmeier, C., and Wegner, M. (2001). The transcription factor Sox10 is a key regulator of peripheral glial development. *Genes Dev.* **15,** 66–78.

Buckley, C. E., Marguerie, A., Roach, A. G., Goldsmith, P., Fleming, A., Alderton, W. K., and Franklin, R. J. (2010). Drug reprofiling using zebrafish identifies novel compounds with potential pro-myelination effects. *Neuropharmacology* **59,** 149–159.

Bunge, R. P. (1968). Glial cells and the central myelin sheath. *Physiol. Rev.* **48,** 197–251.

Chen, Z.-L., and Strickland, S. (2003). Laminin gamma1 is critical for Schwann cell differentiation, axon myelination, and regeneration in the peripheral nerve. *J. Cell Biol.* **163,** 889–899.

Chen, S., Velardez, M. O., Warot, X., Yu, Z.-X., Miller, S. J., Cros, D., and Corfas, G. (2006). Neuregulin 1-erbB signaling is necessary for normal myelination and sensory function. *J. Neurosci.* **26,** 3079–3086.

Citri, A., Skaria, K. B., and Yarden, Y. (2003). The deaf and the dumb: The biology of ErbB-2 and ErbB-3. *Exp. Cell Res.* **284,** 54–65.

Cotter, L., Ozcelik, M., Jacob, C., Pereira, J. A., Locher, V., Baumann, R., Relvas, J. B., Suter, U., and Tricaud, N. (2010). Dlg1-PTEN interaction regulates myelin thickness to prevent damaging peripheral nerve overmyelination. *Science* **328,** 1415–1418.

Coulpier, F., Decker, L., Funalot, B., Vallat, J. M., Garcia-Bragado, F., Charnay, P., and Topilko, P. (2010). CNS/PNS boundary transgression by central glia in the absence of Schwann cells or Krox20/Egr2 function. *J. Neurosci.* **30,** 5958–5967.

Coulpier, F., Decker, L., Funalot, B., Vallat, J. M., Garcia-Bragado, F., Charnay, P., and Topilko, P. (2011). Krox20 inactivation in the PNS leads to CNS/PNS boundary transgression by central glia. *Rev. Neurol.* **167,** 51–56.

Darbas, A., Jaegle, M., Walbeehm, E., van den Burg, H., Driegen, S., Broos, L., Uyl, M., Visser, P., Grosveld, F., and Meijer, D. (2004). Cell autonomy of the mouse claw paw mutation. *Dev. Biol.* **272,** 470–482.

Dong, Z., Brennan, A., Liu, N., Yarden, Y., Lefkowitz, G., Mirsky, R., and Jessen, K. R. (1995). Neu differentiation factor is a neuron-glia signal and regulates survival, proliferation, and maturation of rat Schwann cell precursors. *Neuron* **15,** 585–596.

Dong, Z., Sinanan, A., Parkinson, D., Parmantier, E., Mirsky, R., and Jessen, K. R. (1999). Schwann cell development in embryonic mouse nerves. *J. Neurosci. Res.* **56,** 334–348.

Doyon, Y., McCammon, J. M., Miller, J. C., Faraji, F., Ngo, C., Katibah, G. E., Amora, R., Hocking, T. D., Zhang, L., Rebar, E. J., Gregory, P. D., Urnov, F. D., *et al.* (2008). Heritable targeted gene disruption in zebrafish using designed zinc-finger nucleases. *Nat. Biotechnol.* **26,** 702–708.

Emery, B. (2010). Regulation of oligodendrocyte differentiation and myelination. *Science* **330,** 779–782.

Falls, D. L. (2003). Neuregulins: Functions, forms, and signaling strategies. *Exp. Cell Res.* **284,** 14–30.

Feltri, M. L., Graus Porta, D., Previtali, S. C., Nodari, A., Migliavacca, B., Cassetti, A., Littlewood-Evans, A., Reichardt, L. F., Messing, A., Quattrini, A., Mueller, U., and Wrabetz, L. (2002). Conditional disruption of beta 1 integrin in Schwann cells impedes interactions with axons. *J. Cell Biol.* **156,** 199–209.

Fernandez, C., Goldberg, J. M., and Baird, R. A. (1990). The vestibular nerve of the chinchilla III. Peripheral innervation patterns in the utricular macula. *J. Neurophysiol.* **63,** 767–780.

Finzsch, M., Schreiner, S., Kichko, T., Reeh, P., Tamm, E. R., Bösl, M. R., Meijer, D., and Wegner, M. (2010). Sox10 is required for Schwann cell identity and progression beyond the immature Schwann cell stage. *J. Cell Biol.* **189,** 701–712.

Fraher, J. P. (1999). The transitional zone and CNS regeneration. *J. Anat.* **194**(Pt 2), 161–182.

Fredriksson, R., Gloriam, D. E., Hoglund, P. J., Lagerstrom, M. C., and Schioth, H. B. (2003). There exist at least 30 human G-protein-coupled receptors with long Ser/Thr-rich N-termini. *Biochem. Biophys. Res. Commun.* **301,** 725–734.

Garratt, A. N., Britsch, S., and Birchmeier, C. (2000a). Neuregulin, a factor with many functions in the life of a schwann cell. *Bioessays* **22,** 987–996.

Garratt, A. N., Voiculescu, O., Topilko, P., Charnay, P., and Birchmeier, C. (2000b). A dual role of erbB2 in myelination and in expansion of the Schwann cell precursor pool. *J. Cell Biol.* **148,** 1035–1046.

Geren, B. B., and Raskind, J. (1953). Development of the fine structure of the myelin sheath in sciatic nerves of chick embryos. *Proc. Natl. Acad. Sci. USA* **39,** 880–884.

Ghysen, A., and Dambly-Chaudiere, C. (2007). The lateral line microcosmos. *Genes Dev.* **21,** 2118–2130.

Gilmour, D. T., Maischein, H. M., and Nusslein-Volhard, C. (2002). Migration and function of a glial subtype in the vertebrate peripheral nervous system. *Neuron* **34,** 577–588.

Goebbels, S., Oltrogge, J. H., Kemper, R., Heilmann, I., Bormuth, I., Wolfer, S., Wichert, S. P., Möbius, W., Liu, X., Lappe-Siefke, C., Rossner, M. J., Groszer, M., *et al.* (2010). Elevated phosphatidylinositol 3,4,5-trisphosphate in glia triggers cell-autonomous membrane wrapping and myelination. *J. Neurosci.* **30,** 8953–8964.

Grant, K. A., Raible, D. W., and Piotrowski, T. (2005). Regulation of latent sensory hair cell precursors by glia in the zebrafish lateral line. *Neuron* **45,** 69–80.

Grossmann, K. S., Wende, H., Paul, F. E., Cheret, C., Garratt, A. N., Zurborg, S., Feinberg, K., Besser, D., Schulz, H., Peles, E., Selbach, M., Birchmeier, W., *et al.* (2009). The tyrosine phosphatase Shp2 (PTPN11) directs neuregulin-1/ErbB signaling throughout Schwann cell development. *Proc. Natl. Acad. Sci. USA* **106,** 16704–16709.

Grove, M., Komiyama, N. H., Nave, K.-A., Grant, S. G., Sherman, D. L., and Brophy, P. J. (2007). FAK is required for axonal sorting by Schwann cells. *J. Cell Biol.* **176,** 277–282.

Hahn, A. F., Chang, Y., and Webster, H. D. (1987). Development of myelinated nerve fibers in the sixth cranial nerve of the rat: A quantitative electron microscope study. *J. Comp. Neurol.* **260,** 491–500.

Hartline, D. K., and Colman, D. R. (2007). Rapid conduction and the evolution of giant axons and myelinated fibers. *Curr. Biol.* **17,** R29–R35.

Henry, E. W., Eicher, E. M., and Sidman, R. L. (1991). The mouse mutation claw paw: Forelimb deformity and delayed myelination throughout the peripheral nervous system. *J. Hered.* **82,** 287–294.

Hodgkin, A. L. (1964). The ionic basis of nervous conduction. *Science* **145,** 1148–1154.

Huisken, J., and Stainier, D. Y. (2009). Selective plane illumination microscopy techniques in developmental biology. *Development* **136,** 1963–1975.

Jalink, K., and Moolenaar, W. H. (2010). G protein-coupled receptors: The inside story. *BioEssays* **32,** 13–16.

Jessen, K. R., and Mirsky, R. (2005). The origin and development of glial cells in peripheral nerves. *Nat. Rev. Neurosci.* **6,** 671–682.

Jessen, K. R., Mirsky, R., and Morgan, L. (1991). Role of cyclic AMP and proliferation controls in Schwann cell differentiation. *Ann. N. Y. Acad. Sci.* **633,** 78–89.

Jessen, K. R., Brennan, A., Morgan, L., Mirsky, R., Kent, A., Hashimoto, Y., and Gavrilovic, J. (1994). The Schwann cell precursor and its fate: A study of cell death and differentiation during gliogenesis in rat embryonic nerves. *Neuron* **12,** 509–527.

Kao, S. C., Wu, H., Xie, J., Chang, C. P., Ranish, J. A., Graef, I. A., and Crabtree, G. R. (2009). Calcineurin/NFAT signaling is required for neuregulin-regulated Schwann cell differentiation. *Science* **323,** 651–654.

Kawakami, K. (2007). Tol2: A versatile gene transfer vector in vertebrates. *Genome Biol.* **8**(Suppl. 1), S7.

Kazakova, N., Li, H., Mora, A., Jessen, K. R., Mirsky, R., Richardson, W. D., and Smith, H. K. (2006). A screen for mutations in zebrafish that affect myelin gene expression in Schwann cells and oligodendrocytes. *Dev. Biol.* **297,** 1–13.

Kirby, B. B., Takada, N., Latimer, A. J., Shin, J., Carney, T. J., Kelsh, R. N., and Appel, B. (2006). In vivo time-lapse imaging shows dynamic oligodendrocyte progenitor behavior during zebrafish development. *Nat. Neurosci.* **9,** 1506–1511.

Kucenas, S., Wang, W. D., Knapik, E. W., and Appel, B. (2009). A selective glial barrier at motor axon exit points prevents oligodendrocyte migration from the spinal cord. *J. Neurosci.* **29,** 15187–15194.

Kuhlbrodt, K., Herbarth, B., Sock, E., Hermans-Borgmeyer, I., and Wegner, M. (1998). Sox10, a novel transcriptional modulator in glial cells. *J. Neurosci.* **18,** 237–250.

Larson, T. A., Gordon, T. N., Lau, H. E., and Parichy, D. M. (2010). Defective adult oligodendrocyte and Schwann cell development, pigment pattern, and craniofacial morphology in puma mutant zebrafish having an alpha tubulin mutation. *Dev. Biol.* **346,** 296–309.

Lyons, D. A., Pogoda, H.-M., Voas, M. G., Woods, I. G., Diamond, B., Nix, R., Arana, N., Jacobs, J., and Talbot, W. S. (2005). erbb3 and erbb2 are essential for Schwann cell migration and myelination in zebrafish. *Curr. Biol.* **15,** 513–524.

Lyons, D. A., Naylor, S. G., Scholze, A., and Talbot, W. S. (2009). Kif1b is essential for mRNA localization in oligodendrocytes and development of myelinated axons. *Nat. Genet.* **41,** 854–858.
Martin, J. R., and Webster, H. D. (1973). Mitotic Schwann cells in developing nerve: Their changes in shape, fine structure, and axon relationships. *Dev. Biol.* **32,** 417–431.
McQualter, J. L., and Bernard, C. C. (2007). Multiple sclerosis: A battle between destruction and repair. *J. Neurochem.* **100,** 295–306.
Meng, X., Noyes, M. B., Zhu, L. J., Lawson, N. D., and Wolfe, S. A. (2008). Targeted gene inactivation in zebrafish using engineered zinc-finger nucleases. *Nat. Biotechnol.* **26,** 695–701.
Michailov, G. V., Sereda, M. W., Brinkmann, B. G., Fischer, T. M., Haug, B., Birchmeier, C., Role, L., Lai, C., Schwab, M. H., and Nave, K.-A. (2004). Axonal neuregulin-1 regulates myelin sheath thickness. *Science* **304,** 700–703.
Monk, K. R., Naylor, S. G., Glenn, T. D., Mercurio, S., Perlin, J. R., Dominguez, C., Moens, C. B., and Talbot, W. S. (2009). A G protein-coupled receptor is essential for Schwann cells to initiate myelination. *Science* **325,** 1402–1405.
Monk, K. R., Oshima, K., Jörs, S., Heller, S., and Talbot, W. S. (2011). Gpr126 is essential for peripheral nerve development and myelination in mammals. *Development* **138**(13), 2673–2680.
Monuki, E. S., Weinmaster, G., Kuhn, R., and Lemke, G. (1989). SCIP: A glial POU domain gene regulated by cyclic AMP. *Neuron* **3,** 783–793.
Morrissey, T. K., Levi, A. D., Nuijens, A., Sliwkowski, M. X., and Bunge, R. P. (1995). Axon-induced mitogenesis of human Schwann cells involves heregulin and p185erbB2. *Proc. Natl. Acad. Sci. USA* **92,** 1431–1435.
Nave, K.-A., and Salzer, J. L. (2006). Axonal regulation of myelination by neuregulin 1. *Curr. Opin. Neurobiol.* **16,** 492–500.
Nedelec, S., Dubacq, C., and Trembleau, A. (2005). Morphological and molecular features of the mammalian olfactory sensory neuron axons: What makes these axons so special? *J. Neurocytol.* **34,** 49–64.
Newbern, J., and Birchmeier, C. (2010). Nrg1/ErbB signaling networks in Schwann cell development and myelination. *Semin. Cell Dev. Biol.* **21,** 922–928.
Nishino, J., Saunders, T. L., Sagane, K., and Morrison, S. J. (2010). Lgi4 promotes the proliferation and differentiation of glial lineage cells throughout the developing peripheral nervous system. *J. Neurosci.* **30,** 15228–15240.
Nodari, A., Zambroni, D., Quattrini, A., Court, F. A., D'Urso, A., Recchia, A., Tybulewicz, V. L. J., Wrabetz, L., and Feltri, M. L. (2007). Beta1 integrin activates Rac1 in Schwann cells to generate radial lamellae during axonal sorting and myelination. *J. Cell Biol.* **177,** 1063–1075.
Northcutt, R. G. (2004). Taste buds: Development and evolution. *Brain Behav. Evol.* **64,** 198–206.
Oakley, B., and Witt, M. (2004). Building sensory receptors on the tongue. *J. Neurocytol.* **33,** 631–646.
Ogata, T., Iijima, S., Hoshikawa, S., Miura, T., Yamamoto, S., Oda, H., Nakamura, K., and Tanaka, S. (2004). Opposing extracellular signal-regulated kinase and Akt pathways control Schwann cell myelination. *J. Neurosci.* **24,** 6724–6732.
Ou, H. C., Cunningham, L. L., Francis, S. P., Brandon, C. S., Simon, J. A., Raible, D. W., and Rubel, E. W. (2009). Identification of FDA-approved drugs and bioactives that protect hair cells in the zebrafish (Danio rerio) lateral line and mouse (Mus musculus) utricle. *J. Assoc. Res. Otolaryngol.* **10,** 191–203.
Ozkaynak, E., Abello, G., Jaegle, M., van Berge, L., Hamer, D., Kegel, L., Driegen, S., Sagane, K., Bermingham, J. R., Jr., and Meijer, D. (2010). Adam22 is a major neuronal receptor for Lgi4-mediated Schwann cell signaling. *J. Neurosci.* **30,** 3857–3864.

Pan, Y. A., Livet, J., Sanes, J. R., Lichtman, J. W., and Schier, A. F. (2011). Multicolor brainbow imaging in zebrafish. *Cold Spring Harb. Protoc.* pdb prot5546.

Parichy, D. M., and Turner, J. M. (2003). Zebrafish puma mutant decouples pigment pattern and somatic metamorphosis. *Dev. Biol.* **256,** 242–257.

Parichy, D. M., Turner, J. M., and Parker, N. B. (2003). Essential role for puma in development of postembryonic neural crest-derived cell lineages in zebrafish. *Dev. Biol.* **256,** 221–241.

Pereira, J. A., Benninger, Y., Baumann, R., Gonçalves, A. F., Ozçelik, M., Thurnherr, T., Tricaud, N., Meijer, D., Fässler, R., Suter, U., and Relvas, J. B. (2009). Integrin-linked kinase is required for radial sorting of axons and Schwann cell remyelination in the peripheral nervous system. *J. Cell Biol.* **185,** 147–161.

Perlin, J. R., Lush, M. E., Stephens, W. Z., Piotrowski, T., and Talbot W. S. (2011). Neuronal Neuregulin type III directs Schwann cell migration. *Development* **138,** 4639–4648.

Peters, A. (1964). Observations on the connexions between myelin sheaths and glial cells in the optic nerves of young rats. *J. Anat.* **98,** 125–134.

Peters, A., and Muir, A. R. (1959). The relationship between axons and Schwann cells during development of peripheral nerves in the rat. *Q. J. Exp. Physiol. Cogn. Med. Sci.* **44,** 117–130.

Peterson, R. T., Shaw, S. Y., Peterson, T. A., Milan, D. J., Zhong, T. P., Schreiber, S. L., MacRae, C. A., and Fishman, M. C. (2004). Chemical suppression of a genetic mutation in a zebrafish model of aortic coarctation. *Nat. Biotechnol.* **22,** 595–599.

Pogoda, H. M., Sternheim, N., Lyons, D. A., Diamond, B., Hawkins, T. A., Woods, I. G., Bhatt, D. H., Franzini-Armstrong, C., Dominguez, C., Arana, N., Jacobs, J., Nix, R., *et al.* (2006). A genetic screen identifies genes essential for development of myelinated axons in zebrafish. *Dev. Biol.* **298,** 118–131.

Purcell, I. M., and Perachio, A. A. (1997). Three-dimensional analysis of vestibular efferent neurons innervating semicircular canals of the gerbil. *J. Neurophysiol.* **78,** 3234–3248.

Raphael, A. R., Perlin, J. R., and Talbot, W. S. (2010). Schwann cells reposition a peripheral nerve to isolate it from postembryonic remodeling of its targets. *Development* **137,** 3643–3649.

Raphael, A. R., Lyons, D. A., and Talbot, W. S. (2011). ErbB signaling has a role in radial sorting independent of Schwann cell number. *Glia* **59**(7), 1047–1055.

Riethmacher, D., Sonnenberg-Riethmacher, E., Brinkmann, V., Yamaai, T., Lewin, G. R., and Birchmeier, C. (1997). Severe neuropathies in mice with targeted mutations in the ErbB3 receptor. *Nature* **389,** 725–730.

Rihel, J., Prober, D. A., Arvanites, A., Lam, K., Zimmerman, S., Jang, S., Haggarty, S. J., Kokel, D., Rubin, L. L., Peterson, R. T., and Schier, A. F. (2010). Zebrafish behavioral profiling links drugs to biological targets and rest/wake regulation. *Science* **327,** 348–351.

Sauka-Spengler, T., and Bronner, M. (2010). Snapshot: Neural crest. *Cell* **143**486–486 e481.

Schreiner, S., Cossais, F., Fischer, K., Scholz, S., Bosl, M. R., Holtmann, B., Sendtner, M., and Wegner, M. (2007). Hypomorphic Sox10 alleles reveal novel protein functions and unravel developmental differences in glial lineages. *Development* **134,** 3271–3281.

Shestopalov, I. A., and Chen, J. K. (2010). Oligonucleotide-based tools for studying zebrafish development. *Zebrafish* **7,** 31–40.

Si, X., Zakir, M. M., and Dickman, J. D. (2003). Afferent innervation of the utricular macula in pigeons. *J. Neurophysiol.* **89,** 1660–1677.

Stampfli, R. (1954). Saltatory conduction in nerve. *Physiol. Rev.* **34,** 101–112.

Stemple, D. L. (2004). TILLING—a high-throughput harvest for functional genomics. *Nat. Rev. Genet.* **5,** 145–150.

Svaren, J., and Meijer, D. (2008). The molecular machinery of myelin gene transcription in Schwann cells. *Glia* **56,** 1541–1551.

Syed, N., Reddy, K., Yang, D. P., Taveggia, C., Salzer, J. L., Maurel, P., and Kim, H. A. (2010). Soluble neuregulin-1 has bifunctional, concentration-dependent effects on Schwann cell myelination. *J. Neurosci.* **30,** 6122–6131.
Takada, N., Kucenas, S., and Appel, B. (2010). Sox10 is necessary for oligodendrocyte survival following axon wrapping. *Glia* **58,** 996–1006.
Tasaki, I. (1959). Physiologic properties of the myelin sheath and of the node of Ranvier. *Prog. Neurobiol.* **4,** 159–172.
Taveggia, C., Zanazzi, G., Petrylak, A., Yano, H., Rosenbluth, J., Einheber, S., Xu, X., Esper, R. M., Loeb, J. A., Shrager, P., Chao, M. V., Falls, D. L., *et al.* (2005). Neuregulin-1 type III determines the ensheathment fate of axons. *Neuron* **47,** 681–694.
Topilko, P., Schneider-Maunoury, S., Levi, G., Baron-Van Evercooren, A., Chennoufi, A. B., Seitanidou, T., Babinet, C., and Charnay, P. (1994). Krox-20 controls myelination in the peripheral nervous system. *Nature* **371,** 796–799.
Voas, M. G., Glenn, T. D., Raphael, A. R., and Talbot, W. S. (2009). Schwann cells inhibit ectopic clustering of axonal sodium channels. *J. Neurosci.* **29,** 14408–14414.
Webster, H. D., Martin, R., and O'Connell, M. F. (1973). The relationships between interphase Schwann cells and axons before myelination: A quantitative electron microscopic study. *Dev. Biol.* **32,** 401–416.
Winklbauer, R. (1989). Development of the lateral line system in Xenopus. *Prog. Neurobiol.* **32,** 181–206.
Woldeyesus, M. T., Britsch, S., Riethmacher, D., Xu, L., Sonnenberg-Riethmacher, E., Abou-Rebyeh, F., Harvey, R., Caroni, P., and Birchmeier, C. (1999). Peripheral nervous system defects in erbB2 mutants following genetic rescue of heart development. *Genes Dev.* **13,** 2538–2548.
Wyart, C., and Del Bene, F. (2011). Let there be light: Zebrafish neurobiology and the optogenetic revolution. *Rev. Neurosci.* **22,** 121–130.
Wyart, C., Del Bene, F., Warp, E., Scott, E. K., Trauner, D., Baier, H., and Isacoff, E. Y. (2009). Optogenetic dissection of a behavioural module in the vertebrate spinal cord. *Nature* **461,** 407–410.
Xu, Q., Stemple, D., and Joubin, K. (2008). Microinjection and cell transplantation in zebrafish embryos. *Methods Mol. Biol.* **461,** 513–520.
Yoo, S. K., Deng, Q., Cavnar, P. J., Wu, Y. I., Hahn, K. M., and Huttenlocher, A. (2010). Differential regulation of protrusion and polarity by PI3K during neutrophil motility in live zebrafish. *Dev. Cell* **18,** 226–236.
Yu, W. M., Feltri, M. L., Wrabetz, L., Strickland, S., and Chen, Z. L. (2005). Schwann cell-specific ablation of laminin gamma1 causes apoptosis and prevents proliferation. *J. Neurosci.* **25,** 4463–4472.
Yu, W.-M., Chen, Z.-L., North, A. J., and Strickland, S. (2009). Laminin is required for Schwann cell morphogenesis. *J. Cell Sci.* **122,** 929–936.
Zanazzi, G., Einheber, S., Westreich, R., Hannocks, M. J., Bedell-Hogan, D., Marchionni, M. A., and Salzer, J. L. (2001). Glial growth factor/neuregulin inhibits Schwann cell myelination and induces demyelination. *J. Cell Biol.* **152,** 1289–1299.

CHAPTER TWO

Wnt Signaling: Signaling at and Above the Receptor Level

Tina Buechling* *and* Michael Boutros*

Contents

Abstract

Wnt signaling is one of the most important developmental signaling pathways that controls cell fate decisions and tissue patterning during early embryonic and later development. It is activated by highly conserved Wnt proteins that are secreted as palmitoylated glycoproteins and act as morphogens to form a concentration gradient across a developing tissue. Wnt proteins regulate transcriptional and posttranscriptional processes depending on the distance of their origin and activate distinct intracellular cascades, commonly referred to as canonical (β-catenin-dependent) and noncanonical (β-catenin-independent) pathways. Therefore, the secretion and the diffusion of Wnt proteins needs to be tightly regulated to induce short- and long-range downstream signaling. Even though the Wnt signaling cascade has been studied intensively, key aspects and principle mechanisms, such as transport of Wnt growth factors or regulation of signaling specificity between different Wnt pathways, remain unresolved. Here, we introduce basic principles of Wnt/Wg signal transduction

* German Cancer Research Center (DKFZ), Division of Signaling and Functional Genomics, and Faculty of Medicine Mannheim, Department of Cell and Molecular Biology, University of Heidelberg, Heidelberg, Germany

Current Topics in Developmental Biology, Volume 97
ISSN 0070-2153, DOI: 10.1016/B978-0-12-385975-4.00008-5

and highlight recent discoveries, such as the involvement of vacuolar ATPases and vesicular acidification in Wnt signaling. We also discuss recent findings regarding posttranslational modifications of Wnts, trafficking through the secretory pathway and developmental consequences of impaired Wnt secretion. Understanding the detailed mechanism and regulation of Wnt protein secretion will provide valuable insights into many human diseases based on overactivated Wnt signaling.

1. Overview of Wnt Signal Transduction Cascades

Wnt signaling plays a central role in many developmental processes, such as cell fate specification during early embryonic development, body axis patterning, and tissue homeostasis (reviewed in Croce and McClay, 2008). It has also been implicated in the self-renewing state of stem cells in the skin, the hematopoetic system, and the intestinal epithelium (reviewed in Gu *et al.*, 2010; Nusse, 2008; Staal and Luis, 2010; Wend *et al.*, 2010).

To date, three different branches of Wnt signaling have been identified that are activated upon binding of highly conserved Wnt proteins to different members of the Frizzled (Fz) receptor family, atypical G-protein-coupled receptors (GPCRs): the canonical Wnt pathway which involves the stabilization of the proto-oncogene β-catenin or *Drosophila* Armadillo (Arm) (reviewed in Angers and Moon, 2009; Cadigan and Peifer, 2009; MacDonald *et al.*, 2009; Mosimann *et al.*, 2009; van Amerongen and Nusse, 2009), noncanonical Wnt signaling or planar cell polarity (PCP) pathways which are required to establish tissue polarity of many epithelia (reviewed in Axelrod, 2009; Roszko *et al.*, 2009; Segalen and Bellaiche, 2009; Strutt and Strutt, 2009; Wu and Mlodzik, 2009) and the Wnt/Ca^{2+} pathway which stimulates the intracellular release of Ca^{2+} and activates calcium-dependent mediators (reviewed in Kohn and Moon, 2005; Kuhl *et al.*, 2000). More recently and mainly in the context of neuronal development, Wnt ligands have also been shown to signal via the receptor tyrosine kinase-like orphan receptor (ROR) and the receptor tyrosine kinase RYK, the vertebrate ortholog of *Drosophila* Derailed (Drl) and *Caenorhabditis elegans* LIN-18, which in some cases antagonize Wnt/β-catenin signaling through the activation of the protein kinase c-Jun N-terminal kinase (JNK) (reviewed in Angers and Moon, 2009; van Amerongen *et al.*, 2008).

Most core components of the Wnt transduction cascade are evolutionary conserved from *C. elegans* to humans. For instance, Wnt proteins are a highly conserved class of secreted signaling molecules that can activate the different intracellular signaling branches in a tissue-specific manner. They have also been shown to act as morphogens that are secreted from a restricted source and form a concentration gradient across a developing tissue (Neumann and Cohen, 1997; Zecca *et al.*, 1996). Numerous Wnt proteins have been identified in vertebrates, *Drosophila*, and *C. elegans*. The founding members of the Wnt protein family are *Drosophila* Wingless (Wg) and mammalian Wingless-type mouse mammary tumor virus (MMTV) integration site family member 1 (Wnt1). Wg was identified as a recessive mutation affecting wing development and embryonic segmentation in *Drosophila* (Nusslein-Volhard and Wieschaus, 1980; Sharma and Chopra, 1976). Its mammalian counterpart, Int-1 (later renamed to Wnt-1) was discovered in a retroviral insertion mutagenesis screen (Nusse *et al.*, 1984; van Ooyen and Nusse, 1984). Since then, two classes of Wnt proteins have been identified: canonical Wnt proteins defined by their transforming or axis-inducing ability and noncanonical, nontransforming Wnts that are not able to induce a secondary axis (Du *et al.*, 1995; Wong *et al.*, 1994). Canonical Wnt proteins, such as Wnt1 and Wnt3A (Roelink and Nusse, 1991) signal through the stabilization of β-catenin/Arm. Overexpression of these Wnts results in a duplication of the dorsal–ventral axis in vertebrates (Shimizu *et al.*, 1997) while overexpression of noncanonical Wnts, such as Wnt5A (Gavin *et al.*, 1990) and Wnt11 (Adamson *et al.*, 1994) results in convergent extension of the body axis in Xenopus and zebrafish (Heisenberg *et al.*, 2000; Kilian *et al.*, 2003; Wallingford *et al.*, 2001), a process which is β-catenin independent. However, making it even more complex, these Wnt proteins can bind to a variety of cell surface receptor, including low-density lipoprotein receptor-related proteins (LRP) 5/6 or *Drosophila* Arrow (Arr) in combination with members of the Fz receptor family, thereby inducing different cellular programs (reviewed in Kikuchi *et al.*, 2009; van Amerongen *et al.*, 2008). In recent years, it has emerged that Wnt proteins appear not to be intrinsically canonical or noncanonical, but that the same Wnt can activate multiple pathways depending on the particular Wnt receptor combination present at the cell surface that determines downstream signaling events (reviewed in Kikuchi *et al.*, 2009).

2. Wnt Proteins: Different Branches of Signaling

2.1. Canonical Wnt signal transduction cascades

Canonical Wnt signaling stabilizes the transcriptional activator β-catenin/ Arm, leading to the formation of a complex with the DNA-binding factor TCF/LEF (T-cell lymphoid enhancer factor) or *Drosophila* pangolin (pan) resulting in activation of tissue-specific target genes. While invertebrates have a single TCF member, vertebrates have four TCFs (TCF1, TFC2/ LEF1, TCF3, and TCF4) which are expressed in a tissue-dependent manner and function as bi-modular regulators both in an activating and repressing state (reviewed in Hoppler and Kavanagh, 2007). While TCF1 and TCF2/LEF1 are transcriptional activators, TCF3 has been identified as a repressor that is replaced by TCF1 upon Wnt stimulation. The TCF switch to activate Wnt target genes seems to be mediated through the homeodomain-interacting protein kinase 2 (HIPK2) which has been identified in Xenopus to regulate different TCFs (Hikasa and Sokol, 2011).

In the absence of Wnt ligands, β-catenin/Arm is bound by the destruction complex composed of the scaffolding proteins Axis inhibition protein-1 (Axin1) (Willert *et al.*, 1999; Zeng *et al.*, 1997) and adenomatous polyposis coli (APC), and the serine/threonine kinases, Casein kinase 1 alpha (CK1α) and glycogen synthase kinase 3β (GSK3β) or shaggy (sgg) in *Drosophila.* Axin itself is phosphorylated by CK1α providing binding sites for GSK3β which then phosphorylates APC. Axin and APC form a scaffold for phosphorylation of β-catenin/Arm by Axin-bound CK1α and GSK3β (Amit *et al.*, 2002; Liu *et al.*, 2002; Yanagawa *et al.*, 2002). β-Catenin/Arm is first phosphorylated on Ser45 by CK1α and subsequently on Ser33, Ser37, and Thr41 by cytoplasmic GSK3β to provide adapter sites for the β-transductin-repeat-containing protein (βTrCP), an E3 ubiquitin ligase subunit, which targets ubiquitination and proteasomal degradation of cytoplasmic β-catenin (Aberle *et al.*, 1997). As a consequence, cytoplasmic levels of β-catenin/Arm remain low which in turn leads to binding of TCF/Pan to transcriptional corepressors, such as the C-terminal binding protein CtBP and Groucho (Gro), the Drosophila ortholog of TLE (transducin-like enhancer protein). TLE/Gro in turn binds to histone deacetylases (HDAC), such as *Drosophila* Rpd3 (Chen *et al.*, 1999), to suppress target gene expression through chromatin compression (reviewed in Stadeli *et al.*, 2006) (Fig. 2.1).

In contrast, binding of canonical Wnt ligands, such as Wnt1, Wnt3A, or Wg, to the cystein rich domain (CRD) of Fz receptors and to LRP6/Arr coreceptors results in dissociation of the destruction complex. The serine/threonine phosphatases PP1 and PP2A have been shown to counteract CK1α and GSK3β kinase function by dephosphorylating Axin and β-catenin, respectively (Luo *et al.*, 2007; Seeling *et al.*, 1999). Dephosphorylated Axin is subsequently recruited to the plasma membrane (PM), a process that is regulated by Bili, an evolutionary conserved Band4.1-domain containing protein (Kategaya *et al.*, 2009), the trimeric G protein Go (Egger-Adam and Katanaev, 2009), and Dishevelled (Dvl/Dsh). Dsh is activated by multiple phosphorylation through CK1 epsilon (CK1ε) and functions as the branching point between canonical and noncanonical Wnt signaling (reviewed in Gao and Chen, 2010). It is negatively regulated by an E3 ubiquitin ligase complex containing Cullin-3 which promotes polyubiquitination and subsequent degradation of Dsh in a Wnt-dependent manner (Angers *et al.*, 2006). In the current model of Dsh function, DIX domain-mediated polymerization of Dsh and binding to Fz receptors recruits the Axin-GSKβ complex to the

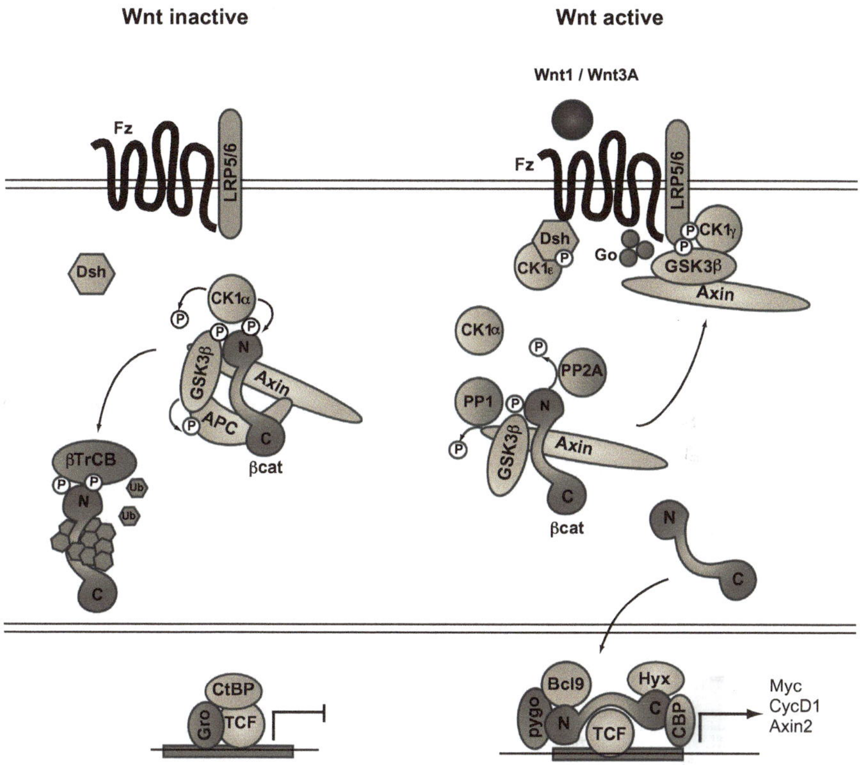

Figure 2.1 The canonical Wnt/β-catenin signaling pathway. Wnt inactive: In the absence of Wnt proteins, β-catenin is bound by the destruction complex composed of Axin, and APC which function as scaffolds for CK1α and GSK3β to phosphorylate β-catenin. Phosphorylated β-catenin is subsequently ubiquitinated by β-TrCP and proteosomally degraded. Groucho (Gro) is recruited to TCF transcription factors to suppress target genes. Wnt active: In the presence of Wnt ligands which bind to Fz receptors and LRP5/6 coreceptors, LRP6 is phosphorylated by CK1γ and GSK3β. Dsh and Axin are recruited to the plasma membrane which results in dissociation of the destruction complex. Stabilized β-catenin translocates to the nucleus and converts TCFs into transcriptional activators through the action of the coactivators Pygopus (Pygo), Bcl9 and Hyrax (Hyx). (For color version of this figure, the reader is referred to the Web version of this chapter.)

cytoplasmic tail of the LRP5/6 coreceptor at the PM. Axin has been shown to bind to phosphorylated LRP6 (reviewed in Niehrs and Shen, 2010). LRP6 is activated through Wnt-induced phosphorylation at conserved PPSP motifs through the action of CK1 gamma (CK1γ) and PM-associated GSK3β (Davidson *et al.*, 2005; Zeng *et al.*, 2005). This process requires the *Drosophila* Cyclin-dependent kinase, Eip63E, or vertebrate PFTK and Cyclin Y (Davidson *et al.*, 2009). As a consequence of Axin-GSK3β recruitment to the

PM, β-catenin is stabilized by a decrease in GSK3β-mediated phosphorylation through PP2A. Stabilized β-catenin translocates to the nucleus to convert TCFs into transcriptional activators. TCF proteins bind to conserved binding motifs in Wnt-responsive elements (WREs) through their high mobility group (HMG) domain and to β-catenin through their catenin binding domain (CBD), providing an adaptor between the DNA and β-catenin/Arm. Bcl9/legless (lgs) and pygopus (pygo) are essential coactivators of Wnt/β catenin signaling that bind to the *N*-terminal transactivation domain (NTD) of β-catenin (Belenkaya *et al.*, 2002; Kramps *et al.*, 2002; Parker *et al.*, 2002; Thompson *et al.*, 2002). In addition, Hyrax (Hyx)/Parafibromin was found to bind to the C-terminal transactivation domain (CTD) of β-catenin which is required for activation of target gene expression (Mosimann *et al.*, 2006). Subsequently, histone acetyltransferases (HATs), such as CBP/p300 and SET1, as well as the chromatin remodeling enzyme brahma-related gene 1 (BRG1) are specifically recruited to WREs to mediate chromatin remodeling of target gene regions (Griffin *et al.*, 2011; Parker *et al.*, 2008; Sierra *et al.*, 2006) (Fig. 2.1). Recently, the histone methylase SET8 has been shown to function as coactivator that is recruited to the β-catenin/TCF complex upon Wnt stimulation linking histone methylation to Wnt target gene expression (Li *et al.*, 2011b). Although canonical Wnt/β-catenin signaling has been studied for more than a decade, novel regulators and links to different cellular processes, such as cell cycle regulation or chromatin remodeling, will continue to be identified, feeding into an even more complex network of core components with multiple, and most likely tissue-specific functions.

2.2. Noncanonical β-catenin-independent Wnt pathways

Noncanonical Wnt signaling is a β-catenin/Arm independent branch that does not require LRP6/Arr or TCF/Pan function. The best described noncanonical Wnt pathways are Wnt/Ca^{2+} (reviewed in Kohn and Moon, 2005; Kuhl *et al.*, 2000) and Fz/PCP signaling (reviewed in Axelrod, 2009; Roszko *et al.*, 2009; Segalen and Bellaiche, 2009; Strutt and Strutt, 2009). Both pathways mediate positional signals important for cell adhesion, polarity, and migration (Fig. 2.2). In vertebrates, these pathways mediate convergent extension movements which are necessary for body axis elongation during gastrulation (reviewed in Roszko *et al.*, 2009). In *Drosophila*, they control the orientation of cuticular structures and bristles in the wing and notum and omatidial formation in the compound eye (reviewed in Wu and Mlodzik, 2009). Wnt/Ca^{2+} signaling has been proposed to involve the action of Wnt proteins and Fz receptors and functions through the activation of G proteins leading to a transient release of calcium from intracellular stores, which in turn activates Ca^{2+}-dependent modulators, such as protein kinase C (PKC) and calcium calmodulin mediated kinase II (CAMKII).

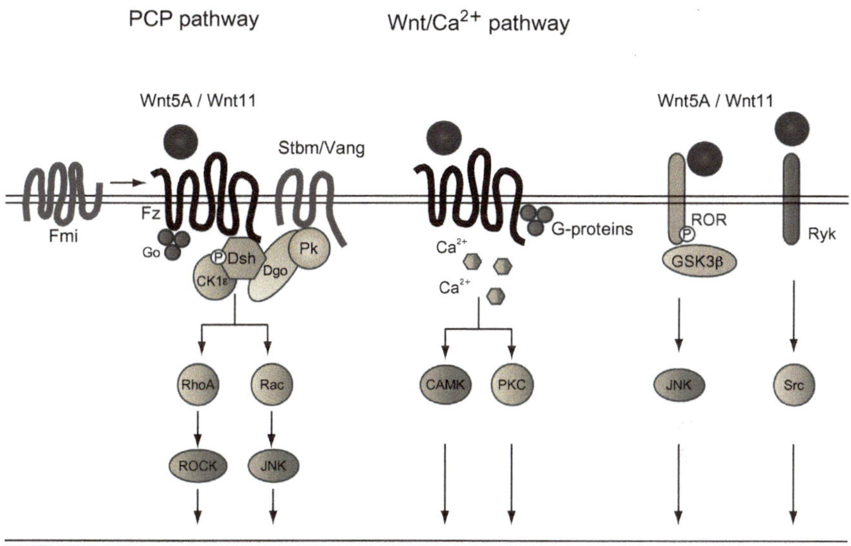

Figure 2.2 Noncanonical β-catenin independent Wnt signaling. Noncanonical Wnt signaling pathways are planar cell polarity (PCP) signaling and the Wnt/Ca^{2+} pathway, activated through nontransforming Wnts, such as Wnt5A or Wnt11. PCP signaling is mediated through recruitment of the PCP complex of Dishevelled (Dsh), the transmembrane proteins Flamingo (Fmi) and Strabismus (Stbm)/VanGogh (Vang), and the cytoplasmic components Diego (Dg) and Prickle (Pk) to the cell surface. This translocation activates the GTPases RhoA and Rac which activate ROCK and the JNK stress response pathway, respectively. Wnt/Ca^{2+} is mediated through G-proteins leading to an increase of Ca^{2+} levels which in turn activates the Ca^{2+}-dependent enzymes PKC and CAMK. In addition, Wnt proteins have been shown to activate distinct signaling branches by binding to the tyrosine kinase-like orphan receptor (ROR) and the receptor tyrosine kinase (RYK) to activate JNK and Src, respectively. (For color version of this figure, the reader is referred to the Web version of this chapter.)

The Wnt/Ca^{2+}-pathway has been implicated in cancer via the induction of epithelial to mesenchymal transition (EMT) through PKC activation (Dissanayake *et al.*, 2007).

While initial studies in *Drosophila* suggested that none of the Wnt proteins might be required for Fz/PCP signal transduction (Chen *et al.*, 2008), PCP signaling in vertebrates has been shown to be initiated by Wnt5A and Wnt11 (He *et al.*, 1997; Heisenberg *et al.*, 2000; Wallingford *et al.*, 2001) through binding to Fz receptors. This binding recruits Dsh to the PM, a process that has been linked to the trimeric G protein Go. Go has been implicated in direct signal transduction of Fz signals from the membrane to downstream components by initiating the recruitment of Dsh to the PM (Katanaev *et al.*, 2005). In *Drosophila*, it has also been shown that Dsh recruitment is dependent on the sodium proton exchanger (Nhe2), the ortholog of vertebrate Nhe3, suggesting that electrochemical cues regulate Fz/Dsh complex formation at the PM during the establishment of PCP (Simons *et al.*, 2009). As PM recruitment of Dsh is required for downstream PCP signaling, Dsh functions as a branching point between canonical and noncanonical Wnt pathways (reviewed in Gao and Chen, 2010).

Besides Fz and Dsh, conserved core components of the Fz/PCP pathway are the four-pass transmembrane protein Strabismus (Stbm)/Van Gogh (Vang) (Wolff and Rubin, 1998), the atypical cadherin and, adhesion-GPCR Flamingo (Fmi) (Usui *et al.*, 1999), and the cytosolic proteins Prickle (Pk) (Gubb *et al.*, 1999) and Diego (Dgo) (Feiguin *et al.*, 2001). These core components are recruited to the apical membrane of polarized cells and subsequently segregate into complementary apical domains (Strutt, 2001). Although, the detailed mechanism of this recruitment and rearrangement remains to be elucidated, the segregation of PCP core components has been shown to activate the small GTPases RhoA (Ras homolog gene family member A) and Rac1, which in turn activate the JNK stress response pathway (Boutros *et al.*, 1998) and ROCK (Rho-associated coiled-coil containing protein kinase) leading to cytoskeletal remodeling (Fig. 2.2). Recently, Wnt5A and Wnt11 have also been shown to trigger downstream signaling by binding to the PM proteins ROR and RYK leading to activation of JNK and Src family members, respectively (reviewed in Kikuchi *et al.*, 2009; van Amerongen and Nusse, 2009) (Fig. 2.2).

While the downstream events triggered by binding of canonical Wnt proteins, such as Wnt1 or Wnt3A, to LRP5/6 and Fz receptors have been characterized in depth, the detailed mechanism of noncanonical signaling is poorly understood. It remains a challenge to understand how positional information creates cell reorientation during the establishment of a polarized tissue and how Wnt signaling leads to remodeling of the cytoskeleton in many vertebrates.

2.3. Wnt signaling in human diseases

As a consequence of aberrant activation, for instance through mutations in core components, Wnt signaling causes severe developmental defects during embryogenesis and has been linked to various human diseases, most notably to cancer (reviewed in Lucero *et al.*, 2010). It has been associated with the maintenance of cancer initiating cells (or cancer stem cells) in the skin, the mammary gland and the intestine (reviewed in Wend *et al.*, 2010). In a majority of colorectal cancers, transformation of epithelial cells is initiated by activating mutations in core components of the Wnt transduction process (reviewed in Klaus and Birchmeier, 2008) indicating that Wnt signals play a crucial role in intestinal progenitor and stem cell maintenance. Originally discovered in adenomatous polyposis (AP), the tumor suppressor protein APC, a negative regulator of Wnt signaling, has been linked to approximately 90% of all sporadic colon cancers (reviewed in Klaus and Birchmeier, 2008). Colon epithelia cells transform as a consequence of loss-of-function mutations within APC leading to an inappropriate stabilization of β-catenin which translocates to the nucleus to form a constitutively active complex with TCF4, one of the TCFs involved in β-catenin dependent transcription. As a consequence, Wnt target genes, such as the proto-oncogene c-Myc and the cell cycle regulator Cyclin D1, are upregulated, directly linking Wnt signaling to mitotic processes, such as cell cycle progression (reviewed in Davidson and Niehrs, 2010).

Less frequently, stabilizing oncogenic mutations in β-catenin removing the N-terminal serine/threonine degradation motif have been identified in colon cancer cells (reviewed in Klaus and Birchmeier, 2008). These mutations result in an accumulation of nuclear β-catenin leading to a constitutive activation of the TCF-mediated transcriptional response in a Wnt-independent manner. In past years, developmental consequences of loss and gain of function of β-catenin have been studied intensively in various mouse models (reviewed in Grigoryan *et al.*, 2008). Further, β-catenin dependent Wnt signaling has been dissected in various cancer cell lines using genome-wide RNA interference (RNAi) approaches which identified numerous components that regulate β-catenin mediated transcription, including components of the Mediator complex. Mediator is a multiprotein complex consisting of at least 18 proteins in mammals that functions as a transcriptional coactivator (reviewed in Biddick and Young, 2005; Bjorklund and Gustafsson, 2005; Conaway *et al.*, 2005). In *Drosophila*, Med12 and Med13 have been shown to be required for Wg target gene expression through interaction with Pygo (Carrera *et al.*, 2008). In human colorectal cancer cells, the cyclin-dependent kinase member of the Mediator complex CDK8 and AGGF1, a nuclear chromatin-associated protein, were found to be required for β-catenin-driven transformation and transcription (Firestein *et al.*, 2008; Major *et al.*, 2008).

In addition, HEF1/NEDD9, the human enhancer of filamentation 1, was recently identified as a novel Wnt target gene. Elevated levels of HEF1 in colorectal cancer cells correlate with increased cell proliferation and migration (Li *et al.*, 2011a). In contrast, TCF7L2/TCF4 was found as a transcriptional repressor that restricts growth of colon cancer cells (Tang *et al.*, 2008). Numerous primary colorectal cancers have been shown to contain mutations in the CtBP binding motifs at the C-terminus of TCF4 which most likely affect the constitutive repressor function of TCF4/CtBP in the absence of Wnt signals in these cells (reviewed in Chinnadurai, 2002). In addition, Bruton's Tyrosine Kinase (BTK) negatively regulates Wnt/β-catenin signaling by binding to CDC73 to repress β-catenin-mediated transcription (James *et al.*, 2009). Loss of BTK function in human colorectal cancer cells therefore correlates with an increased abundance of nuclear β-catenin and elevated target gene expression. Even though the detailed mechanisms of these lately identified Wnt regulators remains vague, they might be suitable entry points for the development of therapeutics in β-catenin-driven malignancies.

3. Regulation of Wnt Signaling Above the Receptor Level

3.1. Wnt proteins and their posttranslational modifications

Wnt proteins constitute a large family of structurally related cysteine-rich glycoproteins with an approximate molecular weight of 38–42 kDa. To date, 19 different Wnts have been identified in mouse and human, 15 in zebrafish, 5 in *C. elegans,* and 7 in *Drosophila*. All of these Wnt proteins share a common signature of 23 highly conserved cysteines (Fig. 2.3). They do not contain any conserved protein domains except an N-terminal signal peptide. Wnt proteins undergo several posttranslational modifications, including *N*-glycosylation and acylation, both of which are required for transport through the secretory pathway and the proper activity.

N-Glycosylation at conserved asparagine (Asn) residues is mediated by the oligosaccharyl transferase (OST) complex within the endoplasmatic reticulum (ER) (Yan and Lennarz, 1999) and has been implicated in the secretion of Wnt proteins (Komekado *et al.*, 2007; Kurayoshi *et al.*, 2007). *Drosophila* Wg is glycosylated at two Asn residues (N103, N414), two additional predicted *N*-glycosylation sites seem not to be modified in S2 cells (Tanaka *et al.*, 2002). Mouse Wnt1 was shown to be glycosylated at three Asn residues (N29, N316, and N359), a predicted forth site is only inefficiently modified in cultured cells (Mason *et al.*, 1992). Mouse Wnt3A has been found to be glycosylated at two Asn residues (N87, N298) (Komekado *et al.*, 2007). A mutant Wnt3A lacking all glycosylation sites is retained within the ER, suggesting that glycosylation is required for proper secretion (Komekado *et al.*, 2007). Likewise, secretion of a nonglycosylated Wnt5A lacking its four glycosylated Asn residues (N114, N120, N311, N325) was significantly impaired whereas its ability to activate Wnt signaling was not affected (Kurayoshi *et al.*, 2007).

S-Palmitoylation (reviewed in Charollais and Van Der Goot, 2009), the covalent attachment of palmitic acid to the most amino-terminal cysteine residue of Wnts, has been found to be responsible for the hydrophobic nature of Wnt proteins (Willert *et al.*, 2003). This palmitoylated cysteine residue is absolutely conserved among all Wnt family members (Takada *et al.*, 2006; Willert *et al.*, 2003). Mouse Wnt3A is *S*-palmitoylated at cysteine 77 (C77), corresponding to C93 in mouse Wnt1 and *Drosophila* Wg (Fig. 2.3). A mutant Wnt3A with a single amino acid substitution of the palmitoylated cysteine to alanine (C77A) is secreted comparable to the

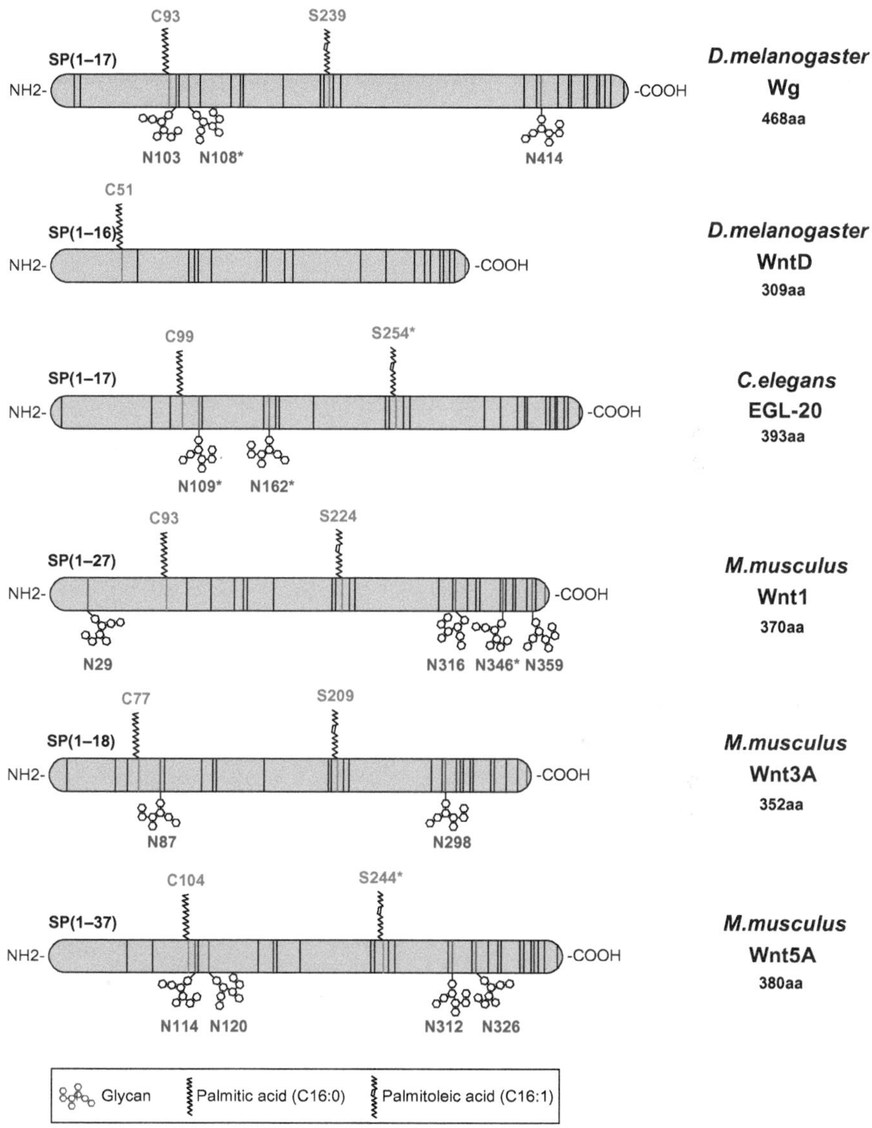

Figure 2.3 Structure and modifications of Wnt proteins. The canonical Wnt proteins Wg, mWnt1, and mWnt3A as well as the noncanonical mWnt5A are secreted glycoproteins with an N-terminal signal peptide (SP). All Wnt proteins share a signature of highly conserved cysteines (black lines), one of which is palmitoylated (C16:0, in blue). In addition to *S*-palmitoylation, Wg, mWnt1, and mWnt3A contain an unsaturated palmitoleate residue at a conserved serine (C16:1, in blue). According serine residues have been predicted for other Wnt proteins due to the highly conserved amino acid motif surrounding the palmitoleated serine residue. *Drosophila* WntD/Wnt8 is thought to be the only nonlipid modified Wnt protein as it lacks the conserved palmitoleated

wild-type protein but its activity is diminished for both autocrine and paracrine signaling due to a decreased affinity for Wnt receptors (Willert *et al.*, 2003). This observation is consistent with the finding that palmitoylation of C104 in Wnt5A is required for binding to Frizzled 5 (Fzd5) (Kurayoshi *et al.*, 2007). A C93A variant of Wg was found to be secreted normally in cultured S2 cells (Franch-Marro *et al.*, 2008a) indicating that palmitoylation is required for signaling activity rather than secretion. Likewise, a cysteine to threonine substitution (C93Y) in the *Drosophila* Wg loss-of-function allele wg^{S21} (Couso and Martinez Arias, 1994), as well as a natural loss-of-function allele of *C. elegans* egl-20(n585) (Maloof *et al.*, 1999) containing a cysteine to serine replacement (C99S) confirmed the requirement of lipid modifications at this most amino-terminal cysteine for Wnt activity (Willert *et al.*, 2003). Interestingly, Wnt3A (C77A) was found to be glycosylated in a wild-type manner, while nonglycosylated Wnt3A was not palmitoylated at C77, indicating that glycosylation precedes palmitoylation (Komekado *et al.*, 2007).

In addition to *S*-palmitoylation at C77, mouse Wnt3A is acylated with an unsaturated fatty acid, palmitoleic acid, at a conserved serine residue (S209) (Takada *et al.*, 2006). A mutant form of Wnt3A preventing the palmitoleoyl modification (S209A) is retained in Wnt-secreting cells in cell culture and Xenopus embryos. However, no glycosylation defects compared to wild-type Wnt3A were detected (Takada *et al.*, 2006), once more indicating that lipid modifications of Wnts are not required for glycosylation. The amino acid sequence surrounding S209 in mouse Wnt3A is highly conserved, suggesting that other Wnt proteins are similarily acylated at this conserved motif (Fig. 2.3). Indeed, the presence of *O*-acyl moieties at residues corresponding to S209 in Wnt3A was confirmed for Wnt1 (S224) and *Drosophila* Wg (S239) (Doubravska *et al.*, 2011; Franch-Marro *et al.*, 2008a). However, in contrast to mouse Wnt3A, S239 in *Drosophila* Wg appears not to be essential for secretion whereas nonpalmitoylated *Drosophila* Wg (C93A) abrogates signaling *in vivo* as it is retained within the ER (Franch-Marro *et al.*, 2008a) indicating that lipidation of Wnts is required for ER exit.

Genetic evidence suggests that Wnt secreting cells require the multipass transmembrane protein Porcupine (Porc) or its ortholog *mom1* in *C. elegans*, a predicted membrane-bound *O*-acyltransferase (MBOAT) localized to the ER (Caricasole *et al.*, 2002; Kadowaki *et al.*, 1996; Tanaka *et al.*, 2000; van den Heuvel *et al.*, 1993). Porc has been found to bind the N-terminal

serine residue, but has been shown to contain a palmitoylated cysteine (C51). *N*-Linked glycosylation sites are indicated in red. Predicted glycosylation or acylation sites are indicated by asterisks. *N*-glycosylation sites were predicted using NetNGlyc1.0, signal peptides were determined using SignalP 3.0. Note that for Wnt5A, N312, and N326 correspond to N311 and N325 determined by Kurayoshi *et al.*, 2007, respectively. (See Color Insert.)

domain of Wnt proteins suggesting that it anchors Wnts to the ER membrane possibly through acylation (Tanaka *et al.*, 2002). Overexpression of Porc promotes the activity of Wnt1 and Wnt3A in cultured cells and steepens the Wnt-mediated proliferation gradient in the chick dorsal neural tube, indicating that Porc limits the distribution of Wnts by an increased affinity for Wnt receptors through membrane anchoring (Galli *et al.*, 2007). In contrast, loss of Porc reduces hydrophobicity and membrane association of Wnt proteins and results in an accumulation of Wnts within the ER (Zhai *et al.*, 2004). In particular, Porc has been found to be required for S209-dependent acylation, an essential modification for Wnt3A ER export (Takada *et al.*, 2006). This *O*-linked acylation of serine in both, Wnt1 and Wnt3A, is essential for subsequent *S*-palmitoylation at the N-terminal cysteine residue (Doubravska *et al.*, 2011). These findings highlight the requirement of Porc-mediated lipid modifications of Wnt proteins for subsequent ER exit and entry into the secretory pathway. The only Wnt protein that has so far been found to be secreted in a Porc-independent manner is *Drosophila* WntD/Wnt8 (Wnt inhibitor of Dorsal) a distant member of the *Drosophila* Wnt family (Ching *et al.*, 2008; Ganguly *et al.*, 2005; Gordon *et al.*, 2005). WntD shares the conserved cysteine signature of Wnt proteins and seems to be *S*-palmitoylated at its most amino-terminal cysteine (C51) (Willert *et al.*, 2003) but lacks the serine corresponding to S209 in murine Wnt3A (Fig. 2.3) raising the question whether there are other Wnt modifying proteins in the ER.

Recently, another ER-resident protein named Oto, the mouse ortholog of the glycosylphosphatidylinisotol (GPI)-inositol-deacylase PGAP1, has been implicated in Wnt maturation with a novel, deacylase-independent function (Zoltewicz *et al.*, 2009). GPI deacylation plays an important role in transport of GPI-anchored proteins from the ER to the Golgi (Tanaka *et al.*, 2004). Oto was found to retain Wnt1 and Wnt3A within the ER of Wnt producing cells by increasing their hydrophobicity through a GPI-anchor to a yet undefined residue. ER retention could fully be reverted by a GPI-specific phospholipase D (GPI-PLD) which dramatically increased secretion of Wnt1 and Wnt3A (Zoltewicz *et al.*, 2009). These findings suggest that Oto and Porc have competitive or synergistic effects on Wnt maturation by altering posttranslational modifications. However, it remains to be investigated whether these proteins are the only Wnt-modifying factors in the ER.

In summary, both glycosylation and lipid modifications are necessary for proper trafficking through the secretory pathway and activity of Wnt proteins. While glycosylation seems to be required for further processing of Wnt proteins in the ER and proper secretion, palmitoylation promotes binding to Fz receptors and membrane anchoring of Wnts. However, it will require further studies to reconcile conflicting data for different Wnts to show which specific posttranslational modifications are required for secretion and Wnt activity.

3.2. Secretion of Wnt proteins

3.2.1. Post-Golgi transport of Wnt proteins

In recent years, progress has been made to identify components required for processing and secretion of active Wnt proteins (reviewed in Bartscherer and Boutros, 2008; Lorenowicz and Korswagen, 2009; Port and Basler, 2010). Wnt secretion has to be tightly regulated as Wnt proteins regulate cellular processes in surrounding tissues and determine cell fate in a concentration dependent manner. All Wnt proteins contain an N-terminal hydrophobic signal peptide for transport through the secretory pathway (Fig. 2.4). Modified Wnt proteins are transported from the ER to the Golgi network by a yet unknown mechanism. They are escorted from the Golgi to the PM by the multipass transmembrane protein Evenness interrupted (Evi)/Wntless (Wls), also known as Sprinter (Srt) (Banziger *et al.*, 2006; Bartscherer *et al.*, 2006; Goodman *et al.*, 2006). Evi/Wls is an evolutionary highly conserved protein and a predicted orphan G-protein coupled receptor (GPCR) with an even number of transmembrane elements (Bartscherer and Boutros, 2008; Jin *et al.*, 2010b; Korkut *et al.*, 2009). It is localized to the Golgi, the PM and the endocytic vesicles (Banziger *et al.*, 2006; Bartscherer *et al.*, 2006; Belenkaya *et al.*, 2008; Franch-Marro *et al.*, 2008b; Yang *et al.*, 2008) (Fig. 2.4). GPR177, the mammalian orthologue of Evi/Wls, has been predicted to be glycosylated at multiple Asn residues. Disturbance of *N*-linked glycosylation prevents Golgi localization of GPR177 in Wnt-producing cells suggesting that this posttranslational modification is necessary for transport through the secretory pathway (Fu *et al.*, 2009; Yu *et al.*, 2010).

Originally identified in *Drosophila*, Evi/Wls is highly specific for Wnt proteins and required for early embryonic patterning in *Drosophila* (Banziger *et al.*, 2006; Bartscherer *et al.*, 2006; Goodman *et al.*, 2006). Its specific requirement for transport of Wnt proteins to the PM is conserved across species. Depletion of Evi/Wls in the flour beetle *Tribolium castaneum* (*Tc*) causes severe segmentation phenotypes similar to those observed upon Tc-Wnt1 RNAi (Bolognesi *et al.*, 2008). The *C. elegans* homolog of Evi/Wls, MIG-14, also known as MOM-3, was originally identified with defects in left Q neuroblasts (QL) migration (Maloof *et al.*, 1999), a Wnt/EGL-20-dependent process. MIG-14 was found to be expressed in Wnt/EGL-20 producing cells in the tail region of the nematode where EGL-20 forms a concentration gradient along the anterior–posterior axis (Coudreuse *et al.*, 2006). Loss of MIG-14 results in mutant phenotypes similar to those caused by mutation of *egl-20* indicating that MIG-14 is required for Wnt secretion (Pan *et al.*, 2008; Yang *et al.*, 2008). A requirement of Evi/Wls for Wnt secretion has recently also been confirmed in mammals. Expression analysis of the vertebrate GPR177 revealed that Evi/Wls is ubiquitously expressed in a variety of embryonic and adult tissues in zebrafish, rat, and mouse confirming a global role in Wnt signaling and organogenesis (Jin *et al.*, 2010b;

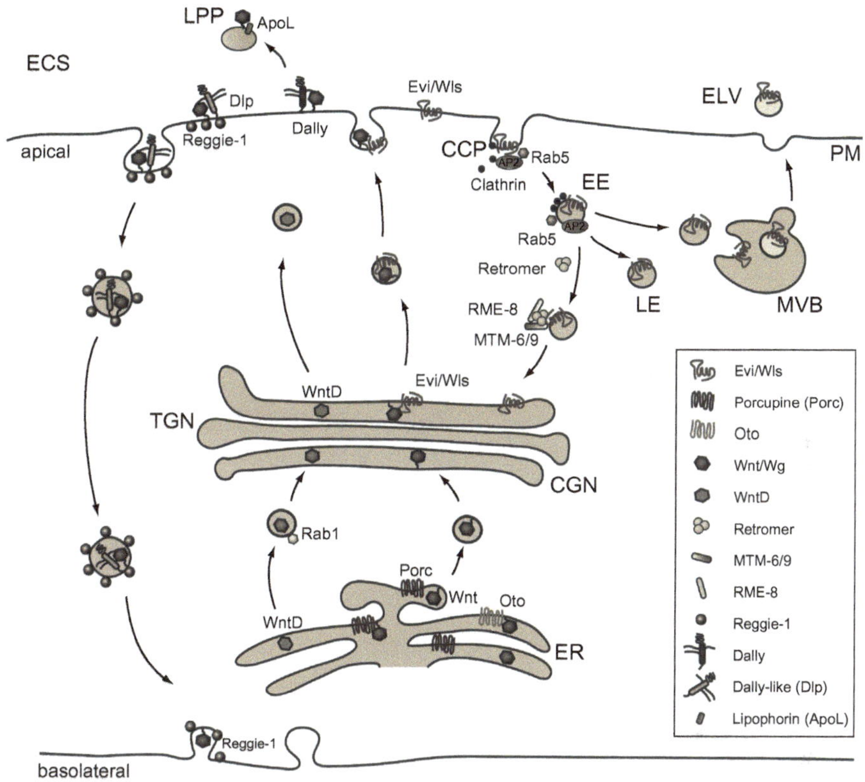

Figure 2.4 Current model of Wnt protein secretion. In the endoplasmatic reticulum (ER), Wnt proteins are palmitoylated through the *N*-acyltransferase Porcupine (Porc) and transported to the cis-Golgi network (CGN). The ER-resident (GPI)-inositol-deacylase Oto/PGAP-1 has been implicated in GPI-modifications of Wnt proteins. The multipass transmembrane protein Evi/Wls functions as a carrier protein for transport of Wnt proteins to the plasma membrane (PM). Evi/Wls itself is internalized to early endosomes (EE) through clathrin-coated pits (CCP), a process which depends on the μ subunit of AP-2 and the GTPase Rab5, and is recycled back to the trans-Golgi network (TGN) through the Retromer complex. RME-8 and MTM-6/9 have been found to bind to Retromer for Evi/Wls recycling. In the absence of Retromer, RME-8 or MTMs, Evi/Wls is targeted to late endosomes (LE) for lysosomal degradation. Wnt proteins are apically secreted from producing cells and associate with lipophorin (ApoL) on lipoprotein particles (LPPs) for short-range signaling and trafficking through the extracellular space (ECS). The glypican-type heparan sulfate proteoglycan (HSPG) Dally functions as a Wnt coreceptor promoting short-range signaling. In contrast, Wnts associate with the noncaveolin microdomain component Reggie-1 and the glypican Dally-like (Dlp) for transcytosis to the basolateral membrane to mediate long-range signaling. Evi/Wls itself has been shown to be secreted across the synaptic cleft in neuromuscular junctions through exosome-like vesicles (ELV) generated via inward budding of multivesicular bodies (MVB). To date, WntD is the only Wnt protein secreted in a Porc- and Evi/Wls-independent manner. However, WntD secretion maintains a requirement for the small GTPase Rab1 for ER to cis-Golgi (CGN) trafficking suggesting an alternative route of nonlipid modified Wnts through the secretory pathway. (See Color Insert.)

Yu *et al.*, 2010). Loss of GPR177 disturbs the patterning of the anterior–posterior body axis due to impaired Wnt secretion, phenocopying a loss of Wnt3 signaling during early embryonic development in mouse (Fu *et al.*, 2009). Conditional deletion of GPR177 leads to early fetal lethality due to severe defects in body axis establishment and mimicks Wnt1 null abnormalities in mid- and hindbrain (Carpenter *et al.*, 2010). Further, ablation of GPR177 in Wnt1-secreting cells causes severe defects in brain and craniofacial development resembling a Wnt1 and Wnt3 double knockout phenotype (Fu *et al.*, 2011). These findings confirm that mammalian GPR177 is necessary for Wnt protein secretion.

As the Evi/Wls protein lacks any conserved protein domains, it has been proposed that Evi/Wls might function as a chaperone for post-Golgi transfer of Wnt proteins to the PM. Indeed, Evi/Wls was found to bind to numerous Wnts, including Wg, Wnt1, and Wnt3A (Banziger *et al.*, 2006), suggesting that it coordinates posttranslational modifications of Wnt proteins. However, Evi/Wls was found to bind both, nonglycosylated Wnt3A and Wnt3A C77A indicating that glycosylation and *S*-palmitoylation are not required for interaction with Evi/Wls (Banziger *et al.*, 2006; Komekado *et al.*, 2007). In contrast, acylation of Wnt3A at S209 was reported to be necessary for binding to Evi/Wls (Coombs *et al.*, 2010) confirming the requirement of palmitoleoyl modification of Wnt proteins for proper secretion.

The interaction of Evi/Wls with multiple Wnt proteins led to the "pan-Wnt secretion model" of Evi/Wls implicating a role in canonical β-catenin-dependent as well as noncanonical processes. Consistently, *S. mediterranea* Evi (Smed-Evi) has been shown to be required for Smed-β-catenin 1-dependent processes, such as posterior identity in regenerating planaria, but also for β-catenin-independent processes, such as signaling through the noncanonical Wnt protein Smed-Wnt5 during nervous system regeneration (Adell *et al.*, 2009). In contrast to a pan-Wnt requirement of Evi/Wls in *Drosophila* and *C. elegans*, *X. laevis* Wntless (XWls) has been suggested to specifically regulate the secretion of XWnt4 during eye development (Kim *et al.*, 2009). In this study, XWls seems not to be required for XWnt1, XWnt3A, XWnt5A, XWnt8, and XWnt11 secretion during frog development (Kim *et al.*, 2009). Even though it has been found to be associated with Wnt5A in neuronal stem cells (Fu *et al.*, 2009) Evi/Wls has so far not been linked to the secretion of noncanonical Wnt proteins which remains to be investigated. To date, *Drosophila* WntD/Wnt8 and *T. castaneum* WntD/8 (Tc-WntD/8) are the only Wnt proteins known to be secreted in an Evi/Wls-independent manner (Bolognesi *et al.*, 2008; Ching *et al.*, 2008), most likely as they lack the conserved palmitoleated serine residue shown to be required for Evi/Wls binding (Coombs *et al.*, 2010). Interestingly, the only trafficking protein known to be involved in WntD secretion is Rab1, the small GTPase necessary for trafficking from the

ER to the cis-Golgi (Ching *et al.*., 2008). Rab1 has so far not been linked to lipid modified Wnt proteins, pointing toward an alternative secretion route of WntD (Fig. 2.4).

Interestingly, Evi/Wls itself has been found to be released in exosome like vesicles in *Drosophila* larval neuromuscular junction (NMJ) and Schneider cells (Korkut *et al.*, 2009). It is endocytosed in a clathrin-dependent process and recycled back to the Golgi network through retromer, a highly conserved multiprotein sorting complex composed of the sorting nexins SNX1/2 and SNX5/6 and several subunits of the vacuolar protein sorting (Vps) protein family (Vps26, Vps29, and Vps35) (reviewed in Bonifacino and Hurley, 2008; Collins, 2008). Retromer has been found to be required for endosome to Golgi recycling of Wnt proteins (Coudreuse *et al.*, 2006; Prasad and Clark, 2006). Evi/Wls is dependent on retromer for endosomal sorting to the trans-Golgi network (TGN) (Belenkaya *et al.*, 2008; Franch-Marro *et al.*, 2008b; Pan *et al.*, 2008; Port *et al.*, 2008; Yang *et al.*, 2008) (Fig. 2.4). Retrieval of Evi/Wls to the Golgi network is thought to be mediated by interaction with the retromer subunit Vps35 (Franch-Marro *et al.*, 2008b). In the absence of retromer, Evi/Wls is targeted for lysosomal degradation leading to retention of Wnts within the Golgi network of Wnt-secreting cells (Belenkaya *et al.*, 2008; Franch-Marro *et al.*, 2008b; Pan *et al.*, 2008; Port *et al.*, 2008; Yang *et al.*, 2008). Mutations affecting the retromer subunit Vps35 result in Evi/Wls accumulation in multivesicular bodies (MVB) targeting Evi/Wls for degradation (Pan *et al.*, 2008; Yang *et al.*, 2008). Interestingly, MIG-14 overexpression in EGL-20 producing cells in Vps35 mutants rescues the neuronal migration defects indicating that MIG-14 is limited in the absence of retromer (Yang *et al.*, 2008).

The J-domain protein RME-8 has been found to interact with the retromer component SNX1 and to be required for proper endosomal sorting of Evi/Wls (Shi *et al.*, 2009). Phenotypes of *rme-8* and *snx-1* mutants phenocopy a loss of other retromer components indicating a requirement of those components for endosomal sorting of internalized Evi/Wls. In addition, MTM-6 and MTM-9, members of the myotubularin lipid phosphatase family, are required in Wnt-producing cells in *C. elegans* and *Drosophila* as part of the Evi/Wls recycling pathway (Silhankova *et al.*, 2010). MTMs are central regulators of endosomal trafficking through dephosphorylation of phosphatidylinositol-3-phosphate (PI3P) at endosomal membrane. The MTM-6/9 myotubularin complex regulates PI3P levels by controlling the recruitment of the retromer subunit SNX-3 which binds PI3P. In MTM-6 mutants, SNX-3 accumulates at endosomal membranes as a consequence of excessive PI3P resulting in a reduced efficiency of retromer-mediated Evi/Wls recycling (Silhankova *et al.*, 2010). These findings suggest that a balance between PI3P synthesis and degradation is required for efficient recycling of Evi/Wls and Wnt signaling. Recycling of Evi/Wls has also been reported to be essential during central nervous system (CNS) development in rodents where GPR177 was identified as a μ-opioid receptor (MOR) interacting protein (Jin *et al.*, 2010a; Reyes *et al.*, 2010). In rodent brain, morphine treatment induced an increased MOR/GPR177 complex formation at the PM due to inefficient internalization of MORs. This results in reduced retromer-mediated retrieval of GPR177 and subsequent inhibition of Wnt secretion in the CNS (Jin *et al.*, 2010a).

Prior to endosomal sorting via retromer, Evi/Wls has been found to be internalized from the PM via clathrin-mediated endocytosis which involves the small GTPase Rab5 (Shi *et al.*, 2009), DPY-23, the *C. elegans* μ subunit of adaptor protein 2 (AP2) (Pan *et al.*, 2008; Yang *et al.*, 2008) and Shibire (Shi), the *Drosophila* Dynamin (Belenkaya *et al.*, 2008) (Fig. 2.4). In the absence of these proteins, Evi/Wls accumulates at the PM leading to impaired Wnt secretion and defective Wnt signaling. Recently, there was evidence that expression and activity of Evi/Wls itself is tightly regulated and under the control of Wnt signaling. Evi/Wls has been found to be a direct target of miR-8, a conserved negative regulator of Wnt/Wg signaling (Kennell *et al.*, 2008). Further, GPR177 has been shown to be a transcriptional target of Wnt signaling, activated by β-catenin/ TCF/LEF transcription to promote Wnt secretion in a positive feedback loop (Fu *et al.*, 2009). Depending on this feedback regulation, a differential compartmentalization of Evi/Wls protein has been reported for Wnt-secreting versus Wnt-receiving cells (Fu *et al.*, 2009) indicating additional functions of Evi/Wls in receiving cells. In accordance with this, the Evi/Wls protein is ubiquitously expressed in the developing *Drosophila* larval wing disc, but is upregulated and stabilized by Wnt proteins in Wg-producing cells along the dorsal–ventral boundary (Port *et al.*, 2008). Likewise, Evi/Wls has been detected at the postsynaptic membrane of Wnt-target cells in synaptic Wnt signaling (Korkut *et al.*, 2009), implicating that Evi/Wls might play a role in signal receiving cells in addition to its function in Wnt/Wg secretion, both of which need to be tightly regulated.

One such regulation is lipidation of membranes and acidification of secretory vesicles (reviewed in Niehrs and Boutros, 2010). Lipidation by Porc is necessary for membrane-anchoring of Wnts and subsequent sorting into lipid raft microdomains, a process which might be facilitated by Evi/Wls within the Golgi as GPCR-like proteins have been found to be associated with these microdomains (Insel *et al.*, 2005). It has been proposed that Evi/Wls binds to Wnt proteins via a lipid-binding barrel and that acidification of secretory vesicles is necessary for Wnt3A release from Evi/Wls (Coombs *et al.*, 2010). In particular, it has been shown that upon inhibition of acidification by blocking V-ATPase function the Wnt3A-Evi/Wls complex accumulates inside cells and at the PM resulting in impaired Wnt3A secretion and Evi/Wls recycling (Coombs *et al.*, 2010). However, it remains to be investigated how Wnt proteins are normally released from Evi/Wls and whether vesicular acidification contributes to this process.

3.2.2. Spreading of Wnt proteins and gradient formation

Once Wnt proteins reach the PM, they associate with LPPs that serve as extracellular transporters to achieve long-range signaling (Katanaev *et al.*, 2008; Neumann *et al.*, 2009; Panakova *et al.*, 2005). Lipoproteins are composed of an outer layer of polar phospholipids, cholesterol, and embedded proteins called apolipoproteins and an inner layer of neutral lipids. Wg has been found to be associated with lipophorin, the scaffold apoprotein of LPPs in *Drosophila*, most likely through insertion of its lipid moieties into the outer phospholipids layer of LPPs (Panakova *et al.*, 2005) (Fig. 2.4). Likewise, Wnt3A seems to be associated with high density lipoprotein particles (HDLs). A palmitate-deficient Wnt3A was released in a HDL-independent manner, confirming a dual role of palmitoylation in membrane anchoring and lipoprotein binding (Neumann *et al.*, 2009).

Spreading of Wnts throughout the extracellular matrix is promoted by heparan sulfate proteoglycans (HSPGs) consisting of a core protein to which heparan sulfate (HS) glycosaminoglycan (GAG) chains are attached. HSPGs are involved in various developmental processes by regulating the activities of numerous growth factors, cell adhesion molecules, and lipoproteins (reviewed in Yan and Lin, 2009).

The involvement of HSPGs in Wg signaling was elucidated by the identification of *Drosophila* mutants encoding HS GAG biosynthesis enzymes, such as sugarless (sgl), an UDP-glucose 6-dehydrogenase (NDST) (Hacker *et al.*, 1997), and sulfateless (sfl), an *N*-deacetylase/*N*-sulfotransferase (Lin and Perrimon, 1999). Purified HSPGs have been found to stabilize the activity of purified Wnt proteins in solution by preventing aggregating in aqueous environments (Fuerer *et al.*, 2009).

Sulfated (Sulf1), a *Drosophila* member of the HS 6-*O*-endosulfatase class of HS modifying enzymes, has recently been identified as a negative regulator of Wg gradient formation (Kleinschmit *et al.*, 2011; You *et al.*, 2011). Sulf1 is part of a regulatory feedback loop specifically upregulated in the wing disc upon Wg signaling and influences the stability and distribution of Wg at the cell surface (Kleinschmit *et al.*, 2011; You *et al.*, 2011). Overexpression of HS modifying enzymes, such as Sulf1, has been linked to compromised Wg activity *in vivo* as 6-*O*-endosulfatases remove 6-*O*-sulfate groups from internal sulfated domains of extracellular HS thereby destabilizing Wg at the cell surface (Kamimura *et al.*, 2010; Kleinschmit *et al.*, 2010). Likewise, extracellular QSulf1, an avian heparan-specific *N*-acetyl glucosamine sulfatase, regulates Wnt signaling through desulfation of cell surface HSPGs (Dhoot *et al.*, 2001). Wnt3A was shown to bind to a specific disaccharide structures of GAG chains with a specific sulfation pattern (4-*O*-sulfate, 6-*O*-sulfate) which concentrates Wnt3A on the cell surface of Wnt-secreting cells (Nadanaka *et al.*, 2008). The enzyme responsible for sulfation of GAG chains, chondroitin 4-*O*-sulfotransferase-1 (C4ST-1) has been demonstrated to be a downstream target of Wnt signaling. Its expression is negatively regulated in Wnt-secreting cells in an autocrine manner triggering diffusion of Wnt molecules from these cells (Nadanaka *et al.*, 2010).

Drosophila members of the glypican family of HSPGs, Dally and Dally-like (Dlp) are anchored to the cell surface through a GPI-modification and can bind a variety of extracellular ligands, including Wg. They have been proposed to control Wg gradient formation by a restricted diffusion mechanism (reviewed in Wu *et al.*, 2010). While Dally has been identified as a positive regulator of Wg gradient formation by acting as a classical coreceptor, a biphasic activity has been proposed for Dlp in the regulation of Wg movement and stability (Yan *et al.*, 2009). In *Drosophila* wing disc, Dlp seems to capture Wg at the cell surface of secreting cells to pass it on to neighboring cells to promote long-range signaling while inhibiting Wg activity near the Wg source (Franch-Marro *et al.*, 2005; Han *et al.*, 2005). The GPI-anchor of Dlp is required for Wg internalization from the apical surface and subsequent transport to the basolateral membrane, hence, Dlp is required for long-range signaling of Wg by promoting transcytosis (Gallet *et al.*, 2008) (Fig. 2.4). The core protein of Dlp has been shown to exhibit a similar biphasic activity as wild-type Dlp by directly interacting with Wg, while the attached HS chains can enhance Dlp's affinity for Wg binding (Yan *et al.*, 2009). In addition to Dally and Dlp, Reggie-1/Flottilin-2, a major component of noncaveolin microdomains has been shown to play an important role in Wg-secreting cells by promoting secretion and long-range diffusion of Wg (Katanaev *et al.*, 2008).

In summary, an increasing amount of data suggests that not only Wnt secretion is tightly regulated but also the regulation of Wnt/Wg protein stability, extracellular diffusion and gradient formation. A large variety of cofactors are involved in shaping the Wnt/Wg gradient and fine-tuning the range and robustness of Wnt/Wg signaling during development. However, in vertebrates, the situation is even more complex and the precise regulatory mechanisms remain elusive.

4. Regulation of Wnt Signaling Pathways at the Receptor Level

As many other signaling pathways, Wnt signaling is a tightly regulated process that is mainly mediated by alterations of phosphorylation states of numerous core components (reviewed in Kikuchi *et al.*, 2006), such as LRP6/Arr (reviewed in Niehrs and Shen, 2010) but also through controlled intracellular transport of Wnt ligands. In addition to, controlled trafficking and defined diffusion of Wnt proteins through the ECS (reviewed in Port and Basler, 2010), Wnt-associated chaperones have been identified to be required for trafficking of Wnt receptors. For instance, Boca, an evolutionary conserved ER protein, has been implicated in trafficking of LDL receptors, such as LRP/Arr in *Drosophila* (Culi and Mann, 2003).

As Fz receptors are required for both, Wnt/β-catenin signaling and the establishment of PCP, the maturation of these receptors, signaling strength and activity need to be tightly regulated. A specific ER-retention mechanism mediated by the ER protein Shisa has been identified that controls the amount of mature Fz receptor at the PM, a process which is required for vertebrate head formation during gastrulation (Yamamoto *et al.*, 2005).

In past years, numerous soluble antagonists of Wnt signaling, such as secreted Fz related proteins (sFRPs) (reviewed in Bovolenta *et al.*, 2008), Wnt inhibitory factor (WIF) or the *Drosophila* ortholog Shifted (Shf) (Gorfinkiel *et al.*, 2005) and Dickkopf homolog-1(DKK-1) (reviewed in Niehrs, 2006) have been identified that either antagonize Wnt signaling by blocking binding of Wnt ligands to the cell surface receptors or by direct interaction with the receptor complex (reviewed in MacDonald *et al.*, 2009). Shf, for instance, has been found to interact with HSPGs thereby preventing Wnt/Wg binding (Gorfinkiel *et al.*, 2005).

In addition, Wnt signaling is regulated by a complex network of intracellular feedback loops, such as expression of the canonical Wnt target gene naked cuticle (nkd), which blocks activity of Dvl/Dsh, or by dual function of the kinases CK1 and GSK3, which act as antagonists at the level of the β-catenin destruction complex, but as agonists at the PM by sequentially phosphorylating LRP6.

Recent studies have also linked microRNAs (miRs) to the responsiveness of cells to Wnt signaling molecules (reviewed in Huang *et al.*, 2010). In particular, miR-135a/b and miR-315 have been reported to upregulate Wnt signaling by suppressing expression of negative regulators, such as APC or Axin, respectively (Nagel *et al.*, 2008; Silver *et al.*, 2007). In contrast, miR-8, the *Drosophila* homolog of the vertebrate miR-200 family, downregulates signaling by targeting positive regulators of Wnt/Wg signaling, such as TCF and Evi/Wls (Kennell *et al.*, 2008; Korpal *et al.*, 2008; Saydam *et al.*, 2009). Likewise, zebrafish lymphoid enhancer-binding factor 1 (leF1) was found to be a conserved target of miR-203 which inhibits lef1-mediated fin regeneration (Thatcher *et al.*, 2008).

In addition to a complex miR-mediated regulation of Wnt signaling components, there is increasing evidence that endocytosis of Wnt receptors is required for selective activation of downstream signaling events (reviewed in Kikuchi *et al.*, 2009). It has been proposed that caveolin-mediated endocytosis is mainly involved in β-catenin-dependent signaling while clathrin-mediated endocytosis has been linked to both, β-catenin-dependent and -independent signaling (reviewed in Gagliardi *et al.*, 2008; Kikuchi *et al.*, 2009). More recently, it has also been proposed that vacuolar ATPase (V-ATPase)-mediated acidification of endosomal vesicles is involved in

Wnt signaling (reviewed in Niehrs and Boutros, 2010). Vesicular acidification was found to be required for Wnt secretion, most likely through the regulation of Evi/Wls recycling (Coombs *et al.*, 2010). Further, it was shown that V-ATPases are required at receptor level for both, canonical and noncanonical Wnt/Wg signaling (Buechling *et al.*, 2010; Cruciat *et al.*, 2010; Hermle *et al.*, 2010). In Drosophila, the V-ATPase accessory protein dPRR (Drosophila homolog of ProRenin receptor) is required for PM localization of both Fz receptors, thereby regulating canonical Wnt/Wg and PCP signaling (Buechling *et al.*, 2010). Similarly, vertebrate PRR seems to be a component of the Wnt receptor complex by interacting with Fz and LRP6 and is required for anterior–posterior patterning of the CNS in Xenopus (Cruciat *et al.*, 2010). As the phosphorylation of LRP6 seems to be dependent on V-ATPase function, it is tempting to speculate that PRR functions as an adapter molecule between Fz/LRP6 and V-ATPases. However, similar to its role in Wnt secretion and Evi/Wls recycling, the detailed mechanism of V-ATPase function on trafficking of Wnt receptor complexes remains elusive.

5. Conclusions and Perspective

Even though much progress has been made in understanding the detailed mechanistic and physiological basis of Wnt signaling, and the downstream cascades triggered by Wnt proteins, it has only emerged recently, that there are multiple additional layers of regulation, such as posttranslational modifications of Wnts, the tightly regulated secretion and surface trafficking of Wnts, and acidification of Wnt receptor containing vesicles that regulate endocytosis and in turn, signaling in receiving cells.

To date, several components have been identified that regulate Wnt protein activity and secretion. However, the current model of Wnt trafficking through the secretory pathway is probably incomplete as multiple steps, for instance, the transport of Wnts from the ER to the Golgi network, are not yet clear. The following questions arise from recently published reports: Are there Wnt specific transport vesicles? If so, how are they regulated? Are the posttranslational modifications identified in Wnt proteins required for membrane tethering in these transport vesicles and binding to trafficking proteins? Are there Wnt-specific cargo receptors required for packaging into these vesicles and transport to the PM? As many of the recently identified components involved in Wnt binding and stability, such as Dally or Reggie-1/Flottilin-2, are not Wnt specific, it remains an open question whether there are any other Wnt specific trafficking components or cargo receptors except Evi/Wls.

In past years, Evi/Wls has been studied intensively in different model organisms. Nevertheless, it still remains elusive how Wnt proteins are released from Evi/Wls and whether vesicular acidification contributes to Wnt-Evi/Wls complex dissociation as lately suggested. As Evi/Wls was found in exosomal structures in Drosophila, it remains an open question how Evi/Wls itself is transported through the early secretory pathway. To date, no Evi/Wls trafficking factor is known as past studies on Evi/Wls were mainly focused on its recycling pathway. However, if Evi/Wls is secreted on exosomes, do these vesicles also contain Wnt proteins? It will be interesting to study how signaling activity of these vesicles is regulated. This would open up the possibility to interfere with Wnt signaling by blocking Evi/Wls function, applicable to several types of cancer with enhanced Wnt secretion.

Valuable information to restrain Wnt signaling in receiving cells was lately added through the discovery that V-ATPases play a crucial role for signaling downstream of the Wnt receptor complex most likely by regulating Wnt receptor endocytosis. However, as for the involvement of vesicular acidification in Wnt secretion, it requires further studies to unravel the detailed mechanism of this regulation by V-ATPases that are required for both branches of Wnt signaling, canonical and noncanonical. Further studies on V-ATPase-mediated Wnt signaling might allow to use V-ATPases as potential drug targets to restrict Wnt signaling in receiving cells, especially in cells with aberrant Wnt activity, such as colon cancer cells.

Taken together, we believe that novel Wnt components will continue to be identified that will help to address the above questions for a more comprehensive view of the regulatory mechanisms underlying different Wnt signaling cascades, Wnt secretion and the formation of Wnt gradients in developing tissues. This information will be useful to understand the nature of various human diseases associated with aberrant Wnt signaling.

Even though most of the cancer causing mutations in Wnt signaling affect components downstream of the receptor complex and are thought to activate Wnt signaling independent from Wnt secretion, Porc and more recently, Evi/Wls have also been linked to cancer. Porc, for instance, has been shown to be overexpressed in several types of lung cancer, closely linked to tumor growth. Notably, recent small molecule screens in colon cancer cell lines with mutations of downstream signaling components, such as APC, identified potent Porc inhibitors, indicating that modulation of upstream signaling events in these cells might open up new possibilities for therapeutic intervention. However, the role of Evi/Wls and in turn, Wnt secretion, remains to be investigated in colon cancer cells. More detailed information will help to develop new therapeutic strategies to treat human diseases associated with aberrant Wnt signaling.

ACKNOWLEDGMENTS

We would like to thank Varun Chaudhary, Julia Gross, and Christina Falschlehner for critical comments and useful suggestions. We apologize to all colleagues whose work was not cited due to length restrictions. This work was supported by the Deutsche Forschungsgemeinschaft (FOR1036).

REFERENCES

Aberle, H., Bauer, A., Stappert, J., Kispert, A., and Kemler, R. (1997). beta-Catenin is a target for the ubiquitin-proteasome pathway. *EMBO J.* **16**(13), 3797–3804.

Adamson, M. C., Dennis, C., Delaney, S., Christiansen, J., Monkley, S., Kozak, C. A., and Wainwright, B. (1994). Isolation and genetic mapping of two novel members of the murine Wnt gene family, Wnt11 and Wnt12, and the mapping of Wnt5a and Wnt7a. *Genomics* **24**(1), 9–13.

Adell, T., Salo, E., Boutros, M., and Bartscherer, K. (2009). Smed-Evi/Wntless is required for beta-catenin-dependent and -independent processes during planarian regeneration. *Development* **136**(6), 905–910.

Amit, S., Hatzubai, A., Birman, Y., Andersen, J. S., Ben-Shushan, E., Mann, M., Ben-Neriah, Y., and Alkalay, I. (2002). Axin-mediated CKI phosphorylation of beta-catenin at Ser 45: A molecular switch for the Wnt pathway. *Genes Dev.* **16**(9), 1066–1076.

Angers, S., and Moon, R. T. (2009). Proximal events in Wnt signal transduction. *Nat. Rev. Mol. Cell Biol.* **10**(7), 468–477.

Angers, S., Thorpe, C. J., Biechele, T. L., Goldenberg, S. J., Zheng, N., MacCoss, M. J., and Moon, R. T. (2006). The KLHL12-Cullin-3 ubiquitin ligase negatively regulates the Wnt-beta-catenin pathway by targeting Dishevelled for degradation. *Nat. Cell Biol.* **8**(4), 348–357.

Axelrod, J. D. (2009). Progress and challenges in understanding planar cell polarity signaling. *Semin. Cell Dev. Biol.* **20**(8), 964–971.

Banziger, C., Soldini, D., Schutt, C., Zipperlen, P., Hausmann, G., and Basler, K. (2006). Wntless, a conserved membrane protein dedicated to the secretion of Wnt proteins from signaling cells. *Cell* **125**(3), 509–522.

Bartscherer, K., and Boutros, M. (2008). Regulation of Wnt protein secretion and its role in gradient formation. *EMBO Rep.* **9**(10), 977–982.

Bartscherer, K., Pelte, N., Ingelfinger, D., and Boutros, M. (2006). Secretion of Wnt ligands requires Evi, a conserved transmembrane protein. *Cell* **125**(3), 523–533.

Belenkaya, T. Y., Han, C., Standley, H. J., Lin, X., Houston, D. W., and Heasman, J. (2002). pygopus Encodes a nuclear protein essential for wingless/Wnt signaling. *Development* **129**(17), 4089–4101.

Belenkaya, T. Y., Wu, Y., Tang, X., Zhou, B., Cheng, L., Sharma, Y. V., Yan, D., Selva, E. M., and Lin, X. (2008). The retromer complex influences Wnt secretion by recycling wntless from endosomes to the trans-Golgi network. *Dev. Cell* **14**(1), 120–131.

Biddick, R., and Young, E. T. (2005). Yeast mediator and its role in transcriptional regulation. *C. R. Biol.* **328**(9), 773–782.

Bjorklund, S., and Gustafsson, C. M. (2005). The yeast Mediator complex and its regulation. *Trends Biochem. Sci.* **30**(5), 240–244.

Bolognesi, R., Farzana, L., Fischer, T. D., and Brown, S. J. (2008). Multiple Wnt genes are required for segmentation in the short-germ embryo of Tribolium castaneum. *Curr. Biol.* **18**(20), 1624–1629.

Bonifacino, J. S., and Hurley, J. H. (2008). Retromer. *Curr. Opin. Cell Biol.* **20**(4), 427–436.

Boutros, M., Paricio, N., Strutt, D. I., and Mlodzik, M. (1998). Dishevelled activates JNK and discriminates between JNK pathways in planar polarity and wingless signaling. *Cell* **94**(1), 109–118.
Bovolenta, P., Esteve, P., Ruiz, J. M., Cisneros, E., and Lopez-Rios, J. (2008). Beyond Wnt inhibition: New functions of secreted Frizzled-related proteins in development and disease. *J. Cell Sci.* **121**(Pt 6), 737–746.
Buechling, T., Bartscherer, K., Ohkawara, B., Chaudhary, V., Spirohn, K., Niehrs, C., and Boutros, M. (2010). Wnt/Frizzled signaling requires dPRR, the Drosophila homolog of the prorenin receptor. *Curr. Biol.* **20**(14), 1263–1268.
Cadigan, K. M., and Peifer, M. (2009). Wnt signaling from development to disease: Insights from model systems. *Cold Spring Harb. Perspect Biol.* **1**(2), a002881.
Caricasole, A., Ferraro, T., Rimland, J. M., and Terstappen, G. C. (2002). Molecular cloning and initial characterization of the MG61/PORC gene, the human homologue of the Drosophila segment polarity gene Porcupine. *Gene* **288**(1–2), 147–157.
Carpenter, A. C., Rao, S., Wells, J. M., Campbell, K., and Lang, R. A. (2010). Generation of mice with a conditional null allele for Wntless. *Genesis* **48**(9), 554–558.
Carrera, I., Janody, F., Leeds, N., Duveau, F., and Treisman, J. E. (2008). Pygopus activates Wingless target gene transcription through the mediator complex subunits Med12 and Med13. *Proc. Natl. Acad. Sci. USA* **105**(18), 6644–6649.
Charollais, J., and Van Der Goot, F. G. (2009). Palmitoylation of membrane proteins (Review). *Mol. Membr. Biol.* **26**(1), 55–66.
Chen, G., Fernandez, J., Mische, S., and Courey, A. J. (1999). A functional interaction between the histone deacetylase Rpd3 and the corepressor groucho in Drosophila development. *Genes Dev.* **13**(17), 2218–2230.
Chen, W. S., Antic, D., Matis, M., Logan, C. Y., Povelones, M., Anderson, G. A., Nusse, R., and Axelrod, J. D. (2008). Asymmetric homotypic interactions of the atypical cadherin flamingo mediate intercellular polarity signaling. *Cell* **133**(6), 1093–1105.
Ching, W., Hang, H. C., and Nusse, R. (2008). Lipid-independent secretion of a Drosophila Wnt protein. *J. Biol. Chem.* **283**(25), 17092–17098.
Chinnadurai, G. (2002). CtBP, an unconventional transcriptional corepressor in development and oncogenesis. *Mol. Cell* **9**(2), 213–224.
Collins, B. M. (2008). The structure and function of the retromer protein complex. *Traffic* **9**(11), 1811–1822.
Conaway, J. W., Florens, L., Sato, S., Tomomori-Sato, C., Parmely, T. J., Yao, T., Swanson, S. K., Banks, C. A., Washburn, M. P., and Conaway, R. C. (2005). The mammalian Mediator complex. *FEBS Lett.* **579**(4), 904–908.
Coombs, G. S., Yu, J., Canning, C. A., Veltri, C. A., Covey, T. M., Cheong, J. K., Utomo, V., Banerjee, N., Zhang, Z. H., Jadulco, R. C., Concepcion, G. P., Bugni, T. S., *et al.* (2010). WLS-dependent secretion of WNT3A requires Ser209 acylation and vacuolar acidification. *J. Cell Sci.* **123**(Pt 19), 3357–3367.
Coudreuse, D. Y., Roel, G., Betist, M. C., Destree, O., and Korswagen, H. C. (2006). Wnt gradient formation requires retromer function in Wnt-producing cells. *Science* **312**(5775), 921–924.
Couso, J. P., and Martinez Arias, A. (1994). Notch is required for wingless signaling in the epidermis of Drosophila. *Cell* **79**(2), 259–272.
Croce, J. C., and McClay, D. R. (2008). Evolution of the Wnt pathways. *Methods Mol. Biol.* **469,** 3–18.
Cruciat, C. M., Ohkawara, B., Acebron, S. P., Karaulanov, E., Reinhard, C., Ingelfinger, D., Boutros, M., and Niehrs, C. (2010). Requirement of prorenin receptor and vacuolar H+−ATPase-mediated acidification for Wnt signaling. *Science* **327**(5964), 459–463.
Culi, J., and Mann, R. S. (2003). Boca, an endoplasmic reticulum protein required for wingless signaling and trafficking of LDL receptor family members in Drosophila. *Cell* **112**(3), 343–354.

Davidson, G., Wu, W., Shen, J., Bilic, J., Fenger, U., Stannek, P., Glinka, A., and Niehrs, C. (2005). Casein kinase 1 gamma couples Wnt receptor activation to cytoplasmic signal transduction. *Nature* **438,** 867–872.

Davidson, G., and Niehrs, C. (2010). Emerging links between CDK cell cycle regulators and Wnt signaling. *Trends Cell Biol.* **20**(8), 453–460.

Davidson, G., Shen, J., Huang, Y. L., Su, Y., Karaulanov, E., Bartscherer, K., Hassler, C., Stannek, P., Boutros, M., and Niehrs, C. (2009). Cell cycle control of wnt receptor activation. *Dev. Cell* **17**(6), 788–799.

Dhoot, G. K., Gustafsson, M. K., Ai, X., Sun, W., Standiford, D. M., and Emerson, C. P., Jr. (2001). Regulation of Wnt signaling and embryo patterning by an extracellular sulfatase. *Science* **293**(5535), 1663–1666.

Dissanayake, S. K., Wade, M., Johnson, C. E., O'Connell, M. P., Leotlela, P. D., French, A. D., Shah, K. V., Hewitt, K. J., Rosenthal, D. T., Indig, F. E., Jiang, Y., Nickoloff, B. J., *et al.* (2007). The Wnt5A/protein kinase C pathway mediates motility in melanoma cells via the inhibition of metastasis suppressors and initiation of an epithelial to mesenchymal transition. *J. Biol. Chem.* **282**(23), 17259–17271.

Doubravska, L., Krausova, M., Gradl, D., Vojtechova, M., Tumova, L., Lukas, J., Valenta, T., Pospichalova, V., Fafilek, B., Plachy, J., Sebesta, O., and Korinek, V. (2011). Fatty acid modification of Wnt1 and Wnt3a at serine is prerequisite for lipidation at cysteine and is essential for Wnt signalling. *Cell. Signal.* **23**(5), 837–848.

Du, S. J., Purcell, S. M., Christian, J. L., McGrew, L. L., and Moon, R. T. (1995). Identification of distinct classes and functional domains of Wnts through expression of wild-type and chimeric proteins in Xenopus embryos. *Mol. Cell. Biol.* **15**(5), 2625–2634.

Egger-Adam, D., and Katanaev, V. L. (2009). The trimeric G protein Go inflicts a double impact on axin in the Wnt/frizzled signaling pathway. *Dev. Dyn.* **239**(1), 168–183.

Feiguin, F., Hannus, M., Mlodzik, M., and Eaton, S. (2001). The ankyrin repeat protein Diego mediates Frizzled-dependent planar polarization. *Dev. Cell* **1**(1), 93–101.

Firestein, R., Bass, A. J., Kim, S. Y., Dunn, I. F., Silver, S. J., Guney, I., Freed, E., Ligon, A. H., Vena, N., Ogino, S., Chheda, M. G., Tamayo, P., *et al.* (2008). CDK8 is a colorectal cancer oncogene that regulates beta-catenin activity. *Nature* **455**(7212), 547–551.

Franch-Marro, X., Marchand, O., Piddini, E., Ricardo, S., Alexandre, C., and Vincent, J. P. (2005). Glypicans shunt the Wingless signal between local signalling and further transport. *Development* **132**(4), 659–666.

Franch-Marro, X., Wendler, F., Griffith, J., Maurice, M. M., and Vincent, J. P. (2008a). In vivo role of lipid adducts on Wingless. *J. Cell Sci.* **121**(Pt 10), 1587–1592.

Franch-Marro, X., Wendler, F., Guidato, S., Griffith, J., Baena-Lopez, A., Itasaki, N., Maurice, M. M., and Vincent, J. P. (2008b). Wingless secretion requires endosome-to-Golgi retrieval of Wntless/Evi/Sprinter by the retromer complex. *Nat. Cell Biol.* **10**(2), 170–177.

Fu, J., Jiang, M., Mirando, A. J., Yu, H. M., and Hsu, W. (2009). Reciprocal regulation of Wnt and Gpr177/mouse Wntless is required for embryonic axis formation. *Proc. Natl. Acad. Sci. USA* **106**(44), 18598–18603.

Fu, J., Ivy Yu, H. M., Maruyama, T., Mirando, A. J., and Hsu, W. (2011). Gpr177/mouse Wntless is essential for Wnt-mediated craniofacial and brain development. *Dev. Dyn.* **240**(2), 365–371.

Fuerer, C., Habib, S. J., and Nusse, R. (2009). A study on the interactions between heparan sulfate proteoglycans and Wnt proteins. *Dev. Dyn.* **239**(1), 184–190.

Gagliardi, M., Piddini, E., and Vincent, J. P. (2008). Endocytosis: A positive or a negative influence on Wnt signalling? *Traffic* **9**(1), 1–9.

Gallet, A., Staccini-Lavenant, L., and Therond, P. P. (2008). Cellular trafficking of the glypican Dally-like is required for full-strength Hedgehog signaling and wingless transcytosis. *Dev. Cell* **14**(5), 712–725.

Galli, L. M., Barnes, T. L., Secrest, S. S., Kadowaki, T., and Burrus, L. W. (2007). Porcupine-mediated lipid-modification regulates the activity and distribution of Wnt proteins in the chick neural tube. *Development* **134**(18), 3339–3348.

Ganguly, A., Jiang, J., and Ip, Y. T. (2005). Drosophila WntD is a target and an inhibitor of the Dorsal/Twist/Snail network in the gastrulating embryo. *Development* **132**(15), 3419–3429.

Gao, C., and Chen, Y. G. (2010). Dishevelled: The hub of Wnt signaling. *Cell. Signal.* **22**(5), 717–727.

Gavin, B. J., McMahon, J. A., and McMahon, A. P. (1990). Expression of multiple novel Wnt1/int1related genes during fetal and adult mouse development. *Genes Dev.* **4**(12(B)), 2319–2332.

Goodman, R. M., Thombre, S., Firtina, Z., Gray, D., Betts, D., Roebuck, J., Spana, E. P., and Selva, E. M. (2006). Sprinter: A novel transmembrane protein required for Wg secretion and signaling. *Development* **133**(24), 4901–4911.

Gordon, M. D., Dionne, M. S., Schneider, D. S., and Nusse, R. (2005). WntD is a feedback inhibitor of Dorsal/NF-kappaB in Drosophila development and immunity. *Nature* **437** (7059), 746–749.

Gorfinkiel, N., Sierra, J., Callejo, A., Ibanez, C., and Guerrero, I. (2005). The Drosophila ortholog of the human Wnt inhibitor factor Shifted controls the diffusion of lipid-modified Hedgehog. *Dev. Cell* **8**(2), 241–253.

Griffin, C. T., Curtis, C. D., Davis, R. B., Muthukumar, V., and Magnuson, T. (2011). The chromatin-remodeling enzyme BRG1 modulates vascular Wnt signaling at two levels. *Proc. Natl. Acad. Sci. USA* **108**(6), 2282–2287.

Grigoryan, T., Wend, P., Klaus, A., and Birchmeier, W. (2008). Deciphering the function of canonical Wnt signals in development and disease: Conditional loss- and gain-of-function mutations of beta-catenin in mice. *Genes Dev.* **22**(17), 2308–2341.

Gu, B., Watanabe, K., and Dai, X. (2010). Epithelial stem cells: An epigenetic and Wnt-centric perspective. *J. Cell. Biochem.* **110**(6), 1279–1287.

Gubb, D., Green, C., Huen, D., Coulson, D., Johnson, G., Tree, D., Collier, S., and Roote, J. (1999). The balance between isoforms of the prickle LIM domain protein is critical for planar polarity in Drosophila imaginal discs. *Genes Dev.* **13**(17), 2315–2327.

Hacker, U., Lin, X., and Perrimon, N. (1997). The Drosophila sugarless gene modulates Wingless signaling and encodes an enzyme involved in polysaccharide biosynthesis. *Development* **124**(18), 3565–3573.

Han, C., Yan, D., Belenkaya, T. Y., and Lin, X. (2005). Drosophila glypicans Dally and Dally-like shape the extracellular Wingless morphogen gradient in the wing disc. *Development* **132**(4), 667–679.

He, X., Saint-Jeannet, J. P., Wang, Y., Nathans, J., Dawid, I., and Varmus, H. (1997). A member of the Frizzled protein family mediating axis induction by Wnt-5A. *Science* **275**(5306), 1652–1654.

Heisenberg, C. P., Tada, M., Rauch, G. J., Saude, L., Concha, M. L., Geisler, R., Stemple, D. L., Smith, J. C., and Wilson, S. W. (2000). Silberblick/Wnt11 mediates convergent extension movements during zebrafish gastrulation. *Nature* **405**(6782), 76–81.

Hermle, T., Saltukoglu, D., Grunewald, J., Walz, G., and Simons, M. (2010). Regulation of Frizzled-dependent planar polarity signaling by a V-ATPase subunit. *Curr. Biol.* **20**(14), 1269–1276.

Hikasa, H., and Sokol, S. Y. (2011). Phosphorylation of TCF proteins by homeodomain-interacting protein kinase 2. *J. Biol. Chem.* **286**(14), 12093–12100.

Hoppler, S., and Kavanagh, C. L. (2007). Wnt signalling: Variety at the core. *J. Cell Sci.* **120**(Pt 3), 385–393.

Huang, K., Zhang, J. X., Han, L., You, Y. P., Jiang, T., Pu, P. Y., and Kang, C. S. (2010). MicroRNA roles in beta-catenin pathway. *Mol. Cancer* **9,** 252.

Insel, P. A., Head, B. P., Ostrom, R. S., Patel, H. H., Swaney, J. S., Tang, C. M., and Roth, D. M. (2005). Caveolae and lipid rafts: G protein-coupled receptor signaling microdomains in cardiac myocytes. *Ann. NY. Acad. Sci.* **1047,** 166–172.

James, R. G., Biechele, T. L., Conrad, W. H., Camp, N. D., Fass, D. M., Major, M. B., Sommer, K., Yi, X., Roberts, B. S., Cleary, M. A., Arthur, W. T., MacCoss, M., *et al.* (2009). Bruton's tyrosine kinase revealed as a negative regulator of Wnt-beta-catenin signaling. *Sci. Signal.* **2**(72), ra25.

Jin, J., Kittanakom, S., Wong, V., Reyes, B. A., Van Bockstaele, E. J., Stagljar, I., Berrettini, W., and Levenson, R. (2010a). Interaction of the mu-opioid receptor with GPR177 (Wntless) inhibits Wnt secretion: Potential implications for opioid dependence. *BMC Neurosci.* **11,** 33.

Jin, J., Morse, M., Frey, C., Petko, J., and Levenson, R. (2010b). Expression of GPR177 (Wntless/Evi/Sprinter), a highly conserved Wnt-transport protein, in rat tissues, zebrafish embryos, and cultured human cells. *Dev. Dyn.* **239**(9), 2426–2434.

Kadowaki, T., Wilder, E., Klingensmith, J., Zachary, K., and Perrimon, N. (1996). The segment polarity gene porcupine encodes a putative multitransmembrane protein involved in Wingless processing. *Genes Dev.* **10**(24), 3116–3128.

Kamimura, K., Maeda, N., and Nakato, H. (2010). In vivo manipulation of heparan sulfate structure and its effect on Drosophila development. *Glycobiology* **21**(5), 607–618.

Katanaev, V. L., Ponzielli, R., Semeriva, M., and Tomlinson, A. (2005). Trimeric G protein-dependent frizzled signaling in Drosophila. *Cell* **120**(1), 111–122.

Katanaev, V. L., Solis, G. P., Hausmann, G., Buestorf, S., Katanayeva, N., Schrock, Y., Stuermer, C. A., and Basler, K. (2008). Reggie-1/fl otillin-2 promotes secretion of the long-range signalling forms of Wingless and Hedgehog in Drosophila. *EMBO J.* **27,** 509–521.

Kategaya, L. S., Changkakoty, B., Biechele, T., Conrad, W. H., Kaykas, A., Dasgupta, R., and Moon, R. T. (2009). Bili inhibits Wnt/beta-catenin signaling by regulating the recruitment of axin to LRP6. *PLoS One* **4**(7), e6129.

Kennell, J. A., Gerin, I., MacDougald, O. A., and Cadigan, K. M. (2008). The microRNA miR-8 is a conserved negative regulator of Wnt signaling. *Proc. Natl. Acad. Sci. USA* **105** (40), 15417–15422.

Kikuchi, A., Kishida, S., and Yamamoto, H. (2006). Regulation of Wnt signaling by protein-protein interaction and post-translational modifications. *Exp. Mol. Med.* **38**(1), 1–10.

Kikuchi, A., Yamamoto, H., and Sato, A. (2009). Selective activation mechanisms of Wnt signaling pathways. *Trends Cell Biol.* **19**(3), 119–129.

Kilian, B., Mansukoski, H., Barbosa, F. C., Ulrich, F., Tada, M., and Heisenberg, C. P. (2003). The role of Ppt/Wnt5 in regulating cell shape and movement during zebrafish gastrulation. *Mech. Dev.* **120**(4), 467–476.

Kim, H., Cheong, S. M., Ryu, J., Jung, H. J., Jho, E. H., and Han, J. K. (2009). Xenopus Wntless and the retromer complex cooperate to regulate XWnt4 secretion. *Mol. Cell. Biol.* **29**(8), 2118–2128.

Klaus, A., and Birchmeier, W. (2008). Wnt signalling and its impact on development and cancer. *Nat. Rev. Cancer* **8**(5), 387–398.

Kleinschmit, A., Koyama, T., Dejima, K., Hayashi, Y., Kamimura, K., and Nakato, H. (2010). Drosophila heparan sulfate 6-O endosulfatase regulates Wingless morphogen gradient formation. *Dev. Biol.* **345**(2), 204–214.

Kleinschmit, A., Koyama, T., Dejima, K., Hayashi, Y., Kamimura, K., and Nakato, H. (2011). Drosophila heparan sulfate 6-O endosulfatase regulates Wingless morphogen gradient formation. *Dev. Biol.* **345**(2), 204–214.

Kohn, A. D., and Moon, R. T. (2005). Wnt and calcium signaling: beta-Catenin-independent pathways. *Cell Calcium* **38**(3–4), 439–446.

Komekado, H., Yamamoto, H., Chiba, T., and Kikuchi, A. (2007). Glycosylation and palmitoylation of Wnt-3a are coupled to produce an active form of Wnt-3a. *Genes Cells* **12**(4), 521–534.

Korkut, C., Ataman, B., Ramachandran, P., Ashley, J., Barria, R., Gherbesi, N., and Budnik, V. (2009). Trans-synaptic transmission of vesicular Wnt signals through Evi/Wntless. *Cell* **139**(2), 393–404.

Korpal, M., Lee, E. S., Hu, G., and Kang, Y. (2008). The miR-200 family inhibits epithelial-mesenchymal transition and cancer cell migration by direct targeting of E-cadherin transcriptional repressors ZEB1 and ZEB2. *J. Biol. Chem.* **283**(22), 14910–14914.

Kramps, T., Peter, O., Brunner, E., Nellen, D., Froesch, B., Chatterjee, S., Murone, M., Zullig, S., and Basler, K. (2002). Wnt/wingless signaling requires BCL9/legless-mediated recruitment of pygopus to the nuclear beta-catenin-TCF complex. *Cell* **109**(1), 47–60.

Kuhl, M., Sheldahl, L. C., Park, M., Miller, J. R., and Moon, R. T. (2000). The Wnt/Ca2+ pathway: A new vertebrate Wnt signaling pathway takes shape. *Trends Genet.* **16**(7), 279–283.

Kurayoshi, M., Yamamoto, H., Izumi, S., and Kikuchi, A. (2007). Post-translational palmitoylation and glycosylation of Wnt-5a are necessary for its signalling. *Biochem. J.* **402**(3), 515–523.

Li, Y., Bavarva, J. H., Wang, Z., Guo, J., Qian, C., Thibodeau, S. N., Golemis, E. A., and Liu, W. (2011a). HEF1, a novel target of Wnt signaling, promotes colonic cell migration and cancer progression. *Oncogene* **30**(23), 2633–2643.

Li, Z., Nie, F., Wang, S., and Li, L. (2011b). From the Cover: Feature Article: Histone H4 Lys 20 monomethylation by histone methylase SET8 mediates Wnt target gene activation. *Proc. Natl. Acad. Sci. USA* **108**(8), 3116–3123.

Lin, X., and Perrimon, N. (1999). Dally cooperates with Drosophila Frizzled 2 to transduce Wingless signalling. *Nature* **400**(6741), 281–284.

Liu, C., Li, Y., Semenov, M., Han, C., Baeg, G. H., Tan, Y., Zhang, Z., Lin, X., and He, X. (2002). Control of beta-catenin phosphorylation/degradation by a dual-kinase mechanism. *Cell* **108**(6), 837–847.

Lorenowicz, M. J., and Korswagen, H. C. (2009). Sailing with the Wnt: Charting the Wnt processing and secretion route. *Exp. Cell Res.* **315**(16), 2683–2689.

Lucero, O. M., Dawson, D. W., Moon, R. T., and Chien, A. J. (2010). A re-evaluation of the "oncogenic" nature of Wnt/beta-catenin signaling in melanoma and other cancers. *Curr. Oncol. Rep.* **12**(5), 314–318.

Luo, W., Peterson, A., Garcia, B. A., Coombs, G., Kofahl, B., Heinrich, R., Shabanowitz, J., Hunt, D. F., Yost, H. J., and Virshup, D. M. (2007). Protein phosphatase 1 regulates assembly and function of the beta-catenin degradation complex. *EMBO J.* **26**(6), 1511–1521.

MacDonald, B. T., Tamai, K., and He, X. (2009). Wnt/beta-catenin signaling: Components, mechanisms, and diseases. *Dev. Cell* **17**(1), 9–26.

Major, M. B., Roberts, B. S., Berndt, J. D., Marine, S., Anastas, J., Chung, N., Ferrer, M., Yi, X., Stoick-Cooper, C. L., von Haller, P. D., Kategaya, L., Chien, A., *et al.* (2008). New regulators of Wnt/beta-catenin signaling revealed by integrative molecular screening. *Sci. Signal.* **1**(45), ra12.

Maloof, J. N., Whangbo, J., Harris, J. M., Jongeward, G. D., and Kenyon, C. (1999). A Wnt signaling pathway controls hox gene expression and neuroblast migration in C. elegans. *Development* **126**(1), 37–49.

Mason, J. O., Kitajewski, J., and Varmus, H. E. (1992). Mutational analysis of mouse Wnt-1 identifies two temperature-sensitive alleles and attributes of Wnt-1 protein essential for transformation of a mammary cell line. *Mol. Biol. Cell* **3**(5), 521–533.

Mosimann, C., Hausmann, G., and Basler, K. (2006). Parafibromin/Hyrax activates Wnt/Wg target gene transcription by direct association with beta-catenin/Armadillo. *Cell* **125** (2), 327–341.

Mosimann, C., Hausmann, G., and Basler, K. (2009). Beta-catenin hits chromatin: Regulation of Wnt target gene activation. *Nat. Rev. Mol. Cell Biol.* **10**(4), 276–286.

Nadanaka, S., Ishida, M., Ikegami, M., and Kitagawa, H. (2008). Chondroitin 4-O-sulfotransferase-1 modulates Wnt-3a signaling through control of E disaccharide expression of chondroitin sulfate. *J. Biol. Chem.* **283**(40), 27333–27343.

Nadanaka, S., Kinouchi, H., Taniguchi-Morita, K., Tamura, J. I., and Kitagawa, H. (2010). Down-regulation of chondroitin 4-O-sulfotransferase-1 by Wnt signaling triggers diffusion of Wnt-3a. *J. Biol. Chem.* **286,** 4199–4208.

Nagel, R., le Sage, C., Diosdado, B., van der Waal, M., Oude Vrielink, J. A., Bolijn, A., Meijer, G. A., and Agami, R. (2008). Regulation of the adenomatous polyposis coli gene by the miR-135 family in colorectal cancer. *Cancer Res.* **68**(14), 5795–5802.

Neumann, C., and Cohen, S. (1997). Morphogens and pattern formation. *Bioessays* **19**(8), 721–729.

Neumann, S., Coudreuse, D. Y., van der Westhuyzen, D. R., Eckhardt, E. R., Korswagen, H. C., Schmitz, G., and Sprong, H. (2009). Mammalian Wnt3a is released on lipoprotein particles. *Traffic* **10**(3), 334–343.

Niehrs, C. (2006). Function and biological roles of the Dickkopf family of Wnt modulators. *Oncogene* **25**(57), 7469–7481.

Niehrs, C., and Boutros, M. (2010). Trafficking, acidification, and growth factor signaling. *Sci. Signal.* **3**(134), pe26.

Niehrs, C., and Shen, J. (2010). Regulation of Lrp6 phosphorylation. *Cell. Mol. Life Sci.* **67** (15), 2551–2562.

Nusse, R. (2008). Wnt signaling and stem cell control. *Cell Res.* **18**(5), 523–527.

Nusse, R., van Ooyen, A., Cox, D., Fung, Y. K., and Varmus, H. (1984). Mode of proviral activation of a putative mammary oncogene (int-1) on mouse chromosome 15. *Nature* **307**(5947), 131–136.

Nusslein-Volhard, C., and Wieschaus, E. (1980). Mutations affecting segment number and polarity in Drosophila. *Nature* **287**(5785), 795–801.

Pan, C. L., Baum, P. D., Gu, M., Jorgensen, E. M., Clark, S. G., and Garriga, G. (2008). C. elegans AP-2 and retromer control Wnt signaling by regulating mig-14/Wntless. *Dev. Cell* **14**(1), 132–139.

Panakova, D., Sprong, H., Marois, E., Thiele, C., and Eaton, S. (2005). Lipoprotein particles are required for Hedgehog and Wingless signalling. *Nature* **435**(7038), 58–65.

Parker, D. S., Jemison, J., and Cadigan, K. M. (2002). Pygopus, a nuclear PHD-finger protein required for Wingless signaling in Drosophila. *Development* **129**(11), 2565–2576.

Parker, D. S., Ni, Y. Y., Chang, J. L., Li, J., and Cadigan, K. M. (2008). Wingless signaling induces widespread chromatin remodeling of target loci. *Mol. Cell. Biol.* **28**(5), 1815–1828.

Port, F., and Basler, K. (2010). Wnt trafficking: New insights into Wnt maturation, secretion and spreading. *Traffic* **11**(10), 1265–1271.

Port, F., Kuster, M., Herr, P., Furger, E., Banziger, C., Hausmann, G., and Basler, K. (2008). Wingless secretion promotes and requires retromer-dependent cycling of Wntless. *Nat. Cell Biol.* **10**(2), 178–185.

Prasad, B. C., and Clark, S. G. (2006). Wnt signaling establishes anteroposterior neuronal polarity and requires retromer in C. elegans. *Development* **133**(9), 1757–1766.

Reyes, A. R., Levenson, R., Berrettini, W., and Van Bockstaele, E. J. (2010). Ultrastructural relationship between the mu opioid receptor and its interacting protein, GPR177, in striatal neurons. *Brain Res.* **1358,** 71–80.
Roelink, H., and Nusse, R. (1991). Expression of two members of the Wnt family during mouse development–restricted temporal and spatial patterns in the developing neural tube. *Genes Dev.* **5**(3), 381–388.
Roszko, I., Sawada, A., and Solnica-Krezel, L. (2009). Regulation of convergence and extension movements during vertebrate gastrulation by the Wnt/PCP pathway. *Semin. Cell Dev. Biol.* **20**(8), 986–997.
Saydam, O., Shen, Y., Wurdinger, T., Senol, O., Boke, E., James, M. F., Tannous, B. A., Stemmer-Rachamimov, A. O., Yi, M., Stephens, R. M., Fraefel, C., Gusella, J. F., *et al.* (2009). Downregulated microRNA-200a in meningiomas promotes tumor growth by reducing E-cadherin and activating the Wnt/beta-catenin signaling pathway. *Mol. Cell. Biol.* **29**(21), 5923–5940.
Seeling, J. M., Miller, J. R., Gil, R., Moon, R. T., White, R., and Virshup, D. M. (1999). Regulation of beta-catenin signaling by the B56 subunit of protein phosphatase 2A. *Science* **283**(5410), 2089–2091.
Segalen, M., and Bellaiche, Y. (2009). Cell division orientation and planar cell polarity pathways. *Semin. Cell Dev. Biol.* **20**(8), 972–977.
Sharma, R. P., and Chopra, V. L. (1976). Effect of the Wingless (wg1) mutation on wing and haltere development in Drosophila melanogaster. *Dev. Biol.* **48**(2), 461–465.
Shi, A., Sun, L., Banerjee, R., Tobin, M., Zhang, Y., and Grant, B. D. (2009). Regulation of endosomal clathrin and retromer-mediated endosome to Golgi retrograde transport by the J-domain protein RME-8. *EMBO J.* **28**(21), 3290–3302.
Shimizu, H., Julius, M. A., Giarre, M., Zheng, Z., Brown, A. M., and Kitajewski, J. (1997). Transformation by Wnt family proteins correlates with regulation of beta-catenin. *Cell Growth Differ.* **8**(12), 1349–1358.
Sierra, J., Yoshida, T., Joazeiro, C. A., and Jones, K. A. (2006). The APC tumor suppressor counteracts beta-catenin activation and H3K4 methylation at Wnt target genes. *Genes Dev.* **20**(5), 586–600.
Silhankova, M., Port, F., Harterink, M., Basler, K., and Korswagen, H. C. (2010). Wnt signalling requires MTM-6 and MTM-9 myotubularin lipid-phosphatase function in Wnt-producing cells. *EMBO J.* **29**(24), 4094–4105.
Silver, S. J., Hagen, J. W., Okamura, K., Perrimon, N., and Lai, E. C. (2007). Functional screening identifies miR-315 as a potent activator of Wingless signaling. *Proc. Natl. Acad. Sci. USA* **104**(46), 18151–18156.
Simons, M., Gault, W. J., Gotthardt, D., Rohatgi, R., Klein, T. J., Shao, Y., Lee, H. J., Wu, A. L., Fang, Y., Satlin, L. M., Dow, J. T., Chen, J., *et al.* (2009). Electrochemical cues regulate assembly of the Frizzled/Dishevelled complex at the plasma membrane during planar epithelial polarization. *Nat. Cell Biol.* **11**(3), 286–294.
Staal, F. J., and Luis, T. C. (2010). Wnt signaling in hematopoiesis: Crucial factors for self-renewal, proliferation, and cell fate decisions. *J. Cell. Biochem.* **109**(5), 844–849.
Stadeli, R., Hoffmans, R., and Basler, K. (2006). Transcription under the control of nuclear Arm/beta-catenin. *Curr. Biol.* **16**(10), R378–R385.
Strutt, D. I. (2001). Asymmetric localization of frizzled and the establishment of cell polarity in the Drosophila wing. *Mol. Cell* **7**(2), 367–375.
Strutt, H., and Strutt, D. (2009). Asymmetric localisation of planar polarity proteins: Mechanisms and consequences. *Semin. Cell Dev. Biol.* **20**(8), 957–963.
Takada, R., Satomi, Y., Kurata, T., Ueno, N., Norioka, S., Kondoh, H., Takao, T., and Takada, S. (2006). Monounsaturated fatty acid modification of Wnt protein: Its role in Wnt secretion. *Dev. Cell* **11**(6), 791–801.

Tanaka, K., Okabayashi, K., Asashima, M., Perrimon, N., and Kadowaki, T. (2000). The evolutionarily conserved porcupine gene family is involved in the processing of the Wnt family. *Eur. J. Biochem.* **267**(13), 4300–4311.

Tanaka, K., Kitagawa, Y., and Kadowaki, T. (2002). Drosophila segment polarity gene product porcupine stimulates the posttranslational N-glycosylation of wingless in the endoplasmic reticulum. *J. Biol. Chem.* **277**(15), 12816–12823.

Tanaka, S., Maeda, Y., Tashima, Y., and Kinoshita, T. (2004). Inositol deacylation of glycosylphosphatidylinositol-anchored proteins is mediated by mammalian PGAP1 and yeast Bst1p. *J. Biol. Chem.* **279**(14), 14256–14263.

Tang, W., Dodge, M., Gundapaneni, D., Michnoff, C., Roth, M., and Lum, L. (2008). A genome-wide RNAi screen for Wnt/beta-catenin pathway components identifies unexpected roles for TCF transcription factors in cancer. *Proc. Natl. Acad. Sci. USA* **105**(28), 9697–9702.

Thatcher, E. J., Paydar, I., Anderson, K. K., and Patton, J. G. (2008). Regulation of zebrafish fin regeneration by microRNAs. *Proc. Natl. Acad. Sci. USA* **105**(47), 18384–18389.

Thompson, B., Townsley, F., Rosin-Arbesfeld, R., Musisi, H., and Bienz, M. (2002). A new nuclear component of the Wnt signalling pathway. *Nat. Cell Biol.* **4**(5), 367–373.

Usui, T., Shima, Y., Shimada, Y., Hirano, S., Burgess, R. W., Schwarz, T. L., Takeichi, M., and Uemura, T. (1999). Flamingo, a seven-pass transmembrane cadherin, regulates planar cell polarity under the control of Frizzled. *Cell* **98**(5), 585–595.

van Amerongen, R., and Nusse, R. (2009). Towards an integrated view of Wnt signaling in development. *Development* **136**(19), 3205–3214.

van Amerongen, R., Mikels, A., and Nusse, R. (2008). Alternative wnt signaling is initiated by distinct receptors. *Sci. Signal.* **1**(35), re9.

van den Heuvel, M., Harryman-Samos, C., Klingensmith, J., Perrimon, N., and Nusse, R. (1993). Mutations in the segment polarity genes wingless and porcupine impair secretion of the wingless protein. *EMBO J.* **12**(13), 5293–5302.

van Ooyen, A., and Nusse, R. (1984). Structure and nucleotide sequence of the putative mammary oncogene int-1; proviral insertions leave the protein-encoding domain intact. *Cell* **39**(1), 233–240.

Wallingford, J. B., Vogeli, K. M., and Harland, R. M. (2001). Regulation of convergent extension in Xenopus by Wnt5a and Frizzled-8 is independent of the canonical Wnt pathway. *Int. J. Dev. Biol.* **45**(1), 225–227.

Wend, P., Holland, J. D., Ziebold, U., and Birchmeier, W. (2010). Wnt signaling in stem and cancer stem cells. *Semin. Cell Dev. Biol.* **21**(8), 855–863.

Willert, K., Logan, C. Y., Arora, A., Fish, M., and Nusse, R. (1999). A Drosophila Axin homolog, Daxin, inhibits Wnt signaling. *Development* **126**(18), 4165–4173.

Willert, K., Brown, J. D., Danenberg, E., Duncan, A. W., Weissman, I. L., Reya, T., Yates, J. R., 3rd, and Nusse, R. (2003). Wnt proteins are lipid-modified and can act as stem cell growth factors. *Nature* **423**(6938), 448–452.

Wolff, T., and Rubin, G. M. (1998). Strabismus, a novel gene that regulates tissue polarity and cell fate decisions in Drosophila. *Development* **125**(6), 1149–1159.

Wong, G. T., Gavin, B. J., and McMahon, A. P. (1994). Differential transformation of mammary epithelial cells by Wnt genes. *Mol. Cell. Biol.* **14**(9), 6278–6286.

Wu, J., and Mlodzik, M. (2009). A quest for the mechanism regulating global planar cell polarity of tissues. *Trends Cell Biol.* **19**(7), 295–305.

Wu, Y., Belenkaya, T. Y., and Lin, X. (2010). Dual roles of Drosophila glypican Dally-like in Wingless/Wnt signaling and distribution. *Methods Enzymol.* **480,** 33–50.

Yamamoto, A., Nagano, T., Takehara, S., Hibi, M., and Aizawa, S. (2005). Shisa promotes head formation through the inhibition of receptor protein maturation for the caudalizing factors, Wnt and FGF. *Cell* **120**(2), 223–235.

Yan, Q., and Lennarz, W. J. (1999). Oligosaccharyltransferase: A complex multisubunit enzyme of the endoplasmic reticulum. *Biochem. Biophys. Res. Commun.* **266**(3), 684–689.
Yan, D., and Lin, X. (2009). Shaping morphogen gradients by proteoglycans. *Cold Spring Harb. Perspect. Biol.* **1**(3), a002493.
Yan, D., Wu, Y., Feng, Y., Lin, S. C., and Lin, X. (2009). The core protein of glypican Dally-like determines its biphasic activity in wingless morphogen signaling. *Dev. Cell* **17** (4), 470–481.
Yanagawa, S., Matsuda, Y., Lee, J. S., Matsubayashi, H., Sese, S., Kadowaki, T., and Ishimoto, A. (2002). Casein kinase I phosphorylates the Armadillo protein and induces its degradation in Drosophila. *EMBO J.* **21**(7), 1733–1742.
Yang, P. T., Lorenowicz, M. J., Silhankova, M., Coudreuse, D. Y., Betist, M. C., and Korswagen, H. C. (2008). Wnt signaling requires retromer-dependent recycling of MIG-14/Wntless in Wnt-producing cells. *Dev. Cell* **14**(1), 140–147.
You, J., Belenkaya, T., and Lin, X. (2011). Sulfated is a negative feedback regulator of wingless in Drosophila. *Dev. Dyn.* **240**(3), 640–648.
Yu, H. M., Jin, Y., Fu, J., and Hsu, W. (2010). Expression of Gpr177, a Wnt trafficking regulator, in mouse embryogenesis. *Dev. Dyn.* **239**(7), 2102–2109.
Zecca, M., Basler, K., and Struhl, G. (1996). Direct and long-range action of a wingless morphogen gradient. *Cell* **87**(5), 833–844.
Zeng, L., Fagotto, F., Zhang, T., Hsu, W., Vasicek, T. J., Perry, W. L., 3rd, Lee, J. J., Tilghman, S. M., Gumbiner, B. M., and Costantini, F. (1997). The mouse Fused locus encodes Axin, an inhibitor of the Wnt signaling pathway that regulates embryonic axis formation. *Cell* **90**(1), 181–192.
Zeng, X., Tamai, K., Doble, B., Li, S., Huang, H., Habas, R., Okamura, H., Woodgett, J., and He, X. (2005). A dual-kinase mechanism for Wnt co-receptor phosphorylation and activation. *Nature* **438,** 873–877.
Zhai, L., Chaturvedi, D., and Cumberledge, S. (2004). Drosophila wnt-1 undergoes a hydrophobic modification and is targeted to lipid rafts, a process that requires porcupine. *J. Biol. Chem.* **279**(32), 33220–33227.
Zoltewicz, J. S., Ashique, A. M., Choe, Y., Lee, G., Taylor, S., Phamluong, K., Solloway, M., and Peterson, A. S. (2009). Wnt signaling is regulated by endoplasmic reticulum retention. *PLoS One* **4**(7), e6191.

CHAPTER THREE

Notch-Independent Functions of CSL

Jane E. Johnson* *and* Raymond J. MacDonald†

Contents

Abstract

Notch-dependent CSL transcription complexes control essential biological processes such as cell proliferation, differentiation, and cell-fate decisions in diverse developmental systems. The orthologous proteins CBF1/Rbpj (mammalian), Su(H) (*Drosophila*), and Lag-1 (*Caenorhabditis elegans*) compose the CSL family of sequence-specific DNA-binding transcription factors. The CSL proteins are best known for their role in canonical Notch signaling. However, CSL factors also form transcription complexes that can function independent of Notch signaling and include repression and activation of target gene transcription. Because the different complexes share CSL as a DNA-binding subunit, they can control overlapping sets of genes; but they can also control distinct sets when

* Department of Neuroscience, UT Southwestern Medical Center, Dallas, Texas, USA
† Department of Molecular Biology, UT Southwestern Medical Center, Dallas, Texas, USA

Current Topics in Developmental Biology, Volume 97

ISSN 0070-2153, DOI: 10.1016/B978-0-12-385975-4.00009-7

partnered with tissue-specific cofactors that restrict DNA-sequence recognition or stability of the DNA-bound complex. The Notch-independent functions of CSL and the processes they regulate will be reviewed here with a particular emphasis on the tissue-specific CSL-activator complex with the bHLH factor Ptf1a.

1. Introduction

The transcription factor CSL is a key component in the phylogenetically conserved Notch-signaling pathway, which is fundamental to many aspects of development including cell-fate decisions, proliferation, and pattern formation. Indeed, dysfunction in this pathway can result in a variety of human pathologies from developmental syndromes to cancer (Garg *et al.*, 2005; van Es *et al.*, 2005; Weng *et al.*, 2004). In addition, CSL is co-opted by viral coactivators such as EBNA2 from the Epstein–Barr virus and 13SE1A from adenovirus (Ansieau and Leutz, 2002; Henkel *et al.*, 1994; Zimber-Strobl *et al.*, 1994). CSL is the sequence-specific DNA-binding subunit of transcription complexes that binds a high-affinity DNA site with the consensus YGTGRGAA (where Y = C or T and R = A or G) (Bailey and Posakony, 1995; Lecourtois and Schweisguth, 1995; Tun *et al.*, 1994). The CSL–DNA interactions are likely dynamic, with the activator and repressor complexes forming off DNA and possibly competing with each other for CSL and for CSL–DNA-binding sites (Bray and Bernard, 2010; Krejci and Bray, 2007; Zhou and Hayward, 2001). CSL is the term used when referring generally to the orthologous family of proteins; but when describing findings in a particular species, the gene name for that species is used: Rbpj (Recombination binding protein Jκ) for mammals, Su(H) (Suppressor of Hairless) for *Drosophila*, and Lag-1 for *Caenorhabditis elegans*.

Although the main focus of this review is on Notch-independent activities of CSL, one cannot fully understand the function of CSL without knowing its central role in the Notch-signaling pathway. As there are many outstanding recent reviews on Notch signaling (e.g., Bray and Bernard, 2010; Kopan and Ilagan, 2009), we summarize only the basics here.

The Notch proteins are transmembrane receptors that undergo complex processing when bound and activated by any of the family of Delta and Jagged extracellular ligands. The cleaved Notch intracellular domain (NICD) translocates to the nucleus, where it activates transcription through CSL (Fig. 3.1). This was the main context in which CSL factors were thought to function. However, even soon after the discovery of these factors, a simple role for CSL as the transcriptional activator of Notch signaling was not sufficient to explain the variety of phenotypes being reported for CSL mutants. For example, whereas the *Drosophila* orthologue,

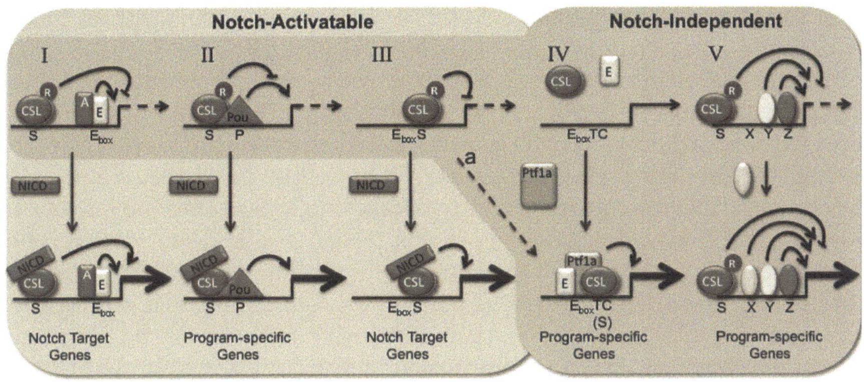

Figure 3.1 Notch-independent and Notch-activatable CSL transcription complexes. CSL forms multiple complexes (I–V), some of which can be activated by Notch signaling through the NICD (I–III) and some that are not responsive to NICD (IV and V). (I) In *Drosophila*, En(Spl) gene transcript m8 is repressed by CSL binding its DNA-recognition motif (S) and its corepressor(s). Notch signaling results in NICD translocating to the nucleus and forming an activator complex with CSL. The collaboration of the proneural bHLH heterodimer achaete/daughterless (A/E) located on a nearby E-box motif (E_{box}) is also required for active gene transcription in this example (Castro *et al.*, 2005; Cave *et al.*, 2005). (II) *Sox15* is repressed by CSL in mechanosensory neuron generation in *Drosophila*. In a cell with Notch signaling, NICD relieves the repression and permits the POU homeodomain factor Vvl to activate transcription. CSL and the POU factor interact with each other and cooperatively recognize a specific S + POU binding motif (P) (Miller *et al.*, 2009). (III) CSL repressor complex can bind the S motif to repress transcription. Notch signaling switches CSL to an activator complex that can also function through this site. However, during pancreas and neural development in a cell expressing Ptf1a and its E-protein heterodimeric partner (E), there may be competition between Ptf1a and NICD for CSL at those DNA elements containing an S plus an E_{box} (constrained with a one or two helical turn spacing) (IIIa, dashed arrow). Both complexes are activators. (IV) The PTF1 heterotrimer with Ptf1a and E-protein can also form with CSL on a DNA motif with an E-box plus a TC-box variant representing a low-affinity CSL site. However, in the absence of Ptf1a, CSL does not bind this site. Only in the presence of Ptf1a, E-protein, and CSL, a transcription activator complex is formed (Beres *et al.*, 2006). (V) In *C. elegans*, *bhlh6* transcription in gland cells requires multiple collaborating activators to overcome repression by CSL. CSL repressor complex is required to suppress inappropriate transcription of *bhlh6* by activators present in nongland cells. Notch signaling does not function in regulating *bhlh6* expression (Ghai and Gaudet, 2008). These models illustrate the different complexes that can form based on the protein components present in a cell, and the specific combination of sites in a given DNA *cis*-regulatory motif. Blue shading indicates the complexes that can be activated by NICD. Tan shading highlights the complexes that can form independent of NICD. The specific repressors used in each case are not necessarily the same and are not defined here. (For interpretation of the references to color in this figure legend, the reader is referred to the Web version of this chapter.)

Su(H), was identified genetically as an activator of Notch target genes (Furukawa *et al.*, 1995; Schweisguth and Posakony, 1992), Rbpj (CBF1, Rbpjκ, Rbpsuh) was first described as a transcriptional repressor (Dou *et al.*, 1994). In addition, loss-of-function mutants for Su(H) and Notch did not have the phenotype expected if Su(H) was simply the transcriptional activator for Notch signaling (Lecourtois and Schweisguth, 1995).

Insight into many of these discrepancies was provided on multiple fronts. In mammals, it was found that viral proteins, in particular the Epstein–Barr virus EBNA2, could switch the transcriptional repressor function of Rbpj to activation (Hsieh and Hayward, 1995). This led to the discovery that the processed NICD functioned similarly by binding to Su(H) in the nucleus to form a transcriptional activator complex (Hsieh *et al.*, 1996). Conversely, transcriptional corepressors were found in flies and in mammals that directly bound CSL (for review, see Borggrefe and Oswald, 2009). It is now widely accepted that CSL is the sole transcriptional mediator of the canonical Notch pathway and acts as a repression-to-activation switch. In the absence of Notch signaling, CSL is a default repressor for Notch targets; upon Notch signaling, the processed NICD switches CSL to an activating complex that induces transcription of those same Notch target genes (Fig. 3.1).

More recently, studies with a variety of organisms are providing evidence supporting diverse CSL functions separate from its role in Notch signaling. Due to the widespread importance of Notch signaling in development, Notch-independent actions of CSL have been difficult to uncover by genetic analysis alone because they are often coincident with and therefore hidden by Notch-dependent actions. Further, Notch-independent CSL activities can regulate a gene in one cell, but in other cells, its activity in regulating that gene can be Notch-dependent, blurring the boundaries of this distinction. However, evidence for Notch-independent CSL repressor complexes and CSL-activator complexes is emerging from combining detailed analysis of *cis*-regulatory sequences controlling specific gene transcription with gain and loss of function of CSL, Notch receptors, and tissue-specific cofactors.

Notch-independent and -dependent CSL activities are defined by whether Notch participates directly in the action of a CSL-mediated event (Fig. 3.1). We distinguish three types of CSL functions that are independent of a direct action by the Notch receptor. (1) *Default or constitutive repression* by CSL is independent of Notch because the NICD does not establish or participate in the repressor complex (Fig. 3.1, tan boxed complexes I–III), even though the regulatory sites have the potential to be activated (or derepressed) by Notch signaling (Fig. 3.1, blue boxed complexes I–III). (2) CSL also utilizes *cooperative interactions* with program-specific transcription factors that add selectivity to target gene recognition. These interactions may reverse the repression by CSL (Fig. 3.1, tan boxed complex IIIa) or recruit CSL to a regulatory site without its corepressor complex (Fig. 3.1, complex IV). (3) CSL-mediated repression can also suppress target genes until the repression is overcome by *collaborative activity* of transcription activators (Fig. 3.1, complex V). Most of the Notch-independent activities for CSL involve transcriptional repression and are required to prevent lineage- or field-specific transcriptional activators from inducing gene expression programs inappropriately (Fig. 3.1, tan boxed complexes I–III, V). An important exception is the cooperative interaction with a program-specific transcription factor required in pancreas and nervous system development. The studies supporting Notch-independent activities of CSL are discussed next.

2. CSL AS A REPRESSOR

2.1. Default repressor

Default repression by CSL provides a critical mechanism for strict control of program-specific genes by keeping them off until signaled, and is a common mechanism shared by multiple signaling pathways in development that was discussed in an insightful review almost 10 years ago (Barolo and Posakony, 2002). Default repression occurs in the absence of an activating signal. Although CSL forms a transcriptional activation complex with NICD, in the absence of NICD, CSL is a repressor and this repressor is defined here as a Notch-independent complex for CSL (Fig. 3.1, tan boxed complexes I–III).

The *in vivo* importance of a default repression function of CSL has been shown most clearly in *Drosophila*. Since Notch signaling mediates transcription through the single factor Su(H), mutants in Notch and Su(H) were predicted to have similar phenotypes. However, the loss of Su(H) had a weaker phenotype than the loss of Notch in many circumstances. This difference is now explained by the default repression function of Su(H) such that in Su(H) mutants, derepression of target genes can be sufficiently strong to partly compensate for the loss of Notch activation through Su(H). This has been carefully studied in the three developmental processes of bristle formation, wing development, and in dorsal–ventral boundary formation where in each case the contribution of the Su(H) default repression to the Notch phenotype varies (Castro *et al.*, 2005; Koelzer and Klein, 2003, 2006; Morel and Schweisguth, 2000).

Multiple corepressors have been identified in *Drosophila* and mammals. A critical component of the Su(H) repressor complex in *Drosophila* is the adapter protein Hairless. Hairless binds Su(H) and recruits at least two distinct global-type corepressors, Groucho and dCtBP (Barolo *et al.*, 2002; Morel *et al.*, 2001; Nagel *et al.*, 2005). Biochemical experiments demonstrated that Hairless directly interacts with Groucho and dCtBP through distinct sites (Barolo *et al.*, 2002; Morel *et al.*, 2001). Genetic studies provided *in vivo* evidence for these corepressors as components of Su(H) complexes and demonstrated that Groucho and dCtBP can have distinct contributions in different contexts of Su(H) function (Barolo *et al.*, 2002; Nagel *et al.*, 2005). For example, Groucho is a player in both the shaft/socket cell-fate decision and bristle number regulation during mechanosensory organ development, whereas dCtBP function was only detected in the shaft/socket cell-fate decision (Barolo *et al.*, 2002). In contrast, in Hairless overexpression phenotypes in wing development and in transcriptional assays in cell culture, the combination of Groucho and dCtBP was required for proper Hairless function (Nagel *et al.*, 2005). It is not known why one repressor complex is used over another, or why in some contexts both are required, but Groucho and dCtBP can have distinct mechanistic repressor properties, with Groucho functioning in long-range repression and CtBP functioning in more proximal promoter regions (Courey and Jia, 2001).

Although the importance of default repression by Rbpj in mammals is not clear, the repression activity of Rbpj was demonstrated when it was identified in human cells bound to the promoter of an adenovirus gene (Dou *et al.*, 1994). Since then, multiple components of the Rbpj repressor complex have been identified. Rbpj interacts with corepressors nCor1/nCor2 (previously SMRT) and Cir1, which can recruit class 1 histone deacetylases (HDACs) (Hsieh *et al.*, 1999; Kao *et al.*, 1998). An adapter protein Spen (previously Sharp, Mint) was identified that binds Rbpj as well as the corepressors (Oswald *et al.*, 2002, 2005; Tsuji *et al.*, 2007). Somewhat surprisingly, although both Spen and Hairless are adapters for CSL in mammals and *Drosophila*, respectively, and appear to serve analogous function in CSL repressor complexes, they are not orthologs. More recently, a distinct Rbpj repressor complex was defined when the adapter Spen was shown to recruit members of the ETO family of repressors to an Rbpj complex (Engel *et al.*, 2010; Salat *et al.*, 2008). An additional complexity in Rbpj repressor complexes involves Fhl1 (KyoT2). Fhl1 is a lim-only protein that binds in the same pocket as NICD (as does the other cofactor, Ptf1a; see below) and, thus, renders Rbpj repressor complexes resistant to activation (Taniguchi *et al.*, 1998). Recently, Fhl1 was also shown to serve as an adapter to bring polycomb-type repressors to Rbpj, creating another form for the Rbpj transcriptional repressor (Qin *et al.*, 2004, 2005). Thus, at least three distinct Rbpj repressor complexes have been observed, but the *in vivo* distinctions between and functional relevance of the different Rbpj repressor complexes remain to be discovered.

Although Su(H) and Rbpj can both function in repressor complexes, the specific components of the complexes in the two species are unique as detailed above. Nonetheless, the corepressors recruit HDACs, which alter the acetylation of the chromatin to suppress transcription (for review, see Borggrefe and Oswald, 2009). Demethylation of H3K4me3, a chromatin activation mark, was also recently shown to be a part of the mechanism for CSL repression when the histone demethylase KDM5A (previously Jarid1a) was shown to be a part of the CSL repressor complex in mammals and *Drosophila* (Liefke *et al.*, 2010). These results suggest that CSL repressor complexes involve multiple mechanisms for chromatin modifications for repressing transcription.

2.2. Su(H) suppresses the activity of local transcriptional activators

Examples of CSL repressor action other than default repression for Notch signaling are uncommon. One case that has been looked at carefully involves the socket-versus-shaft cell-fate decision during mechanosensory organ development in *Drosophila*. Here, a *cis*-regulatory module for the socket specific gene *Sox15* is required for directing expression of *Sox15* in the socket cell and repressing it in the shaft cell (Miller *et al.*, 2009). This regulatory module requires Su(H) at a site adjacent to a site bound by the Pou homeodomain transcriptional activator Vvl, ventral veins lacking (Fig. 3.1, complex II). Su(H) represses transcription directed by Vvl in the shaft cell, shown by mutations in the Su(H) binding site that result in ectopic expression of *Sox15* in that cell. In contrast, in the socket cell where there is active Notch signaling, repression from Su(H) is relieved and Vvl is free to activate transcription of *Sox15*. This regulation is thought to be "Notch-permissive" rather than "Notch-activated" because mutations in the Su(H) site do not alter Sox15 levels in the socket cell. The Notch-independent Su (H) component of this regulation (Fig. 3.1, tan boxed complex II) is critical for suppressing gene expression in the alternate lineage, a common biological function reported for Su(H) activity.

In the example of the *Sox15* enhancer, Su(H) and Vvl bind cooperatively to DNA to sites separated by two nucleotides (Miller *et al.*, 2009; Fig. 3.1, complex II). DNA-binding proteins that interact with transcription factors to increase the specificity of site selection or to strengthen DNA binding are generally termed cofactors. Thus, Vvl acts as a Su(H) cofactor. However, in other instances where Su(H) is a default repressor, an activator is not bound cooperatively but is bound nearby (Fig. 3.1, complexes I, V). DNA-binding proteins that modulate the activity of the other but do not bind cooperatively with the other factor have been called collaborators, and we retain that term. Multiple examples of collaboration have been reported for *Drosophila* genes and include (1) the presence of a proneural bHLH heterodimer binding site near the Su(H) site in the *cis*-regulatory module for the E(spl) m8 gene that is required in the development of sensory organs (Bailey and Posakony, 1995; Cave *et al.*, 2005; Nellesen *et al.*, 1999; Fig. 3.1, complex I), (2) the presence and requirement for the bHLH factor Twist with Su(H) in many targets in muscle progenitors (Bernard *et al.*, 2010), and (3) the combined requirement for Ets and Runx transcription factors with Su(H) in cone photoreceptor specification (Flores *et al.*, 2000). Thus, cooperative binding of Su(H) with program-specific transcriptional activators as well as collaboration but independent binding of Su(H) with other factors provides a common theme in cell-fate specification. In most of these cases, the importance of the repressor function of Su(H) to maintain restricted expression of lineage specific genes has been demonstrated.

2.3. Lag-1 as a constitutive repressor in *C. elegans*

In contrast to *Drosophila* and mammals, corepressors for Lag-1 in *C. elegans* have been more elusive. Up until recently, there has been little evidence that Lag-1 has any repressor activity, while its activity in Notch signaling is undisputable (Greenwald, 2005). Evidence for Lag-1 acting as a repressor was reported in studies of the *cis*-regulatory module controlling *bhlh6* expression in pharyngeal gland cell development (Ghai and Gaudet, 2008). Three sites within the *cis*-regulatory module can be bound by transcriptional activators that include Fkd, Peb-1, and an unknown activator. Another site acts as a silencer element, because when it was mutated, ectopic reporter gene expression was detected in nongland cells. Lag-1 binds the silencer site, and in Lag-1 mutants nongland cells ectopically expressed *bhlh6*, while expression in gland cells was not affected. Notably, repression by Lag-1 in *bhlh6* expression is not relieved by Notch signaling, distinctly identifying a function for Lag-1 independent of Notch signaling. The model proposes that gland cells, where bhlh6 is present, express the tissue-specific collaborators that overcome repression by Lag-1. In the nongland cells, there are not enough tissue-specific activators present to overcome the repression (Fig. 3.1, complex V). Thus, Lag-1 acts as a transcription repressor to restrict bhlh6 to those cells with sufficient levels of the unknown activator (Ghai and Gaudet, 2008).

3. CSL as a Notch-Independent Transcriptional Activator

3.1. Rbpj in a Notch-independent activator complex with the tissue-specific bHLH factor Ptf1a

CSL can function as a transcriptional repressor or as an activator depending on the cofactors recruited to the DNA-bound complex. CSL participation in these distinct complexes depends on specific DNA sequences within a given *cis*-regulatory module and the availability of tissue-specific cofactors. The clearest example of this comes from the identification of the transcription complex PTF1 (pancreas transcription factor 1). Earlier studies examining the regulation of exocrine pancreatic genes uncovered a conserved DNA motif in their promoters that was both necessary and sufficient to direct proper expression in acinar cells of the pancreas (Boulet *et al.*, 1986; Cockell *et al.*, 1989; Davis *et al.*, 1992; Kruse *et al.*, 1995; Meister *et al.*, 1989; Swift *et al.*, 1984). This pancreatic consensus motif contained two parts stereotypically separated by one or two helical turns that are now known as the bHLH consensus E-box (CACCTG) and the TC-box (TTTCCCA) (Cockell *et al.*, 1989). Three cosegregating proteins were biochemically identified that bound the consensus motif, and the resulting complex was termed PTF1. The three subunits are Ptf1a (p48) (Krapp *et al.*, 1996), one member of the E-protein family (E12, Heb, E2.2) (Roux *et al.*, 1989), and Rbpjl, the Notch indifferent paralogue of Rbpj (Beres *et al.*, 2006; Roux *et al.*, 1989). The importance of Ptf1a, in particular, to the maintenance of the acinar program was demonstrated by the severely diminished expression of secretory enzyme genes in an acinar cell line upon reduction of the level of Ptf1a mRNA (Krapp *et al.*, 1996).

Two lines of evidence suggested Ptf1a might also form a complex with Rbpj. It was proposed that Ptf1a might be important for early pancreatic development because Ptf1a is present prior to differentiation of pancreatic acinar cells. Indeed, analysis of mice null for *Ptf1a* revealed that it is required for all pancreatic structures (Krapp *et al.*, 1998), and cells initially specified to become pancreas converted to either bile ductal or duodenal fates (Kawaguchi *et al.*, 2002). In addition, Ptf1a was identified as an Rbpj-interacting protein in a yeast-2-hybrid screen (Obata *et al.*, 2001). Biochemical analyses combined with expression characterization of Ptf1a, Rbpj, and Rbpjl in the early developing pancreas revealed that Rbpj in a PTF1-J (contains Rbpj rather than Rbpjl) trimer complex is responsible for the developmental functions of Ptf1a, including the acinar specific activation of *Rbpjl* (Masui *et al.*, 2007). Taken together, these studies showed that Rbpj participates in a novel transcription activator complex that establishes pancreatic versus duodenal or biliary fates in the developing endoderm.

Ptf1a is the tissue-specific subunit of PTF1-J. Besides expression in the developing pancreas, it is found in discrete regions within the developing nervous system in progenitors to inhibitory neurons in the dorsal spinal cord, cerebellum, and retina (Fujitani *et al.*, 2006; Glasgow *et al.*, 2005; Hoshino *et al.*, 2005). Studies in *Ptf1a* null mice reveal that Ptf1a functions as a specification factor for these inhibitory neurons, and in its absence, cells are *trans*-fated to excitatory neuronal lineages (Dullin *et al.*, 2007; Fujitani *et al.*, 2006; Glasgow *et al.*, 2005; Hori *et al.*, 2008; Hoshino *et al.*, 2005; Nakhai *et al.*, 2007; Pascual *et al.*, 2007). In the dorsal spinal cord, Ptf1a specifies the GABAergic dorsal interneurons dI4/dIL^{A} while suppressing the glutamatergic dorsal interneurons dI5/dIL^{B} (Glasgow *et al.*, 2005; Wildner *et al.*, 2006). In the cerebellum, Ptf1a is required for the generation of multiple populations of GABAergic neurons including Purkinje, stellate, basket, and golgi neurons while suppressing granule neuron cell fates (Hoshino *et al.*, 2005; Pascual *et al.*, 2007). Similarly, in the retina, Ptf1a is required for the inhibitory neurons in the horizontal and amacrine neuronal lineages while suppressing ganglion neuron fates (Dullin *et al.*, 2007; Fujitani *et al.*, 2006; Nakhai *et al.*, 2007). Thus, similar to its function in cell-fate decisions in endodermal lineages, in multiple regions within the nervous system, Ptf1a is a fate-determinant switch controlling whether a progenitor will execute an inhibitory neuronal program rather than one for excitatory neurons.

Importantly, the interaction between Ptf1a and Rbpj is crucial for these functions of Ptf1a (Hori *et al.*, 2008; Masui *et al.*, 2007). A series of biochemical and genetic experiments were used to demonstrate the importance of the Ptf1a–Rbpj complex in pancreas and the dorsal neural tube. First, a tryptophan was identified within a conserved C-terminal peptide in Ptf1a that is required for PTF1-J trimer formation with Rbpj but not for the Ptf1a:E-protein heterodimer. The tryptophan-containing peptide matches the consensus ΦWΦP peptide motif identified for other CSL-interacting proteins including NICD, EBNA2, and Fhl1 (KyoT2) (Beres *et al.*, 2006). Mutations in *Ptf1a* for both humans and mice revealed the importance of this peptide for Ptf1a function. Infants with mutations in *PTF1a* that resulted in truncated proteins that retained the bHLH domain but lacked the conserved ΦWΦP peptide motif have pancreatic and cerebellar agenesis (Sellick *et al.*, 2004). Further, mouse embryos genetically modified to express only a mutant form of Ptf1a with the tryptophan in the ΦWΦP motif mutated to an alanine phenocopied the developmental defects of the null for both the pancreas and the nervous system (Hori *et al.*, 2008; Masui *et al.*, 2007). Moreover, in the dorsal neural tube, the deletion of *Rbpj*, but not inhibition of Notch signaling, produced identical results (Hori *et al.*, 2008). Together, these studies demonstrate that Rbpj, in a Notch-independent complex, is required for cell-fate decisions in the pancreas and spinal cord (Hori *et al.*, 2008; Masui *et al.*, 2007).

The DNA-binding properties of the PTF1-J complex involve cooperative binding of Rbpj to its consensus site and the binding of Ptf1a:E-protein to an E-box (Beres *et al.*, 2006). Whereas the bHLH heterodimer and Rbpj can bind the individual sites separately, DNA binding by the trimeric PTF1 complex requires a combination of these sites. The formation of the PTF1 complex prior to DNA binding creates a synergistic dependence on the presence of both DNA sites spaced appropriately. Moreover, the binding of the complex is highly cooperative, especially for binding sites with divergent E- and TC-boxes. Thus, the cooperativity allows the recognition of variant E- and TC-boxes that cannot be bound individually by bHLH heterodimers or Rbpj and so remain unoccupied and inactive in the absence of the trimeric complex (Fig. 3.1, complex IV). In contrast, the possibility exists that a PTF1 binding site with an optimal sequence for the E- and TC-boxes can be bound by Rbpj individually with its co-repressor. In this case, transcription might be activated by Notch signaling (Fig. 3.1, complex III), a bHLH dimer, or the PTF1 trimer (Fig. 3.1, complex IIIa). Thus, depending on the specifics of the PTF1 binding site in a given enhancer, a regulated gene could collectively be subject to repression by Rbpj and activation by Notch signaling and a bHLH dimer not containing Ptf1a

(Ptf1a:E-protein does not activate transcription (Beres *et al.*, 2006)), or it could be subject only to activation by Ptf1a in the PTF1 complex.

Although no biological function has been described for a PTF1-orthologous complex in *Drosophila*, there are multiple lines of evidence that support the existence of the complex. First, the ΦWΦP peptide motif in Ptf1a required for Rbpj binding is conserved in its orthologue, Fer1 (Beres *et al.*, 2006). Second, Fer1 and Su(H) have a direct interaction in a yeast-2-hybrid screen (Formstecher *et al.*, 2005). Third, the three orthologous *Drosophila* subunits in the PTF1 complex, Fer1, Daughterless, and Su(H), form a complex capable of binding the mammalian bipartite consensus sequence for PTF1 (Beres *et al.*, 2006). Fourth, diagnostic mutations of the E- and TC-boxes that disrupt binding of the mammalian complex also prevent binding of the *Drosophila* trimeric complex. And fifth, all possible trimeric complexes of mixed mammalian and *Drosophila* subunits bind with similar efficiency (Beres *et al.*, 2006). Thus, the Notch-independent Rbpj-containing activator complex, PTF1-J, is an evolutionary conserved transcription complex.

3.2. Su(H) activator function independent of Notch signaling in bristle formation

Other evidence for an activator function for Su(H) comes from detailed analysis of a *cis*-regulatory enhancer for Su(H) expression in socket cells of *Drosophila* external mechanosensory organs (bristles) where Notch-independent Su(H) autoactivation was revealed (Barolo *et al.*, 2000). Socket cells arise from the bipotent progenitor pIIB, which also gives rise to a shaft cell during bristle formation (Hartenstein and Posakony, 1989). An autoregulatory enhancer 3′ of the Su(H) gene contains five Su(H) binding sites within a 372-bp region. Su(H) itself at this *cis*-regulatory module has three distinct activities that depend on the cellular context (Barolo *et al.*, 2000). First, Su(H) acts as a repressor on this site in the shaft cell where there is no Notch signaling, thus maintaining low levels of Su(H) in that cell. This was one of the earliest examples for function of Su(H) as a default repressor in *Drosophila*. Second, in the canonical Notch-signaling pathway, Su(H) is switched to an activator in the presence of NICD in the socket cell and this is required for specifying the socket cell fate. And finally, a Notch-independent Su(H) activator function takes over, dramatically upregulating Su(H) in the socket cell, and this is required in the socket cell for the normal physiological function of the mechanosensory organ. Although this autoactivation of Su(H) has been demonstrated genetically, the cofactor that endows Su(H) with this cell-type specific activation potential has not been identified.

4. Role of *cis*-Regulatory Modules in Determining Target Gene Transcription

Defining the different CSL transcription complexes is important to understand CSL functions; however, equally important is the sequence-specific binding information encoded in the DNA. Understanding which complexes will be stabilized at what DNA target sites involves differing DNA-recognition sites for the different complexes combined with distinct cell-type specific cofactors. The originally defined high-affinity DNA consensus site for CSL, YGTGRGAA (Bailey and Posakony, 1995; Lecourtois and Schweisguth, 1995; Tun *et al.*, 1994), has been confirmed most recently using a protein-binding DNA microarray with Rbpj (Del Bianco *et al.*, 2010). This *in vitro* strategy detected no difference between Rbpj binding alone and in complexes with NICD from the different Notch receptors. However, genome-wide binding studies in *Drosophila* cells demonstrate that not all Su(H) binding sites have similar levels of occupancy, and that occupancy varies with the different Su(H) complexes (Krejci and Bray, 2007). Additional protein–protein interactions with other transcriptional regulators and epigenetic mechanisms *in vivo* could reconcile these findings and are likely critical for controlling the distribution of CSL transcription complexes throughout the genome.

Detailed analysis of cis-regulatory motifs in target genes provides an increasing number of examples in which the function of CSL requires specific combinations of DNA-sequence elements indicative of additional cooperative cofactors and collaborating factors. A compound motif consisting of paired CSL consensus sites (S), often in a head-to-head configuration, separated by 15–22 nucleotides, and associated with a nearby E-box (CANNTG) has been identified as a high-affinity site for Su(H) complexes (called SPS + A) (Bailey and Posakony, 1995; Cave *et al.*, 2005; Nellesen *et al.*, 1999). The E-box is a binding site for proneural bHLH heterodimers and is thought to be important in some Su(H) target genes for activation by Notch signaling (Cooper *et al.*, 2000). At least in the case of E(spl) m8 regulation, Daughterless (the E-protein subunit in the bHLH heterodimer) can directly bind Su(H) to act synergistically in activating transcription (Cave *et al.*, 2009). This is reminiscent of the PTF1 complex, except that the interaction between Rbpj/Su(H) and the bHLH heterodimer is through the common E-protein. Another notable difference is that the SPS + A codes for a collaborative interaction of the transcription factors whereas the PTF1 site has cooperative binding between the components. The SPS + A motif is critical to the regulation of several *Drosophila* E(spl) genes (Castro *et al.*, 2005; Cave *et al.*, 2005), but it is uncertain whether SPS motifs are involved in gene regulation in other organisms.

Detailed dissection of CSL-responsive enhancers has yielded additional motif combinations that provide context-dependent binding activity. Two of these combinatorial motifs provide examples for cooperative binding and include the S + Pou site in the *Sox15* enhancer in *Drosophila* bristle formation (Miller *et al.*, 2009), and the PTF1 site in mammalian pancreas and neural tube development (Beres *et al.*, 2006; Henke *et al.*, 2009; Masui *et al.*, 2008; Meredith *et al.*, 2009). In each case, one component of the complex is tissue specific (Vvl and Ptf1a, respectively), supporting the context-dependent binding of these CSL complexes. For the S + Pou site bound by Su(H) and Vvl, the motifs are two nucleotides apart (CATGGGAAaaATGAAAT) with one-helical DNA turn between the centers of the motifs (Fig. 3.1, complex II; Miller *et al.*, 2009). The spacing requirement for this bipartite binding site was not assessed.

The Ptf1a/E-protein heterodimer binding motif (E-box, CANNTG) and the Rbpj binding motif (TC-box, YGTGRGAA, generally written as the complementary strand TTCYCACR) are required in combination for the PTF1 activator complex to bind and have a spacing requirement of either one or two DNA-helical turns, center to center (Beres *et al.*, 2006). The motif sequences and conserved spacing have been confirmed for multiple regulatory regions in pancreatic and neural specific genes, including those regulating expression of *Ptf1a* itself, *Rbpjl*, and *Neurog2* (Henke *et al.*, 2009; Masui *et al.*, 2007, 2008; Meredith *et al.*, 2009). Notably, since the binding of the three PTF1 subunits is cooperative, either a weak Rbpj binding site or a suboptimal E-box is tolerated. This property establishes the possibility of functional classes of PTF1 binding sites with distinct regulatory properties: targets for Rbpj repressor complexes and activatable by Notch signaling; targets for other bHLH heterodimers; uniquely bound and activated by PTF1; or competition among these possibilities with the outcome dictated by levels of the binding proteins (Fig. 3.1, complexes III/IV). Identification of additional targets of the PTF1 transcriptional activator using genome-wide approaches like chromatin immunoprecipitation followed by sequencing (ChIP-seq) is providing support for the existence of these diverse binding sites in the genome (J.E.J. and R.J.M., unpublished observations). The *in vivo* consequences of such combinatorial regulation and the existence of alternative Rbpj complexes have yet to be determined. It is clear, however, that tissue-specific cofactors alter the DNA-binding recognition and transcriptional activation properties of CSL-containing transcription complexes.

5. Role of Competition Between CSL-Containing Complexes—Is CSL a Limiting Factor?

Because CSL is required in distinct complexes for regulating gene expression, competition between complexes for limiting amounts of CSL is possible. In a cell with high levels of Notch signaling, the NICD activator complex may compete with the CSL repressor complex, and in cells where Ptf1a is present, with the PTF1 activator complex. Thus, although the CSL complexes lacking NICD have by definition Notch-independent function, the level of Notch signaling can modulate their function by competition for CSL. Further, the concentration of any given CSL-containing complex will be affected by the concentration of the competing cofactors. This type of competition for the CSL factor, in this case Rbpj, was proposed to explain a fate-determining step in pancreas (Cras-Meneur *et al.*, 2009). Here, a bHLH heterodimer of Neurog3 plus an E-protein (Ngn3:E12) directs progenitor cells to an endocrine fate, and the PTF1 heterotrimer (Ptf1a:E12:Rbpj) directs cells to the exocrine fate. If Notch signaling is active, there may be decreased availability of Rbpj for the PTF1 complex, which would allow Ngn3:E12 to push cells to the endocrine fate. This type of mechanism may also explain Notch phenotypes in the dorsal neural tube. In this case, PTF1 is required to direct a progenitor cell to the inhibitory neuronal fate, and in the absence of a PTF1 trimer complex, the fate of these progenitors is switched to an excitatory neuronal fate (Glasgow *et al.*, 2005; Hori *et al.*, 2008). In contrast, Notch signaling biases cells to the excitatory neuronal fate (Hori *et al.*, 2008; Mizuguchi *et al.*, 2006). The argument could be made that active Notch signaling decreases the availability of Rbpj for the PTF1 complex, allowing other specification regulators in the bipotent cells to push the cells to the excitatory neuronal fate. The combinatorial reuse of regulatory proteins in complexes with diverse function, added to distinct composition of DNA-binding motifs in enhancers, makes for an elegantly complex transcriptional network.

6. Concluding Remarks

Many questions remain concerning Notch-dependent and -independent functions for CSL. What is the role of DNA sequence and chromatin modifications in constraining which CSL complexes can bind and the consequences to transcription after they bind? What is the primary mechanism for tissue-specific binding of CSL complexes, and how is it determined when a CSL complex is or is not responsive to Notch? How many different combinations of CSL-containing complexes are there? Addressing the latter question, it is notable that multiple CSL-interacting proteins, particularly the activating cofactors, bind to a region of CSL through a related peptide motif in their amino acid sequence (ΦWΦP, where Φ is a hydrophobic residue). These include NICD, Ptf1a, and viral proteins such as EBNA2. A consequence of the shared binding surface is that these activators may compete for available CSL. Indeed, the LIM-only adapter Fhl1 also has this motif and is known to interfere with activation by Notch signaling in overexpression experiments (Taniguchi *et al.*, 1998). Although these experiments demonstrated that Fhl1 can act as a repressor by blocking activation of CSL by NICD, LIM-only proteins similar to Fhl1 can act as docking molecules that assemble transcription factor complexes via their LIM domains (Rabbitts, 1998). Thus, it remains plausible that rather than just blocking NICD action, Fhl1 as an adapter may bring in other tissue-specific transcription activators into CSL complexes. Identifying additional tissue-specific CSL cofactors and how they function in combination with known CSL complexes to regulate gene expression is critical for understanding Notch-dependent and -independent CSL function during the multitude of biological processes requiring these activities.

There are now several examples of CSL function independent of its role in a Notch-containing complex. These examples highlight CSL activities that involve a combination of transcriptional mechanisms including default or constitutive repression, cooperative binding for target selectivity and synergistic activity, and collaboration with other transcription factors to regulate cell-type specific gene transcription. Current models for CSL–DNA interactions suppose dynamic rather than static complex formation in which CSL repressor or activator complexes form off DNA, have different affinity for DNA, and compete for site-specific DNA motifs (Bray and Bernard, 2010). Future experiments detailing mechanistic underpinnings of CSL repressor and activator functions, whether Notch-dependent or -independent will be required to confirm current models and to fully understand the biological consequences of the competing complexes. This next generation of discoveries of CSL functions will be aided by improvements in technology that are allowing the genomic landscape of tissue-specific transcription factors to be defined.

ACKNOWLEDGMENTS

We are grateful to Galvin Swift for helpful comments on the manuscript. J. E. J. is supported by NIH R01 HD037932 and NS032817. R. J. M. is supported by R01 DK061220 and DK089570.

REFERENCES

Ansieau, S., and Leutz, A. (2002). The conserved Mynd domain of BS69 binds cellular and oncoviral proteins through a common PXLXP motif. *J. Biol. Chem.* **277,** 4906–4910.

Bailey, A. M., and Posakony, J. W. (1995). Suppressor of hairless directly activates transcription of enhancer of split complex genes in response to Notch receptor activity. *Genes Dev.* **9,** 2609–2622.

Barolo, S., and Posakony, J. W. (2002). Three habits of highly effective signaling pathways: Principles of transcriptional control by developmental cell signaling. *Genes Dev.* **16,** 1167–1181.

Barolo, S., Walker, R. G., Polyanovsky, A. D., Freschi, G., Keil, T., and Posakony, J. W. (2000). A notch-independent activity of suppressor of hairless is required for normal mechanoreceptor physiology. *Cell* **103,** 957–969.

Barolo, S., Stone, T., Bang, A. G., and Posakony, J. W. (2002). Default repression and Notch signaling: Hairless acts as an adaptor to recruit the corepressors Groucho and dCtBP to Suppressor of Hairless. *Genes Dev.* **16,** 1964–1976.

Beres, T. M., Masui, T., Swift, G. H., Shi, L., Henke, R. M., and MacDonald, R. J. (2006). PTF1 is an organ-specific and Notch-independent basic helix-loop-helix complex containing the mammalian Suppressor of Hairless (RBP-J) or its paralogue, RBP-L. *Mol. Cell. Biol.* **26,** 117–130.

Bernard, F., Krejci, A., Housden, B., Adryan, B., and Bray, S. J. (2010). Specificity of Notch pathway activation: Twist controls the transcriptional output in adult muscle progenitors. *Development* **137,** 2633–2642.

Borggrefe, T., and Oswald, F. (2009). The Notch signaling pathway: Transcriptional regulation at Notch target genes. *Cell. Mol. Life Sci.* **66,** 1631–1646.

Boulet, A. M., Erwin, C. R., and Rutter, W. J. (1986). Cell-specific enhancers in the rat exocrine pancreas. *Proc. Natl. Acad. Sci. USA* **83,** 3599–3603.

Bray, S., and Bernard, F. (2010). Notch targets and their regulation. *Curr. Top. Dev. Biol.* **92,** 253–275.

Castro, B., Barolo, S., Bailey, A. M., and Posakony, J. W. (2005). Lateral inhibition in proneural clusters: Cis-regulatory logic and default repression by Suppressor of Hairless. *Development* **132,** 3333–3344.

Cave, J. W., Loh, F., Surpris, J. W., Xia, L., and Caudy, M. A. (2005). A DNA transcription code for cell-specific gene activation by notch signaling. *Curr. Biol.* **15,** 94–104.

Cave, J. W., Xia, L., and Caudy, M. A. (2009). The Daughterless N-terminus directly mediates synergistic interactions with Notch transcription complexes via the SPS+A DNA transcription code. *BMC Res. Notes* **2,** 65.

Cockell, M., Stevenson, B. J., Strubin, M., Hagenbuchle, O., and Wellauer, P. K. (1989). Identification of a cell-specific DNA-binding activity that interacts with a transcriptional activator of genes expressed in the acinar pancreas. *Mol. Cell. Biol.* **9,** 2464–2476.

Cooper, M. T., Tyler, D. M., Furriols, M., Chalkiadaki, A., Delidakis, C., and Bray, S. (2000). Spatially restricted factors cooperate with notch in the regulation of Enhancer of split genes. *Dev. Biol.* **221,** 390–403.

Courey, A. J., and Jia, S. (2001). Transcriptional repression: The long and the short of it. *Genes Dev.* **15,** 2786–2796.

Cras-Meneur, C., Li, L., Kopan, R., and Permutt, M. A. (2009). Presenilins, Notch dose control the fate of pancreatic endocrine progenitors during a narrow developmental window. *Genes Dev.* **23,** 2088–2101.

Davis, B. P., Hammer, R. E., Messing, A., and MacDonald, R. J. (1992). Selective expression of trypsin fusion genes in acinar cells of the pancreas and stomach of transgenic mice. *J. Biol. Chem.* **267,** 26070–26077.

Del Bianco, C., Vedenko, A., Choi, S. H., Berger, M. F., Shokri, L., Bulyk, M. L., and Blacklow, S. C. (2010). Notch and MAML-1 complexation do not detectably alter the DNA binding specificity of the transcription factor CSL. *PLoS One* **5,** e15034.

Dou, S., Zeng, X., Cortes, P., Erdjument-Bromage, H., Tempst, P., Honjo, T., and Vales, L. D. (1994). The recombination signal sequence-binding protein RBP-2N functions as a transcriptional repressor. *Mol. Cell. Biol.* **14,** 3310–3319.

Dullin, J. P., Locker, M., Robach, M., Henningfeld, K. A., Parain, K., Afelik, S., Pieler, T., and Perron, M. (2007). Ptf1a triggers GABAergic neuronal cell fates in the retina. *BMC Dev. Biol.* **7,** 110.

Engel, M. E., Nguyen, H. N., Mariotti, J., Hunt, A., and Hiebert, S. W. (2010). Myeloid translocation gene 16 (MTG16) interacts with Notch transcription complex components to integrate Notch signaling in hematopoietic cell fate specification. *Mol. Cell. Biol.* **30,** 1852–1863.

Flores, G. V., Duan, H., Yan, H., Nagaraj, R., Fu, W., Zou, Y., Noll, M., and Banerjee, U. (2000). Combinatorial signaling in the specification of unique cell fates. *Cell* **103,** 75–85.

Formstecher, E., *et al.* (2005). Protein interaction mapping: A Drosophila case study. *Genome Res.* **15,** 376–384.

Fujitani, Y., Fujitani, S., Luo, H., Qiu, F., Burlison, J., Long, Q., Kawaguchi, Y., Edlund, H., MacDonald, R. J., Furukawa, T., Fujikado, T., Magnuson, M. A., *et al.* (2006). Ptf1a determines horizontal and amacrine cell fates during mouse retinal development. *Development* **133,** 4439–4450.

Furukawa, T., Kobayakawa, Y., Tamura, K., Kimura, K., Kawaichi, M., Tanimura, T., and Honjo, T. (1995). Suppressor of hairless, the Drosophila homologue of RBP-J kappa, transactivates the neurogenic gene E(spl)m8. *Jpn. J. Genet.* **70,** 505–524.

Garg, V., Muth, A. N., Ransom, J. F., Schluterman, M. K., Barnes, R., King, I. N., Grossfeld, P. D., and Srivastava, D. (2005). Mutations in NOTCH1 cause aortic valve disease. *Nature* **437,** 270–274.

Ghai, V., and Gaudet, J. (2008). The CSL transcription factor LAG-1 directly represses hlh-6 expression in C. elegans. *Dev. Biol.* **322,** 334–344.

Glasgow, S. M., Henke, R. M., Macdonald, R. J., Wright, C. V., and Johnson, J. E. (2005). Ptf1a determines GABAergic over glutamatergic neuronal cell fate in the spinal cord dorsal horn. *Development* **132,** 5461–5469.

Greenwald, I. (2005). LIN-12/Notch signaling in C. elegans. *WormBook* **8,** 1–16.

Hartenstein, V., and Posakony, J. W. (1989). Development of adult sensilla on the wing and notum of Drosophila melanogaster. *Development* **107,** 389–405.

Henke, R. M., Savage, T. K., Meredith, D. M., Glasgow, S. M., Hori, K., Dumas, J., MacDonald, R. J., and Johnson, J. E. (2009). Neurog2 is a direct downstream target of the Ptf1a-Rbpj transcription complex in dorsal spinal cord. *Development* **136,** 2945–2954.

Henkel, T., Ling, P. D., Hayward, S. D., and Peterson, M. G. (1994). Mediation of Epstein-Barr virus EBNA2 transactivation by recombination signal-binding protein J kappa. *Science* **265,** 92–95.

Hori, K., Cholewa-Waclaw, J., Nakada, Y., Glasgow, S. M., Masui, T., Henke, R. M., Wildner, H., Martarelli, B., Beres, T. M., Epstein, J. A., Magnuson, M. A., Macdonald, R. J., *et al.* (2008). A nonclassical bHLH Rbpj transcription factor complex is required for specification of GABAergic neurons independent of Notch signaling. *Genes Dev.* **22,** 166–178.

Hoshino, M., Nakamura, S., Mori, K., Kawauchi, T., Terao, M., Nishimura, Y. V., Fukuda, A., Fuse, T., Matsuo, N., Sone, M., Watanabe, M., Bito, H., *et al.* (2005). Ptf1a, a bHLH transcriptional gene, defines GABAergic neuronal fates in cerebellum. *Neuron* **47,** 201–213.

Hsieh, J. J., and Hayward, S. D. (1995). Masking of the CBF1/RBPJ kappa transcriptional repression domain by Epstein-Barr virus EBNA2. *Science* **268,** 560–563.

Hsieh, J. J., Henkel, T., Salmon, P., Robey, E., Peterson, M. G., and Hayward, S. D. (1996). Truncated mammalian Notch1 activates CBF1/RBPJk-repressed genes by a mechanism resembling that of Epstein-Barr virus EBNA2. *Mol. Cell. Biol.* **16,** 952–959.

Hsieh, J. J., Zhou, S., Chen, L., Young, D. B., and Hayward, S. D. (1999). CIR, a corepressor linking the DNA binding factor CBF1 to the histone deacetylase complex. *Proc. Natl. Acad. Sci. USA* **96,** 23–28.

Kao, H. Y., Ordentlich, P., Koyano-Nakagawa, N., Tang, Z., Downes, M., Kintner, C. R., Evans, R. M., and Kadesch, T. (1998). A histone deacetylase corepressor complex regulates the Notch signal transduction pathway. *Genes Dev.* **12,** 2269–2277.

Kawaguchi, Y., Cooper, B., Gannon, M., Ray, M., MacDonald, R. J., and Wright, C. V. (2002). The role of the transcriptional regulator Ptf1a in converting intestinal to pancreatic progenitors. *Nat. Genet.* **32,** 128–134.
Koelzer, S., and Klein, T. (2003). A Notch-independent function of Suppressor of Hairless during the development of the bristle sensory organ precursor cell of Drosophila. *Development* **130,** 1973–1988.
Koelzer, S., and Klein, T. (2006). Regulation of expression of Vg and establishment of the dorsoventral compartment boundary in the wing imaginal disc by Suppressor of Hairless. *Dev. Biol.* **289,** 77–90.
Kopan, R., and Ilagan, M. X. (2009). The canonical Notch signaling pathway: Unfolding the activation mechanism. *Cell* **137,** 216–233.
Krapp, A., Knofler, M., Frutiger, S., Hughes, G. J., Hagenbuchle, O., and Wellauer, P. K. (1996). The p48 DNA-binding subunit of transcription factor PTF1 is a new exocrine pancreas-specific basic helix-loop-helix protein. *EMBO J.* **15,** 4317–4329.
Krapp, A., Knofler, M., Ledermann, B., Burki, K., Berney, C., Zoerkler, N., Hagenbuchle, O., and Wellauer, P. K. (1998). The bHLH protein PTF1-p48 is essential for the formation of the exocrine and the correct spatial organization of the endocrine pancreas. *Genes Dev.* **12,** 3752–3763.
Krejci, A., and Bray, S. (2007). Notch activation stimulates transient and selective binding of Su(H)/CSL to target enhancers. *Genes Dev.* **21,** 1322–1327.
Kruse, F., Rose, S. D., Swift, G. H., Hammer, R. E., and MacDonald, R. J. (1995). Cooperation between elements of an organ-specific transcriptional enhancer in animals. *Mol. Cell. Biol.* **15,** 4385–4394.
Lecourtois, M., and Schweisguth, F. (1995). The neurogenic suppressor of hairless DNA-binding protein mediates the transcriptional activation of the enhancer of split complex genes triggered by Notch signaling. *Genes Dev.* **9,** 2598–2608.
Liefke, R., Oswald, F., Alvarado, C., Ferres-Marco, D., Mittler, G., Rodriguez, P., Dominguez, M., and Borggrefe, T. (2010). Histone demethylase KDM5A is an integral part of the core Notch-RBP-J repressor complex. *Genes Dev.* **24,** 590–601.
Masui, T., Long, Q., Beres, T. M., Magnuson, M. A., and MacDonald, R. J. (2007). Early pancreatic development requires the vertebrate Suppressor of Hairless (RBPJ) in the PTF1 bHLH complex. *Genes Dev.* **21,** 2629–2643.
Masui, T., Swift, G. H., Hale, M. A., Meredith, D. M., Johnson, J. E., and Macdonald, R. J. (2008). Transcriptional autoregulation controls pancreatic Ptf1a expression during development and adulthood. *Mol. Cell. Biol.* **28,** 5458–5468.
Meister, A., Weinrich, S. L., Nelson, C., and Rutter, W. J. (1989). The chymotrypsin enhancer core. Specific factor binding and biological activity. *J. Biol. Chem.* **264,** 20744–20751.
Meredith, D. M., Masui, T., Swift, G. H., MacDonald, R. J., and Johnson, J. E. (2009). Multiple transcriptional mechanisms control Ptf1a levels during neural development including autoregulation by the PTF1-J complex. *J. Neurosci.* **29,** 11139–11148.
Miller, S. W., Avidor-Reiss, T., Polyanovsky, A., and Posakony, J. W. (2009). Complex interplay of three transcription factors in controlling the tormogen differentiation program of Drosophila mechanoreceptors. *Dev. Biol.* **329,** 386–399.
Mizuguchi, R., Kriks, S., Cordes, R., Gossler, A., Ma, Q., and Goulding, M. (2006). Ascl1 and Gsh1/2 control inhibitory and excitatory cell fate in spinal sensory interneurons. *Nat. Neurosci.* **9,** 770–778.
Morel, V., and Schweisguth, F. (2000). Repression by suppressor of hairless and activation by Notch are required to define a single row of single-minded expressing cells in the Drosophila embryo. *Genes Dev.* **14,** 377–388.
Morel, V., Lecourtois, M., Massiani, O., Maier, D., Preiss, A., and Schweisguth, F. (2001). Transcriptional repression by suppressor of hairless involves the binding of a hairless-dCtBP complex in Drosophila. *Curr. Biol.* **11,** 789–792.

Nagel, A. C., Krejci, A., Tenin, G., Bravo-Patino, A., Bray, S., Maier, D., and Preiss, A. (2005). Hairless-mediated repression of notch target genes requires the combined activity of Groucho and CtBP corepressors. *Mol. Cell. Biol.* **25,** 10433–10441.
Nakhai, H., Sel, S., Favor, J., Mendoza-Torres, L., Paulsen, F., Duncker, G. I., and Schmid, R. M. (2007). Ptf1a is essential for the differentiation of GABAergic and glycinergic amacrine cells and horizontal cells in the mouse retina. *Development* **134,** 1151–1160.
Nellesen, D. T., Lai, E. C., and Posakony, J. W. (1999). Discrete enhancer elements mediate selective responsiveness of enhancer of split complex genes to common transcriptional activators. *Dev. Biol.* **213,** 33–53.
Obata, J., Yano, M., Mimura, H., Goto, T., Nakayama, R., Mibu, Y., Oka, C., and Kawaichi, M. (2001). p48 subunit of mouse PTF1 binds to RBP-Jkappa/CBF-1, the intracellular mediator of Notch signalling, and is expressed in the neural tube of early stage embryos. *Genes Cells* **6,** 345–360.
Oswald, F., Kostezka, U., Astrahantseff, K., Bourteele, S., Dillinger, K., Zechner, U., Ludwig, L., Wilda, M., Hameister, H., Knochel, W., Liptay, S., and Schmid, R. M. (2002). SHARP is a novel component of the Notch/RBP-Jkappa signalling pathway. *EMBO J.* **21,** 5417–5426.
Oswald, F., Winkler, M., Cao, Y., Astrahantseff, K., Bourteele, S., Knochel, W., and Borggrefe, T. (2005). RBP-Jkappa/SHARP recruits CtIP/CtBP corepressors to silence Notch target genes. *Mol. Cell. Biol.* **25,** 10379–10390.
Pascual, M., Abasolo, I., Mingorance-Le Meur, A., Martinez, A., Del Rio, J. A., Wright, C. V., Real, F. X., and Soriano, E. (2007). Cerebellar GABAergic progenitors adopt an external granule cell-like phenotype in the absence of Ptf1a transcription factor expression. *Proc. Natl. Acad. Sci. USA* **104,** 5193–5198.
Qin, H., Wang, J., Liang, Y., Taniguchi, Y., Tanigaki, K., and Han, H. (2004). RING1 inhibits transactivation of RBP-J by Notch through interaction with LIM protein KyoT2. *Nucleic Acids Res.* **32,** 1492–1501.
Qin, H., Du, D., Zhu, Y., Li, J., Feng, L., Liang, Y., and Han, H. (2005). The PcG protein HPC2 inhibits RBP-J-mediated transcription by interacting with LIM protein KyoT2. *FEBS Lett.* **579,** 1220–1226.
Rabbitts, T. H. (1998). LMO T-cell translocation oncogenes typify genes activated by chromosomal translocations that alter transcription and developmental processes. *Genes Dev.* **12,** 2651–2657.
Roux, E., Strubin, M., Hagenbuchle, O., and Wellauer, P. K. (1989). The cell-specific transcription factor PTF1 contains two different subunits that interact with the DNA. *Genes Dev.* **3,** 1613–1624.
Salat, D., Liefke, R., Wiedenmann, J., Borggrefe, T., and Oswald, F. (2008). ETO, but not leukemogenic fusion protein AML1/ETO, augments RBP-Jkappa/SHARP-mediated repression of notch target genes. *Mol. Cell. Biol.* **28,** 3502–3512.
Schweisguth, F., and Posakony, J. W. (1992). Suppressor of Hairless, the Drosophila homolog of the mouse recombination signal-binding protein gene, controls sensory organ cell fates. *Cell* **69,** 1199–1212.
Sellick, G. S., Barker, K. T., Stolte-Dijkstra, I., Fleischmann, C., Coleman, R. J., Garrett, C., Gloyn, A. L., Edghill, E. L., Hattersley, A. T., Wellauer, P. K., Goodwin, G., and Houlston, R. S. (2004). Mutations in PTF1A cause pancreatic and cerebellar agenesis. *Nat. Genet.* **36,** 1301–1305.
Swift, G. H., Hammer, R. E., MacDonald, R. J., and Brinster, R. L. (1984). Tissue-specific expression of the rat pancreatic elastase I gene in transgenic mice. *Cell* **38,** 639–646.
Taniguchi, Y., Furukawa, T., Tun, T., Han, H., and Honjo, T. (1998). LIM protein KyoT2 negatively regulates transcription by association with the RBP-J DNA-binding protein. *Mol. Cell. Biol.* **18,** 644–654.

Tsuji, M., Shinkura, R., Kuroda, K., Yabe, D., and Honjo, T. (2007). Msx2-interacting nuclear target protein (Mint) deficiency reveals negative regulation of early thymocyte differentiation by Notch/RBP-J signaling. *Proc. Natl. Acad. Sci. USA* **104,** 1610–1615.

Tun, T., Hamaguchi, Y., Matsunami, N., Furukawa, T., Honjo, T., and Kawaichi, M. (1994). Recognition sequence of a highly conserved DNA binding protein RBP-J kappa. *Nucleic Acids Res.* **22,** 965–971.

van Es, J. H., van Gijn, M. E., Riccio, O., van den Born, M., Vooijs, M., Begthel, H., Cozijnsen, M., Robine, S., Winton, D. J., Radtke, F., and Clevers, H. (2005). Notch/gamma-secretase inhibition turns proliferative cells in intestinal crypts and adenomas into goblet cells. *Nature* **435,** 959–963.

Weng, A. P., Ferrando, A. A., Lee, W., Morris, J. P., 4th, Silverman, L. B., Sanchez-Irizarry, C., Blacklow, S. C., Look, A. T., and Aster, J. C. (2004). Activating mutations of NOTCH1 in human T cell acute lymphoblastic leukemia. *Science* **306,** 269–271.

Wildner, H., Muller, T., Cho, S. H., Brohl, D., Cepko, C. L., Guillemot, F., and Birchmeier, C. (2006). dILA neurons in the dorsal spinal cord are the product of terminal and non-terminal asymmetric progenitor cell divisions, and require Mash1 for their development. *Development* **133,** 2105–2113.

Zhou, S., and Hayward, S. D. (2001). Nuclear localization of CBF1 is regulated by interactions with the SMRT corepressor complex. *Mol. Cell. Biol.* **21,** 6222–6232.

Zimber-Strobl, U., Strobl, L. J., Meitinger, C., Hinrichs, R., Sakai, T., Furukawa, T., Honjo, T., and Bornkamm, G. W. (1994). Epstein-Barr virus nuclear antigen 2 exerts its transactivating function through interaction with recombination signal binding protein RBP-J kappa, the homologue of Drosophila Suppressor of Hairless. *EMBO J.* **13,** 4973–4982.

CHAPTER FOUR

Sonic Hedgehog Signaling in the Developing CNS: Where It Has Been and Where It Is Going

Michael P. Matise* *and* Hui Wang*

Contents

* UMDNJ/Robert Wood Johnson Medical School, Piscataway, New Jersey, USA

Current Topics in Developmental Biology, Volume 97
ISSN 0070-2153, DOI: 10.1016/B978-0-12-385975-4.00010-3

Abstract

Sonic Hedgehog (Shh) is one of three mammalian orthologs of the Hedgehog (Hh) family of secreted proteins first identified for their role in patterning the *Drosophila* embryo. In this review, we will highlight some of the outstanding questions regarding how Shh signaling controls embryonic development. We will mainly consider its role in the developing mammalian central nervous system (CNS) where the pathway plays a critical role in orchestrating the specification of distinct cell fates within ventral regions, a process of exquisite complexity that is necessary for the proper wiring and hence function of the mature system. Embryonic development is a process that plays out in both the spatial and the temporal dimensions, and it is becoming increasingly clear that our understanding of Shh signaling in the CNS is grounded in an appreciation for the dynamic nature of this process. In addition, any consideration of Hh signaling must by necessity include a consideration of data from many different model organisms and systems. In many cases, the extent to which insights gained from these studies are applicable to the CNS remains to be determined, yet they provide a strong framework in which to explore its role in CNS development. We will also discuss how Shh controls cell fate diversification through the regulation of patterned target gene expression in the spinal cord, a region where our understanding of the morphogenetic action of graded Shh signaling is perhaps the furthest advanced.

1. Prologue: Identification of Shh

Sonic Hedgehog (Shh) is one of three mammalian orthologs of the Hedgehog (Hh) family of secreted proteins that was identified in the famous *Drosophila* screen to identify larval patterning genes in Tübingen over three decades ago (Nusslein-Volhard and Wieschaus, 1980). The subsequent early reports on the cloning and characterization of vertebrate Shh were met with great excitement as it became clear that this protein appeared to fulfill the role of an organizing center-derived morphogen that had been predicted by the elegant work of experimental embryologists nearly half a century prior (Echelard *et al.*, 1993; Krauss *et al.*, 1993; Riddle *et al.*, 1993; Roelink *et al.*, 1994). We now know that Shh plays a critical part in directing limb digit and skeletal polarity and in the patterned formation of cells within the ventral portion of the central nervous system (CNS). More recently, it has been found that Shh is also involved importantly in many other processes, including cell proliferation and axon guidance. Investigations into the cellular transduction cascade downstream of this signaling protein have revealed a surprisingly high degree of interspecies conservation between many core elements of the pathway, which not only greatly benefits our general understanding but also illustrates its central role in patterning biological tissues. Yet there are also some interesting and significant differences in the vertebrate pathway that suggest it has taken on new roles.

The list of biologically important roles for Hh signaling has also grown, revealing diverse functions in both embryos and adults, including many human cancers. Yet its most fundamental function still remains largely a mystery: How does this extracellular, diffusible protein alter the destiny and physiology of cells that it comes in contact with?

2. Setting the Stage: Shh Signaling During Nervous System Regionalization

Of the three vertebrate Hh homologs, only Shh is found in the mammalian CNS. It is expressed both during embryogenesis and, as is becoming more well appreciated, also in adults, where its function may be related to one that is among the most commonly conserved across species and tissues: maintaining reservoirs of stem cells that are located in specific niches (Ahn and Joyner, 2005; Balordi and Fishell, 2007; Han *et al.*, 2008; Palma *et al.*, 2005; Traiffort *et al.*, 2010). The degree to which this latter role may recapitulate aspects of the former is an interesting area of further consideration but will not be addressed in this review. Rather, we will focus on the significance and role of Shh during CNS development, recognizing that knowledge gained from such studies may provide an important framework on which to build a better appreciation of its role in adults and in human disease.

Shh begins to exert an influence over CNS development at the very earliest stages of embryogenesis when neural tissue is first allocated during gastrulation. At that time, expression is seen in the embryonic node, the undifferentiated group of cells that are responsible for producing the signals that induce neural ectoderm and then for ventralizing this tissue. At presumptive spinal cord levels, Shh expression is maintained in the mesodermal notochord underlying the ventral neural plate and then, shortly thereafter, in the floor plate (FP), a small group of cells located at the ventral neural midline that disappear later in development. In the forebrain, midline mesodermal cells expressing Shh form the prechordal plate, which does not coalesce into a rod-shaped structure like the notochord. Here, too Shh expression is initially induced secondarily in the midline, but is soon downregulated as the hypothalamus and pituitary begin to arise, after which time, it is only expressed in two lateral domains in the approximate boundary between the future hypothalamic/thalamic boundary. After brain vesicle formation, Shh expression is also upregulated in the zona limitans intrathalamica (ZLI), a morphological dividing line between emerging thalamic nuclei in the caudal diencephalon that extends dorsally from the FP in the midbrain, and later still in distinct foci in the preoptic area, presumptive amygdala, and the medial ganglionic eminence (MGE), the anlage to the basal ganglia (Sousa and Fishell, 2010). Another region of the CNS where Shh expression has been characterized is in the developing cerebellum. Here, Shh is seen in postmitotic Purkinje cells, and from this source signals to granule cell precursors to support their extensive expansion (Lewis *et al.*, 2004; Wechsler-Reya and Scott, 1999). This relationship has provided a unique opportunity to study the regulation of cell proliferation by Shh, a pursuit that has important clinical implications for understanding childhood brain tumors arising in the cerebellum (Hatten and Roussel, 2011; Vaillant and Monard, 2009).

Shh serves as an important early signal from all of these regions but its specific requirement varies. Early loss of Shh is associated with the most severe consequences, a failure to induce ventral midline neural tissue, resulting in holoprosencephaly and cyclopia, a condition where only a single forebrain vesicle and eye form (Chiang *et al.*, 1996). Shh in the thalamic midline and ZLI is required to generate specific thalamic neuronal subtype (Jeong *et al.*, 2011). Loss of downstream Shh transduction components also results in the misspecification of ventral forebrain neuronal structures and subtypes (Carney *et al.*, 2010; Cocas *et al.*, 2009; Wang *et al.*, 2010b; Xu *et al.*, 2010; Yu *et al.*, 2009).

3. I Am Dying to Hear from You: The Range of Shh Signaling in the CNS

One common theme that emerges from this body of work is that, as development brings about significant, region-specific tissue expansion, Shh seems to take on a progressively more spatially limited role in regulating neuronal cell fates. This observation raises one of the key questions regarding Shh signaling in the CNS: at what distance from the source can it exert a direct influence over cell behavior? The most straightforward, but technically challenging, approach to this question is to visualize the proteins *in situ* using antibody staining or fusion-reporter techniques. In one classic example, Shh protein distribution was examined in tissue fixed to preserve extracellular proteoglycans, which are known to bind Hh ligands (see below; Gritli-Linde *et al.*, 2001). In all cases, Shh is detected far from its source cells, roughly halfway the distance toward the dorsal neural tube in both brain and spinal cord by the time neurogenesis is fully underway. However, this is not as far as where phenotypes are detected in its absence. In *Shh* nulls, there is an overall decrease in the size of the neural tube accompanied by reduced cell proliferation rates and a substantial increase in cell death throughout both ventral and dorsal regions at both brain and spinal cord levels (Chiang *et al.*, 1996; Huang *et al.*, 2007a; Thibert *et al.*, 2003). One fundamental question is whether these phenotypes are a direct or indirect consequence of reduced Shh signaling.

In support of the latter possibility, it has been shown that there is an upregulation of dorsal signaling molecules including Fgf8, Wnt3a, and BMP4 proteins when Shh signaling is reduced (Hayhurst *et al.*, 2008; Himmelstein *et al.*, 2010; Rash and Grove, 2007). However, these phenotypes can be largely reversed if one of the key repressors of Shh signaling that is expressed dorsally, Gli3, is also reduced or absent (Litingtung and Chiang, 2000; Rash and Grove, 2007). Since intracellular Gli3 repressor levels are directly under the control of Shh signaling (discussed in greater detail below), these results indicate a link between dorsal Gli3 activity and long-range Shh activity. Further, they suggest that early cell cycle and cell death genes may be among the most sensitive targets of the Shh–Gli pathway.

Another long-range function associated with changes in Shh signaling is commissural axon guidance (Sanchez-Camacho and Bovolenta, 2009). In mice lacking Shh expression in the FP, ventrally projecting neurons located in the dorsal spinal cord begin to make guidance errors around 2/3 of the way toward their ventral midline target, just after entering the basal plate region (Charron *et al.*, 2003). These observations together suggest complexity in the mechanisms that control the spread and diffusion of Shh at different stages of development, a topic that will be discussed in the section below.

A second approach involves assaying the output of the pathway in responding cells by examining changes in the expression of putative or known target genes or reporters driven by native target gene regulatory elements or constructs generated from multimerized binding sites for Gli transcription factors, the primary genetic mediators of "canonical" Shh signaling. Two of the most commonly employed lines involve *lacZ* targeted into the *Gli1* and *Ptc1* loci (reviewed in Zhu and Scott, 2004). Of the two, monitoring *Gli1* is more straightforward as its expression absolutely depends on Shh signaling (Bai *et al.*, 2004; Lei *et al.*, 2004). In contrast, the link between *Ptc1* expression and Shh signaling levels is more complex. As the receptor for Shh, Ptc is already expressed at basal levels in most tissues prior to the appearance of Shh but is upregulated in response to signaling (Goodrich *et al.*, 1997). In both cases, the major question is whether these target genes have only a limited sensitivity to signaling and therefore would not report activity at the lowest meaningful levels *in vivo*. Therefore, a definitive answer to questions regarding the range of Shh signaling will likely require new sensitive tools that would allow investigators to directly assay Shh diffusion *in vivo*.

4. Character Creation: Production and Secretion of Shh Protein

The distance over which Shh can travel is intimately linked to the biochemical properties of the active signaling species that affect its interaction with components in the tissue environment (Beachy *et al.*, 2011; Eaton,

2008; Gallet, 2011; Guerrero and Chiang, 2007). Active Shh protein is produced from a longer precursor protein via an intramolecular cleavage reaction that involves the addition of a cholesterol adjunct to the C-terminus of the N-terminal peptide that is the active signaling species (Lewis *et al.*, 2001; Porter *et al.*, 1996), and S-palmitoylation of the N-terminus by Skn/Hhat, which occurs in the secretory pathway (referred to as Shh-Np; Buglino and Resh, 2008). Shh-Np is ultimately released at the cell surface in a regulated process mediated by Dispatched-1 (Disp1; Caspary *et al.*, 2002; Ma *et al.*, 2002; Tian *et al.*, 2005), a transmembrane (TM) protein containing sterol-sending domains with homology to Ptc1, the receptor for Hh proteins. The presence of the cholesterol adjunct renders Shh-Np peptides hydrophobic and charge polarized, imposing restrictions on its ability to move away from the cells that produce it (Goetz *et al.*, 2006). Notably, artificially produced Shh proteins lacking the cholesterol moiety (Shh-N) can be released independent of Disp1 and would be predicted to diffuse further in tissues without membrane anchoring (Huang *et al.*, 2007b). More recent evidence shows that Disp1 is required cell-autonomously for contact-dependent spread of Shh-Np protein in cultured embryoid bodies, suggesting this factor could function not only for the proper release of lipid-modified Shh protein but also for its diffusion (Etheridge *et al.*, 2010). Thus, cholesterol modification plays a critical role in determining the range and shape of the extracellular gradient, which are both critical determinants of the response.

The precise form in which Shh is released into the extracellular environment and how it moves away from the cells producing it and between cells in the target field is another open question (Eaton, 2008). The extracellular environment comprises lipid surfaces in the form of cell membranes, cell surface proteins, diffusible proteins, and extracellular matrix (ECM). Movement of signaling proteins such as Shh through this milieu involves both active and passive mechanisms. At least two theoretical signaling/transport configurations are conceivable: one involving monomeric Shh-Np and the other multimeric (Eaton, 2008). If monomers were released, they would likely require the association with additional proteins to neutralize their inherent hydrophobicity. In this form, secreted Shh-Np might remain closely associated with lipid membranes of producing cells and their close neighbors. Another configuration would involve the assemblage of Shh-Np proteins into micelle-like, multimeric structures (e.g., lipoprotein particles, analogous to the way cholesterol itself is transported in circulating blood) exposing only hydrophilic portions of the complex to the extracellular space. Supporting this, recent evidence shows that Hh proteins can be detected in high molecular weight particles *in vivo* and *in vitro* (Gallet, 2011). Notably, Shh protein can only be detected in the ventral midline (notochord and FP) in mouse *Skn* mutants that produce only monomeric Shh-N protein (Chen *et al.*, 2004a). Thus, if both forms were generated *in vivo*, their range of influence would differ significantly: a local (autocrine or limited paracrine) response would be expected for monomeric Shh-N, while long-range effects would be mediated by Shh-Np. One question that remains is whether both forms can bind to the Ptc receptor with similar affinities or if some aspects of the *Skn* mouse mutant phenotype (which resemble *Shh* mutants; Chen *et al.*, 2004a) can be explained by a lower activation strength for monomeric Shh. However, even if the latter possibility is true, direct visualization of protein distribution in Shh–GFP tagged mice shows that it can be detected further from the FP in *Skn* mutants than in wild-type (WT) embryos, suggesting at minimum that the diffusion rate is altered by changes in N-terminal palmitoylation (Chamberlain *et al.*, 2008). Although no experimental evidence has been provided to date supporting the presence of monomeric Shh in tissues, if they are released, they might contribute toward the establishment of the overall Shh extracellular gradient. In this scenario, monomeric or multimeric Shh could play specific roles in organizing different aspects of local or long-range tissue patterning (e.g., progenitor gene expression, cell proliferation, or axon guidance). Finally, recent studies characterizing human SHH alleles associated with holoprosencephaly raise the interesting possibility that full-length, unprocessed Shh protein can be released and is bioactive *in vivo*, although these results contradict earlier studies (Tokhunts *et al.*, 2010). A better understanding of the species of Shh complexes found *in vivo* and how their formation and processing are regulated will be required to resolve these issues.

5. Won't You Take My Hand: Shh-Binding Proteins and Diffusion

A number of proteins have been identified that bind to Hh proteins and play a key role in shaping the extracellular gradient. Several recent reviews have addressed this issue (Beachy *et al.*, 2011; Gallet, 2011; Wilson and Chuang, 2010), so here we will only briefly highlight a few key points. Hh-binding proteins can be generally divided into two categories: those that promote or amplify the signal and those that oppose or antagonize it. Intriguingly, both categories of Hh-binding proteins are directly regulated by Hh signaling; positive cofactors are inhibited, while negative cofactors are induced. This multitiered negative feedback loop is critical to generate the proper intracellular responses to extracellular signaling strength over time, as will be discussed in a later section.

Cdo/Cdon, Boc, and Gas1 are three factors that belong to the category of Shh signal amplifiers. Cdo and Boc are TM proteins with closely related structures, each containing similar numbers of Fibronectin-type repeats and four or five Ig repeats, as well as divergent cytoplasmic domains (Tenzen *et al.*, 2006; Zhang *et al.*, 2006). Cdo is expressed in the FP region of the neural tube, and in *Cdo* mutants, this structure is not fully induced by signaling from the underlying notochord (Tenzen *et al.*, 2006). In addition, *Cdo* mutants exhibit mild holoprosencephaly, a failure to induce forebrain midline CNS structures that typifies many mutations that reduce Hh signaling. In contrast, *Boc* mutants have commissural axon guidance defects, a phenotype that is perhaps more consistent with the predominantly dorsal expression pattern of this gene (Okada *et al.*, 2006). Another Shh-binding factor is Gas1, a GPI-anchored cell surface protein (Lee *et al.*, 2001). Like *Cdo, Gas1* mutants fail to fully induce a FP, indicating a similar positive role in Shh signal transduction (Allen *et al.*, 2007). Further, genetic evidence indicate that Cdo and Gas1 cooperate to promote Shh signaling, suggesting they serve to amplify Shh signaling in specific cells/tissues (Allen *et al.*, 2007). Because Shh signaling also serves to downregulate the expression of Cdo/Boc/Gas1, together these data indicate that they function to transiently "sensitize" cells to the ambient levels of signaling. The exact mechanisms for how these proteins promote Shh signaling have not been definitively established. However, the fly Cdo/Boc ortholog, Ihog, has been shown to function as a necessary coreceptor for Hh binding to Ptc (Zheng *et al.*, 2010). Together, these results suggest the possibility that Cdo (and potentially Boc) promote Shh signaling by enhancing its activity at the receptor. Although the role of Gas1 in this process has remained unclear, genetic and biochemical evidence supports the idea that it too is involved in delivering Shh protein to Ptc1 for signaling. One impediment is that no invertebrate *Gas1* homolog has been identified, so its role will need to be dissected in the more complex vertebrate pathway. Even more of a mystery surrounds the potential function of the intracellular domains of Cdo and Boc, and whether these domains contribute in some way to the Shh ligand-mediated inhibition of Ptc1, which may involve conformational changes in the protein (Wilson and Chuang, 2010).

Another group of Hh-binding factors function in opposition to Cdo/Boc/Gas1 by inhibiting pathway activity. Ptc1, the Hh receptor (discussed in the next section), is the primary member of this group, while another vertebrate-specific protein called Hhip (Hh interacting protein) also antagonizes signaling. The structural similarities in the Shh-binding domains of both Ptc1 and Hhip suggest a model where Hhip competes with Ptc1 for Shh ligand binding, thus inhibiting signaling (Chaung and Kornberg, 2000). The negative role of Ptc1 in the pathway is more complex and multifaceted. On the one hand, the basic function of Ptc1 is block downstream pathway activation (by inhibiting Smo, as will be described below) in the absence of Hh in a process termed "ligand-independent antagonism" (Jeong and McMahon, 2005). However, the internalization of Hh-bound Ptc1 in cells exposed to the signal also serves another role: the sequestration of ligand that limits its extracellular spread. Additionally, both *Ptc1* and *Hhip* expression are positively regulated by Hh signaling such that more Ptc1 and Hhip are eventually made in cells exposed to Hh. The net effect of this regulatory feedback loop is to attenuate the cellular response to signaling over time. This process has been called "ligand-dependent antagonism" (Jeong and McMahon, 2005). Together, both positive and negative feedbacks likely serve to dynamically regulate the shape of the extracellular concentration gradient of Shh outside the cell in a manner that likely contributes buffering and stabilization functions to potential fluctuations in ligand concentrations.

Hh proteins also bind to glycosaminoglycans in the ECM, specifically heparan sulfate proteoglycans (HSPGs) produced by the integral membrane proteins Ttv (in *Drosophila*) or Ext in vertebrates (Varjosalo and Taipale, 2008). Shh has been shown to bind to HSPGs through a conserved 7-peptide C-terminal domain known as the Cardin–Weintraub motif, which is required for binding to HSPGs but not Ptc1 (Rubin *et al.*, 2002). HSPGs play a widespread role in many signaling pathways including Tgf-β and Wnt, and generally function in a similar manner: to restrict ligand spreading in the ECM. Three lines of evidence support this role for Shh in vertebrates. First, mutations in the HSPG-binding motif in Shh result in a proliferative defect but no patterning deficit in the developing mouse spinal cord, a phenotype that resembles that seen in animals that were engineered to produce only noncholesterol-modified Shh protein that has a greater extracellular diffusion rate (Chan *et al.*, 2009). Second, mice mutant for *GPC3*, a member of the GPI-linked glypican HSPG family that appears to compete with Ptc for Shh binding, also display overgrowth defects consistent with enhanced long-range Shh signaling (Capurro *et al.*, 2008). Finally, both *Sulf1*, an HSPG sulfation modifier, and *Ext1* are expressed in chicken ventral midline cells, and upregulation of Sulf1 activity in the ventral spinal cord has been linked to increased Shh protein levels in oligodendrocyte (OL) progenitors prior to their generation *in vivo* (Danesin *et al.*, 2006). It is not clear whether any of these HSPG regulators are themselves targets of Shh and thus could participate in feedback regulation. Together, these results illustrate the important, potentially signaling-independent role that HSPGs have in establishing the extracellular Shh gradient by regulating the ECM diffusion rate.

An additional Shh-binding protein that has been identified is Megalin/Lrp2/gp 330, a member of the low-density lipoprotein (LDL) receptor family (McCarthy and Argraves, 2003). *Megalin* mutant mice develop holoprosencephaly and show abnormal patterning of the ventral telencephalon, phenotypes associated with diminished Shh signaling (Spoelgen *et al.*, 2005). Further, Megalin has been shown to bind to Shh and can mediate ligand internalization. Notably, in the neural tube, Megalin expression is concentrated at the apical surface of ependymal zone (EZ) epithelial cells lining the central canal (Wicher *et al.*, 2005). This localization supports one of the proposed functions for Megalin as a mediator of Shh ligand transcytosis across ventral EZ cells (McCarthy *et al.*, 2002). This potential function could explain the interesting observation that Shh protein can be detected at both the apical and basal surfaces of the early neural tube (Gritli-Linde *et al.*, 2001). Megalin could also function to deliver Shh to intracellular pools of Ptc to promote signaling. In the spinal cord, loss of Megalin leads to a progressive hypoplasia that results in gross morphological abnormalities at later gestational stages (Wicher and Aldskogius, 2008). It is unclear whether this late-arising phenotype is related to a role for Megalin/Shh signaling in promoting ongoing tissue growth or to a Shh-independent requirement; further studies will be necessary to address these and other issues.

6. Character Development: Transduction of Shh Signal in Receiving Cells

Two distinct TM proteins are required for Shh signal transduction: Ptc (Patched) and Smo (Smoothened). Ptc, which functions as the receptor for Shh, is a 12-pass integral membrane protein with homology to bacterial proton-driven TM transporters. Smo is an integral TM protein whose seven-membrane spanning domain is homologous to G-protein-coupled receptors (GPCRs), while the N-terminal extracellular domain contains a cysteine-rich domain (CRD) motif that is structurally related to Frizzled proteins that function as Wnt receptors (Alcedo *et al.*, 1996); however, no natural ligand has been identified that binds to the Smo CRD. The C-terminal intracellular domain of Smo differs significantly between flies and vertebrates and is likely to be involved in signal transduction in both organisms, but its specific role and importance appear to have diverged (Huangfu and Anderson, 2006).

Ptc serves as a constitutive pathway inhibitor in the absence of Shh by blocking the activity of Smo, the primary receptor-like positive signaling element in the transduction pathway. When Shh binds to Ptc1, Smo activity is disinhibited, triggering a cascade of intracellular events that lead to the accumulation of the gene-activating form of the Gli transcription factor proteins. Smo activation ultimately leads to changes in the phosphorylation status of Gli proteins, the primary mediators of Shh target gene expression. There are three Gli proteins in vertebrates, Gli1–3, that are homologous to the *Drosophila* Cubitus interruptus (Ci) protein. In the absence of Hh signaling, Gli2 and Gli3 proteins are constitutively phosphorylated by protein kinase-A (PKA), casein kinase I (CKI), and glycogen synthase kinase 3 (Gsk3; Pan *et al.*, 2006, 2009; Tempe *et al.*, 2006). In vertebrates, *Gli1* is a target of Shh signaling and is only present after signaling has commenced; thus, in the absence of Shh, no Gli1 protein is present. For Gli2, Gli3, phosphorylation occurs at conserved sites located within the C-terminal half of each protein (Pan *et al.*, 2006; Tempe *et al.*, 2006). Hyperphosphorylated Gli2/Gli3 proteins become substrates for ubiquitination and proteosomal targeting via β-Trcp/Cul1 proteins (Tempe *et al.*, 2006; Wang and Li, 2006). In the proteosome, each Gli protein is processed in a distinct manner. Gli3 is proteolyzed partially to form a C-terminally truncated protein that functions as a transcriptional repressor (Gli3R). In contrast, the proteosomal processing of Gli2 into a shorter repressor form is much less efficient than for Gli3, with the majority of the protein being fully degraded (Pan and Wang, 2007; Pan *et al.*, 2006). It is still unclear why Gli2 is preferentially degraded rather than truncated, whereas for Gli3, the opposite situation occurs, but this appears to be related to the divergence of their C-terminal processing determinant domains (PDD; Li *et al.*, 2011; Pan and Wang, 2007). Previous studies have provided genetic evidence for a role for Gli2R in epaxial muscle progenitors in the developing somites, but this is only revealed once the more significant Gli3R influence is removed, consistent with their differential processing (Buttitta *et al.*, 2003; McDermott *et al.*, 2005). It is unclear whether Gli2R plays a similar minor role in CNS development. Thus, of the three Gli proteins, Gli3 plays the most significant role in the CNS absence of Shh signaling.

Upon relief of Smo inhibition by Shh binding to Ptc1, constitutive Gli2 and Gli3 protein phosphorylation is inhibited. As a consequence, Gli2/Gli3 proteins escape from the β-Trcp/Cul1-mediated proteosome degradation/cleavage pathway and instead remain as full-length proteins retaining the C-terminus that function as transcriptional activators (GliA). However, there are also important differences between the subsequent processing of the two proteins in this form. Recent experiments in cultured cell lines provide evidence that both full-length Gli2 and Gli3 proteins are targeted for proteosomal degradation via a Spop/Cul3-dependent proteosome pathway (Wang *et al.*, 2010a; Wen *et al.*, 2010). Interestingly Spop protein can be detected in both the cytoplasm and nucleus in cultured cells, suggesting roles in both compartments (Chen *et al.*, 2009). One interesting question is whether Spop differentially mediates the degradation of full-length Gli2 and Gli3. Notably, it has been shown in cell lines that Shh signaling can promote the Spop-mediated degradation of full-length Gli3 (Wen *et al.*, 2010). In contrast, in addition to inhibiting processing, Shh signaling also blocks full-length Gli2 degradation, although it is unclear whether this is mediated via Spop (Pan *et al.*, 2006). Thus, if full-length Gli3 were degraded more rapidly than Gli2 or before nuclear entry, it would provide another possible explanation for genetic data showing that Gli2 is the more important activator of Shh target genes than Gli3. There is also evidence that Shh may regulate full-length Gli2A activity at an additional level. In mice engineered to produce a modified form of Gli2 that cannot be phosphorylated by PKA, a gain of function phenotype is seen in both the brain and spinal cord consistent with enhanced Shh-pathway activation and Gli2A stability (Pan *et al.*, 2009). This result suggests that, under normal signaling conditions, some small proportion of full-length Gli2 is degraded, and this may contribute to establishing the proper Gli2A levels. Further, these phenotypes are largely reversed in the absence of Shh, indicating that inhibition of PKA-mediated phosphorylation is not sufficient to fully activate Gli2 (Pan *et al.*, 2009). These results together suggest that Shh exerts at least two additional levels of control over Gli2A activity, raising the possibility that this regulatory interaction is sensitive to graded signaling levels.

In contrast to Gli2 and Gli3, Gli1 does not appear to contain a Spop recognition region and is not degraded via the Spop/Cul3-mediated proteosomal pathway (Chen *et al.*, 2009). Notably, two degradation initiation regions "degrons" were identified in the N-terminus and C-terminus of Gli1, and deletion studies have shown that the latter is required for β-TrCP-mediated degradation and its loss can accelerate tumor formation in skin (Huntzicker *et al.*, 2006). Thus, full-length Gli1 and Gli2/Gli3 proteins are differentially processed and regulated by Shh and proteosomal degradation.

7. Cutting in on the Dance: The Interplay Between Ptc and Smo

The limited knowledge of the mechanisms by which Ptc1 activity controls Smo is one of the biggest holes in our current understanding of the Hh transduction pathway in all organisms. Complicating matters somewhat in vertebrates is the fact that there are two *Ptc* genes expressed in overlapping patterns in the mammalian CNS, *Ptc1* and *Ptc2*, whose protein products both bind to Shh with similar affinities (Carpenter *et al.*, 1998; Motoyama *et al.*, 1998a,b). Of the two, *Ptc1* is more important in CNS development as null mouse mutants are lethal at early neural tube stages and exhibit severe patterning defects consistent with its normal role as a constitutive suppressor of the pathway (Goodrich *et al.*, 1997). In contrast, *Ptc2* mutants are viable but develop problems related to skin postnatally (Lee *et al.*, 2006), although it is unclear if these reflect a role in later Hh signaling. Thus, at least during CNS development, *Ptc1* appears to serve as the primary Hh receptor.

8. Taking It to the Extreme: The Central Role of the Primary Cilium in Shh Signaling

A number of recent studies have brought the role of Ptc1 in the pathway into clearer focus. It has been shown that Ptc1 does not interact directly with Smo but rather functions indirectly to maintain it in a signaling-inactive state, which may involve retention in the endoplasmic reticulum (ER) or other intracellular endosomal compartment (Chen *et al.*, 2002a; Taipale *et al.*, 2002). Localization studies reveal that Ptc1 is present in a specialized structure known as the primary cilium in the absence of Hh signaling. Hh ligand binding to Ptc induces its clearance from the cilium and into endosomes for degradation or, potentially, rerelease (Eggenschwiler and Anderson, 2007; Rohatgi *et al.*, 2007; Wong and Reiter, 2008). This is followed by the translocation of Smo into the cilium along microtubules via an active transport mechanism involving at least two adapter proteins, Kif3a and β-arrestin (Corbit *et al.*, 2005; Kovacs *et al.*, 2008). Thus, the primary function for Ptc in the Hh pathway appears to be to prevent Smo from entering the microenvironment at the tip of the primary cilium. Recent work clearly implicates this vertebrate-specific cellular organelle as playing a central role in the transduction of Shh signaling (reviewed in Eggenschwiler and Anderson, 2007; Wong and Reiter, 2008), although some questions remain about whether it serves as the only site of signal transduction within target cells. Indeed, there is evidence from zebrafish that Smo ciliary localization may only be required for full Hh-pathway activation but not low-level signaling (Aanstad *et al.*, 2009). Most vertebrate cells, including those within the CNS of developing embryos, possess a single membrane protrusion known as the primary cilium. This immotile structure is thought to serve as cellular "antenna," allowing the sensing of extracellular molecules or mechanical deflection (Eggenschwiler and Anderson, 2007). Consistent with this, in the neural tube, cells lining the ventricle extend an apical cilium into the central canal, where Shh protein has been shown to accumulate (Caspary *et al.*, 2007; Chamberlain *et al.*, 2008). In all cells, IFT (intraflagellar transport) proteins play a key structural role in cilia formation through retrograde and anterograde intracellular transport of cargo in and out of the cilium along microtubules (Eggenschwiler and Anderson, 2007). The critical role of these transporter proteins, and of the primary cilium, in Shh signal transduction was discovered through mouse mutagenic screens that perturb neural patterning; many of the genes identified were found to encode IFT or associated proteins (Herron *et al.*, 2002; Huangfu *et al.*, 2003; Reiter and Skarnes, 2006). The general model that has emerged from this work indicates that the primary cilium is a key site for localization of receptor components that mediate signal transduction.

Interestingly, although Smo relocation to the cilium appears to be required for its activation (Corbit *et al.*, 2005), it does not appear to be sufficient as translocation still occurs in the presence of cyclopamine, a naturally occurring plant alkaloid that is a potent inhibitor of Hh signaling, as well as several other antagonists (Ayers and Therond, 2010; Rohatgi *et al.*, 2009; Wang *et al.*, 2009). It has been shown that cyclopamine can bind directly to the TM domain of Smo and possibly alter its structure (Chen *et al.*, 2002a; Wilson *et al.*, 2009a). In addition, small molecule agonists have been identified, some of which also bind to the TM Smo domain (Chen *et al.*, 2002b). More recently, naturally occurring molecules such as oxysterols and possibly provitamin D3 have been shown in cell culture assays to bind Smo, and can function to activate or inhibit activity, respectively (Bijlsma *et al.*, 2006; Corcoran and Scott, 2006). Notably, antagonists such as cyclopamine and the related compound jervine, as well as Smo agonists such as purmorphamine, are structurally similar to sterols, suggesting that they function by competing against or inhibiting/mimicking the effects of endogenous Smo ligands in the ciliary membrane. Together, these data support a two-step model where translocation to the cilium is followed by a second event that is required to induce Smo activation, likely involving a ligand-mediated conformation change in the protein (Rohatgi *et al.*, 2009).

Structural studies of Ptc1 have revealed that it has homology to RND (resistance-nodulation-division) family proteins that function in bacteria to transport small sterol-like molecules across cell membranes (Taipale *et al.*, 2002). Ptc1 also contains a sterol-sensing domain (SSD) with homology to NPC1 (Niemann-Pick type C1), a protein that transports lipophilic molecules (Eaton, 2008). These data suggest a mechanism whereby Ptc1 indirectly influences Smo subcellular trafficking by controlling the flow of molecules into and out of the cell or within intracellular vesicles (for review, see Ayers and Therond, 2010; Eaton, 2008). The presence of the RND domain of Ptc1 further suggests that the pH of the cellular compartments is also important. Thus, intracellular lipid trafficking appears to be a key link between Ptc and Smo (Eaton, 2008). Incorporating these data lead to a model where Ptc1 activity constitutively prevents the activation of Smo by either keeping the net levels of positively acting oxysterols low or increasing the overall ratio of negative to positive sterols. Shh binding to Ptc1 relieves this inhibition by allowing positive sterol ligands to accumulate and induce Smo translocation and/or function, a process that may involve conformational changes in Smo. Whether Ptc directly regulates the levels of endogenous membrane-associated sterol ligands or the access of Smo to them, or both, is not clear. The fact that sterol intermediates and derivatives are also normally produced in the cholesterol biosynthetic pathway provides for numerous potential endogenous Smo regulators. Further, it is also conceivable that the sterol-transporting function of Ptc could independently influence the activity of proteins that directly mediate Smo trafficking so that they become inactive when Ptc is localized to the ciliary membrane.

IFT proteins have also been shown to be required to transport pre- and postprocessed Gli proteins in and out of the cilium and are thus involved in the canonical pathway leading to target gene regulation (Eggenschwiler and Anderson, 2007). As described above, the phosphorylation status of Gli proteins determines whether they are targeted to the cytoplasmic proteosome, which occurs near the base of the cilium, or directed into the nucleus to control gene expression. Thus, Gli2 and Gli3 proteins take different routes to the nucleus depending on whether they first visit the proteosome or not. It would be interesting to determine whether there are distinct IFT proteins and/or transport mechanisms serving each of these trajectories.

9. Stepping into the Dark: How Does Smo Control Gli Phosphorylation?

The overall role of Smo as a key activator of Gli/Ci proteins has been conserved between flies and vertebrates, yet significant differences have arisen in the specific mechanisms of action that have added complexity. This can largely be attributed to the fact that vertebrate Hh signaling involves an evolutionarily novel cellular compartment, the primary cilium (as described above). The appearance of this specialized structure in vertebrate cells is likely to have been the driving force for the functional divergence of conserved factors, as well as the novel additions that are interposed in the pathway at this critical juncture.

In flies, activated Smo alters Ci phosphorylation by interfering with its association with a complex of kinase proteins that retain Ci in the cytoplasm in the absence of signaling and are required for its sequential phosphorylation that leads to its processing into a repressor (Hooper and Scott, 2005). The factors in this complex include Cos2, a kinesin-related scaffold protein that functions to tether Ci to a cytoplasmic complex; fused kinase (Fu); suppressor of fused (Su(Fu)); and PKA/Gsk3/CKI kinases (Zhang *et al.*, 2005). When signaling occurs, changes in Smo conformation (brought about by phosphorylation by PKA/CKI) allow it to bind to Cos2 at its C-terminal tail (Jia *et al.*, 2003, 2004; Ogden *et al.*, 2003). This induces the dissociation of Ci from the complex where it escapes phosphorylation and subsequent processing in the proteosome to remain a full-length activator protein (Ruel *et al.*, 2003, 2007).

Although all of these core factors appear to have homologs in mammals, their functional role in signal transduction has not been fully conserved. In flies, Cos2 primarily functions as a negative regulator by promoting Ci processing into a truncated repressor (Smelkinson *et al.*, 2007; Wang *et al.*, 2000b). However, Cos2 also positively regulates Hh signaling by directly interacting with Smo (Lum *et al.*, 2003). At least two factors with homology to Cos2, Kif7, and Kif27 have been identified in mice. Recent mutant mouse data have provided clear evidence that Kif7 plays a conserved role in the Shh pathway. Loss of Kif7 results in polydactyly and a dorsal expansion of ventral spinal cord cell types, a phenotype similar to but milder than *Ptc1* mutants (Endoh-Yamagami *et al.*, 2009; Liem *et al.*, 2009). Consistent with this, biochemical data show that Kif7 is essential for Gli3 repressor processing *in vivo*. Genetic evidence also suggests that Kif7 activity contributes weakly to pathway upregulation in the absence of Ptc1 (Liem *et al.*, 2009). Taken with expression studies that indicate it localizes to the primary cilium, these data together support the idea that Kif7 plays a similar homologous dual role as Cos2 in the mammalian Hh pathway. In contrast, the only function ascribed to Kif27 to date is related to motile ciliogenesis (Wilson *et al.*, 2009b). Thus, it is currently unclear whether Kif27 function is partially redundant with Kif7 in mediating mammalian Shh signaling transduction, although if so, it is only likely to serve a comparatively minor role.

Another factor shared in both fly and vertebrate pathways is cAMP-dependent PKA, a key intracellular factor that has a broad role in phosphorylating many protein targets including Ci/Gli (Chen *et al.*, 1998; Kaesler *et al.*, 2000; Price and Kalderon, 1999; Wang *et al.*, 2000a). In addition to its well-characterized role in promoting full-length Gli/Ci protein degradation/processing into repressors, phosphorylation of other Hh-pathway targets has been shown to positively regulate signaling in both flies and vertebrates, yet likely through different mechanisms. In flies, Smo activation is dependent on the PKA-mediated C-terminal tail phosphorylation (Jia *et al.*, 2004). In mammals, PKA activity appears to be linked to the lateral translocation and activation of Smo into the cilia in cell lines, although this may be through an indirect pathway since vertebrate Smo proteins do not contain conserved PKA sites and evidence that it can be phosphorylated *in vivo* has been lacking (Milenkovic *et al.*, 2009; Varjosalo *et al.*, 2006). Therefore, for both flies and vertebrates, PKA may serve both negative and positive roles in the Hh pathway although the mechanisms for the latter appear to differ significantly.

Current evidence suggests that Fu kinase no longer has a central role in the vertebrate Hh pathway. In flies, Fu positively regulates Hh signaling by phosphorylating Cos2 and Su(Fu), promoting their dissociation from the cytoplasmic Ci-tethering complex (Lum *et al.*, 2003; Ruel *et al.*, 2007; Therond *et al.*, 1996). In addition, Fu activity is intimately linked to Su (Fu) (Alves *et al.*, 1998; Dussillol-Godar *et al.*, 2006; Pham *et al.*, 1995). By contrast, mouse *Fu* mutants do not exhibit any of the classical phenotypes associated with diminished Hh signaling (Chen *et al.*, 2005; Merchant *et al.*, 2005). Further, loss of Fu activity does not alter the spinal cord phenotype seen in *Su(Fu)* mutant mice, suggesting an uncoupling of their function in the vertebrate Hh pathway (Chen *et al.*, 2009). Thus, Fu appears to be dispensable for mammalian Shh signaling transduction. It will be important, however, to determine whether other unidentified mammalian Fu orthologs or related kinases (e.g., Ulk3) have taken the place of Fu in Shh signal pathway (Maloverjan *et al.*, 2010).

Su(Fu) has emerged as an important conserved regulator of Hh signaling that has taken on new and specific roles in mammals. Notably, mammalian Su(Fu) regulates Gli activities at multiple levels including the primary cilium, cytoplasm, and nucleus. In flies, *Su(Fu)* mutants are viable and the repressive function of Su(Fu) is revealed only on a *Fu* mutant background (Dussillol-Godar *et al.*, 2006; Preat, 1992). In contrast, in *Su(Fu)* null mutant mice, ventral gene expression in the neural tube is expanded dorsally, a phenotype that is strikingly similar to *Ptc1−/−* knockouts (Cooper *et al.*, 2005; Svard *et al.*, 2006). These data provide genetic evidence that the primary or major function for Su(Fu) is as a negative regulator of the Shh pathway. However, Su(Fu) may also play a more limited role as a positive regulator of signaling. These seemingly contradictory functions can be best understood by examining the relationship between Su(Fu) and Gli proteins, their primary targets in the pathway. Current data suggest that Su(Fu) modulates Gli activities at several distinct steps, including their stability, nuclear translocation, and transcriptional activity (Ruel and Therond, 2009). Su(Fu) proteins have been shown to directly interact with both N-termini and C-termini of all three Gli proteins (Dunaeva *et al.*, 2003; Merchant *et al.*, 2004; Wang *et al.*, 2010a). In the absence of Shh signaling, Su(Fu) can retain Gli1 and full-length Gli2 (and possibly Gli3) in the cytoplasm to prevent the nuclear entry of activator forms of Gli proteins (Barnfield *et al.*, 2005; Ding *et al.*, 1999; Kogerman *et al.*, 1999). *In vitro* data also suggest the possibility that Su(Fu) can function as a transcriptional cofactor for Gli repressors (GliRs) through the recruitment of specific corepressors (discussed in more detail below; Cheng and Bishop, 2002; Dai *et al.*, 2002). In contrast to these generally negative functions, Su(Fu) has also been shown to antagonize Spop-mediated degradation of full-length Gli2 and Gli3, potentially by masking the Spop-binding sites in these proteins; in this role, Su(Fu) would function as a positive factor in Shh signal transduction to protect full-length Gli2 and Gli3 from degradation and/or prolong target gene activation (Chen *et al.*, 2009; Wang *et al.*, 2010a). Consistent with this, full-length Gli2 and Gli3 protein levels are diminished in *Su(Fu)* null mutant embryos (Chen *et al.*, 2009). How can these data be reconciled with pathway derepression seen in *Su(Fu)* mutants? The selective regulation of Gli2 and Gli3, but not Gli1, by Spop offers one possible explanation. In this scenario, loss of full-length Gli2 and Gli3 proteins in *Su(Fu)* mutants would reduce overall activator *and* repressor levels, with the latter resulting in the derepression of *Gli1*, which does not interact with Spop (Chen *et al.*, 2009). Clearly more work is needed to further elaborate the mechanisms by which Su(Fu) regulates Gli protein levels and activity and whether this interaction can be regulated by Shh signaling.

Among these described factors, most of them (e.g., Sufu and Kif7) dynamically traffic in or out of the primary cilium (Endoh-Yamagami *et al.*, 2009; Liem *et al.*, 2009). This supports the idea that the functional divergence of conserved component likely links to the primacy of this new organelle in mediating signaling. Primary cilia also provide a localized signaling compartment that could sensitize the cellular responses to extracellular Shh ligands. In addition, the regulatory complexity of the primary cilium allows for multiple new levels to control signaling transduction from Smo to Gli. Indeed, many factors that have been recently identified as vertebrate-specific Shh signaling components, including IFT proteins, Rab23, FK 506-binding protein 8 (FKBP8), and coiled-coil protein Iguana (Igu), appear to function at this step (Bulgakov *et al.*, 2004; Cho *et al.*, 2008; Huangfu *et al.*, 2003; Liu *et al.*, 2005; Qin *et al.*, 2011; Wolff *et al.*, 2004). Together, these novel elements elevate the importance of intracellular membrane trafficking in the vertebrate signaling pathway compared to *Drosophila*. Because of this, the regulation of Gli activity by Smo has also become more complex. Recent evidence suggests that Gli proteins are continually cycling through the cilium and that this dynamic translocation is critical to their processing and hence function (Wen *et al.*, 2010). This would suggest that there is a step in the pathway that could be mediated by IFT proteins that was required for Gli activation downstream of Smo. Supporting this idea, epistasis analysis of several IFT proteins suggests that their activity is required for Gli activation downstream (and independent) of Smo activity (Eggenschwiler and Anderson, 2007). Smo could therefore alter Gli phosphorylation indirectly by redirecting their transport to a subcellular compartment in the cilium where they can no longer be phosphorylated by PKA. This raises the possibility that the activity of proteins such as Rab23 might be regulated by Smo. A better understanding of this process will require high-resolution localization studies of pathway components within cellular micro-domains in Shh-responsive cells. It is also conceivable that the phosphorylation status of Gli proteins could be influenced by other inputs that are independent of Smo that converge on these regulators. While it has been shown that Smo is absolutely required for Shh signal transduction, it also remains unclear whether it is involved in all forms of signaling or if there are other ways to control Gli activity (via phosphorylation) independent of Smo.

10. Are You Going to Use That Thing? Smo as a G-Protein-Coupled Receptor

In addition to the mounting evidence that Smo trafficking into the cilia triggers its downstream activity, recent evidence also suggests that some of the functions of Smo could be regulated by a G-protein-coupled pathway (Ayers and Therond, 2010). The close homology of the Smo 7-TM domain to the large family of GPCRs has long suggested a potential role for heterotrimeric G-protein second messengers in mediating aspects of its signaling activity; however, clear evidence supporting this possibility has only been recently provided.

Signaling through GPCRs generally results in the activation of heterotrimeric G-proteins by GDP to GTP exchange, resulting in G-protein complex dissociation. There are four classes of heterotrimeric G-proteins, $G_{i/o}$, G_s, $G_{q/11}$, and $G_{12/13}$, each characterized by distinct $G\alpha$ subunits. There is evidence from cell culture suggesting that Smo can function as a guanine nucleotide exchange factor (GEF) to catalyze the activation of $G_{i/o}$ family proteins but not other classes (Barzi *et al.*, 2011; Riobo *et al.*, 2006). Recent genetic studies with cultured cells provide further evidence that G-protein activation is coupled to Smo through its TM domain but not C-terminal tail. First, although deletion studies suggest that the C-terminal tail is essential for full Smo downstream activity (Varjosalo *et al.*, 2006), a C-terminal tail deleted Smo (SmoΔC) can still activate G proteins, which can be blocked in the presence of the Smo antagonist cyclopamine that binds selectively to the TM domain (Riobo *et al.*, 2006). Further, a SmoM2 mutation containing a single amino acid substitution located in the seventh TM domain that can constitutively activate signaling also enhances G-protein activation (Xie *et al.*, 1998). These data support the idea that G-protein coupling to Smo is via the 7-TM domain like most or all other GPCRs.

The next question is whether Smo GEF activity is linked to Shh signaling. There is evidence from cell culture studies that full activation of Gli transcriptional activity requires $G_{i/o}$ activation (DeCamp *et al.*, 2000; Riobo *et al.*, 2006). However, the *in vivo* relevance of this has been challenged by experiments showing that $G\alpha_i$ activity, which is associated with the $G_{i/o}$ complex, does not seem to be required for Shh–Gli-dependent neuronal patterning in chick embryos (Low *et al.*, 2008). However, more recent evidence has implicated specific $G_{i/o}$ activity to Shh-mediated proliferation in cerebellar granule cells (Barzi *et al.*, 2011).

A potential mechanism for transmitting Smo G-protein activity to Gli phosphorylation is through PKA. It is known that $G\alpha_{i/o}$-GTP activity inhibits adenyl cyclase (AC), leading to a reduction in the overall cytoplasmic cAMP levels (Engelhardt and Rochais, 2007). Further, cAMP can interact with the regulatory subunits of PKA (R-PKA) to mediate dissociation from the catalytic subunits of PKA (C-PKA) to activate PKA activity (Iyer *et al.*, 2005). Therefore, signaling through $G_{i/o}$ family second messengers could inhibit PKA and hence Gli phosphorylation.

It is also possible that Smo G-protein coupling affects aspects of its function that are only indirectly related to Gli phosphorylation. GPCR signaling is typically associated with phospholipase C (PLC) and IP3-mediated Ca^{2+} influx. Using imaging, it has recently been shown that spinal cord explants exposed to Shh exhibit acute transient increases in Ca^{2+} influx that could be blocked by cyclopamine and thus required Smo (Belgacem and Borodinsky, 2011). The potential role for this mechanism in controlling Shh-mediated target gene expression (via Gli activation) remains to be determined. Another function for Smo G-protein coupling could be related to its translocation to the cilium (Ayers and Therond, 2010). It has been shown that phosphorylation by Grk2 (G protein-coupled receptor kinase 2) of Smo promotes its association with β-arrestin2 and shuttling into the cilium (Chen *et al.*, 2004b; Philipp *et al.*, 2008). Moreover, as mentioned above, it has been shown that PKA–cAMP pathway can also function at the proximal region of primary cilium to facilitate translocation of Smo (Barzi *et al.*, 2011; Milenkovic *et al.*, 2009). This relationship could establish a negative feedback loop to limit Smo activation and would indicate that the role of PKA in regulating signaling is more complex.

11. Being in the Right Place at the Right Time: Shh–Gli Signaling and Patterning in the Ventral Neural Tube

There are still many unanswered questions regarding Shh signaling in the CNS that have yet to be fully addressed. One of the major future challenges will be to unravel its role in establishing the cellular organization and patterning in different regions. In rostral regions of the CNS, because of the more complex pattern of Shh expression and tissue growth, even the most straightforward questions regarding which specific cell types require signaling and for what purpose are still being asked. In contrast, the most caudal region of the CNS, the spinal cord, has a comparatively simpler organization that has been well characterized over many decades of investigation. Because of this, the developing spinal cord has served as an important model system in which to study the mechanisms by which Shh controls the generation of pattern in tissues.

One of the most significant contributions from spinal cord studies is in understanding the morphogenetic action of graded Shh signaling. A basic tenet of morphogen signaling is that secreted protein diffusion from a localized group of cells sets up an extracellular concentration gradient that diminishes progressively from the source. This requires that cellular responses are precisely regulated in both space and time to control tissue patterning, a process that depends on the enactment of specific gene expression programs in distinct cell lineages (Ashe and Briscoe, 2006; Gurdon and Bourillot, 2001; Kutejova *et al.*, 2009). There are many good recent reviews that discuss the background organization of gene expression and cell types in the ventral neural tube, so we will outline them only briefly here. Through many studies, it has been shown that Shh signaling is necessary to establish five distinct ventral progenitor domains that give rise to unique classes of neurons and glia in the developing spinal cord (Briscoe and Ericson, 2001). During the early stages following neural tube closure, progenitor gene expression in the spinal cord becomes established in a characteristic pattern. Initially, the entire structure has a "dorsal" character that becomes progressively ventralized under the influence of Shh signaling from the notochord and FP (Goulding *et al.*, 1993). The major neuronal subclasses that form the circuits involved in sensorimotor processing in the spinal cord are arrayed across the dorso-ventral (DV) or transverse axis. The diversification of neuronal fates begins at the onset of neurogenesis, where groups of cells representing a common broad subclass exit the cell cycle and begin their differentiation in specific DV domains. These groups of cells will ultimately share similar properties and general functions, including their migration patterns, initial axon projections, and role in spinal cord circuits (Lee and Pfaff, 2001). Significantly, subtypes can be identified prior to exhibiting these differentiated characteristics by the fact that they express the same transcription factor profiles. On the basis of these features, five cardinal neuronal classes and one nonneuronal class have been identified, termed V0, V1, V2, MN, V3, and FP, from dorsal to ventral (Ericson *et al.*, 1992, 1997b; Jessell, 2000; Matise and Joyner, 1997). All "V" cell classes are interneuron subtypes that differentiate in adjacent territories to motoneurons (MN) and FP cells. Significantly, each of the neuronal subtypes is generated from progenitor cells that also display distinct transcription factor profiles in specific domains in the ventricular zone (VZ; correspondingly termed p0, p1, p2, pMN, p3; Briscoe *et al.*, 2000).

Through elegant work, it has been shown that early neuronal subclass fates are initially specified at the progenitor level and that graded Shh signaling plays an essential role in this process through the establishment of distinct gene expression territories in VZ progenitor cells (Ericson *et al.*, 1997b; Jessell, 2000). However, not all cells are irreversibly committed to specific fates at these early stages; many remain labile for a short period of time after cell cycle withdrawal or longer. This progressive refinement of fate is a common characteristic of both neuronal and glial differentiation and serves to further diversify the specialized cell types in the CNS. Since Shh expression in the spinal cord persists in the FP throughout both neurogenesis and gliogenesis, it is relevant to ask whether it continues to play a role in this process. We can consider this question in the context of an important developmental milestone that is likely to impact on this issue. For a short period following neural tube closure, both the notochord and prechordal plate remain in close contact with the overlying neural ectoderm along the entire anterior–posterior axis (the exception is in the caudal region of the neural tube that arises by secondary neurulation; this will not be addressed in this review). Later, at spinal cord levels, the notochord regresses ventrally/

anteriorly away from the CNS, thus leaving the FP as the only intrinsic source of Shh for most of neurogenesis and all of gliogenesis. What is the specific contribution of each source to Shh signaling in the spinal cord? In embryos lacking Gli2, a key downstream regulator of Shh signaling, FP cells are not induced, and consequently, Shh is never expressed within the ventral spinal cord (Ding *et al.*, 1998). Nevertheless, the overall early neuronal differentiation pattern develops largely normally with the exception of V3 interneurons defined by the expression of Nkx2.2, which are almost entirely absent. One major caveat for interpreting this result is that the loss of Gli2 from all ventral progenitors severely compromises their ability to read out an extracellular Shh gradient as Gli2 is the primary positive mediator of signaling. More recently, it has been shown that GFP-tagged Shh produced by the notochord can be visualized in a gradient up to five cell diameters from the ventral midline into the presumptive p2 domain even before the FP is induced (Chamberlain *et al.*, 2008). In addition, selective loss of Shh expression in the FP results in a failure to maintain distinct Olig2 (pMN) and Nkx2.2 (pMN) progenitor gene expression (Dessaud *et al.*, 2010). These findings together suggest that the early notochord-derived Shh source may be largely if not entirely responsible for providing the spatial cues that set up early neuronal differentiation patterns, at least in the most ventral domains, and that FP-derived Shh may have only a limited, local influence on early neurogenic progenitor gene patterning.

Another important role for Shh signaling in the spinal cord is in the generation of OL, the myelinating CNS glial cell type that arises from a ventral progenitor domain that earlier generates MN during neurogenesis. Shh signaling appears to be required during two phases of OL development: progenitor specification and differentiation. OLs are not generated in *Shh* mutant mice but can be rescued if Gli3 is also absent (in *Shh*;*Gli3* double mutants), suggesting that the primary early function of Shh is to block Gli3R activity that inhibits OL progenitor cell (OPC) gene expression (Oh *et al.*, 2005; Tan *et al.*, 2006). However, OL differentiation does not progress normally in these double mutants (Tan *et al.*, 2006). These findings raise the likely possibility that there is an ongoing, direct requirement for Shh signaling during gliogenesis, when the only intrinsic source is the FP.

In addition to the spatial pattern of ventral progenitor gene expression that was characterized over a decade ago, more recent work has revealed that there is also a temporal sequence of activation that conforms to the following basic rule: earlier induced genes occupy progressively more dorsal positions as more ventral genes are induced. The consistent temporal sequence of progenitor gene activation also underlies the hypothesis that they are core components of a spinal cord "gene regulatory network" or GRN (Dessaud *et al.*, 2008; Nishi *et al.*, 2009; Vokes *et al.*, 2007). Thus, the earliest ventral genes in this GRN to appear, Nkx6.2 and Nkx6.1, will later demarcate the p1 and p2 domains, respectively, followed by Olig2 (pMN),

Nkx2.2 (p3), and finally FoxA2 (FP). Significantly, each of these factors is first induced in the ventral midline region that will ultimately be occupied by FP cells, closest to the notochord source of Shh, and then become displaced dorsally to occupy their characteristic DV position along the axis.

Based on these observations, it is reasonable to suggest that many or most progenitors in the ventral spinal cord are initially exposed to a similar level of Shh. However, as neurogenesis proceeds, their spatial address changes to become more dorsal, so that in theory they will be exposed to progressively lower levels (further from the source) of Shh. However, as discussed in a preceding section, the establishment of the extracellular Shh gradient also develops over time under the influence of both positive and negative feedback mechanisms. Thus, the spatial Shh gradient may be less important than the length of time spent within a particular (putative) concentration range. Because these changes occur in the context of cell division, ventral progenitors may need to retain a record of their prior levels of exposure. The temporal activation of specific transcription factors within the GRN could achieve this by acting in a feed-forward manner to sequentially alter target gene regulation in progenitors. Recent evidence supporting this idea has been provided by studies showing that early, transient midline activation of Nkx2.2 is required for FP induction prior to becoming restricted to p3 progenitors (Lek *et al.*, 2010). These studies also show that a transcriptional feed-forward mechanism could function to intrinsically amplify the Shh–Gli signal to provide a response that is not strictly graded in nature.

12. Walking Down the Runway: A Model for Shh–Gli Control of Early Progenitor Patterning

A recent model has emerged for the role of Shh signaling in organizing distinct progenitor domains in the neural tube (Dessaud *et al.*, 2008). This model incorporates data showing that experimental manipulation of Gli protein activity levels in transfection assays can recapitulate to a degree the graded activation of ventral progenitor genes along the DV axis, extending earlier classic studies demonstrating a similar correlation between Shh protein levels and positional marker gene induction (Lei *et al.*, 2004; Stamataki *et al.*, 2005). These data together lead to the idea that graded Gli protein activities translate extracellular Shh levels/length of exposure into a corresponding transcriptional response to direct selective target gene activation, suggesting that the principal determinant controlling patterned gene expression in ventral progenitors is the level and/or duration of intracellular Gli transcriptional activity (Dessaud *et al.*, 2008; Ribes and Briscoe, 2009; Stamataki *et al.*, 2005). The corollary to this idea is that

Shh–Gli target genes are "tuned" to specific Gli activities that permit their induction only after a particular threshold level has been reached (Dessaud *et al.*, 2008).

The results from this work suggest that temporal adaptation to Shh–Gli signaling shapes progenitor gene expression over time. For Olig2 and Nkx2.2 expression in the pMN and p3 lineages, negative feedback inhibition of the pathway by Ptc1 has been implicated (Dessaud *et al.*, 2007). This model suggests that an initial pulse of Shh activity functions first to induce Olig2 and Nkx2.2 in ventral cells, but that with continued Shh signaling ligand-dependent antagonisms via *Ptc1* upregulation results in decreased Gli activity levels in pMN cells below the theoretical threshold necessary to induce Nkx2.2 here, thereby creating a differential spatiotemporal profile of Gli activity between the two domains. It is unclear whether the net changes in Gli activity levels would be the result of a reduction in Gli transcriptional activator (GliA) and/or increased GliR levels. This interesting model provides a mechanistic explanation for how a morphogenetic gradient can be shaped over time through negative feedback, and how temporal adaptation correlates to the progressive changes in gene expression in the expanding progenitor pool in the ventral neural tube. One of the central questions that remain is how activation thresholds are regulated for each Shh–Gli target gene. The answer to this must take into account the observation that both time and concentration are important parameters in the response. This could conceivably be achieved by regulating the inherent sensitivity of each gene to absolute Shh–Gli levels by differential inhibition. Addressing this question requires that the regulatory elements for each gene are identified and characterized, something that has been done for only a few ventral progenitor Shh–Gli targets that will be discussed in more detail below.

In addition, the model is consistent with expression and lineage data suggesting that ventral progenitors in the pMN/p3/FP domains sequentially activate progenitor gene expression as a function of their time of exposure to Shh. Recent work has sought to extend this model to cells arising in the more dorsal p0/p1/p2 territories (Dessaud *et al.*, 2010). Using Cre-loxP-mediated lineage tracing, it was shown that progenitors that activate the *Dbx1* gene, which ultimately marks p0 cells that arise in the most distant part of the ventral progenitor territory, can give rise to progeny located more ventrally (and more dorsally), including a few as far away as the pMN domain (although the majority appear to give rise to p1 and pD5/6 progeny; Dessaud *et al.*, 2010). The unique activation profile of genes in this territory, which have been previously shown to be under the control of retinoid signaling from the adjacent somatic mesoderm (Novitch *et al.*, 2003; Pierani *et al.*, 1999), suggests the possibility that these progenitors may not follow a strictly dorsal-to-ventral progression like the pMN/p3/FP domains. These observations suggest that not all ventral progenitors undergo the same progression and that distinct sublineages exist, creating in effect a pattern within a pattern. Indeed, FP induction may be another example of the latter possibility. The origin of this nonneuronal ventral midline tissue is unique among other ventral progenitor territories. First, there are interesting species-specific differences in the cellular origin of FP cells, specifically whether they are induced in distinct segregated neural tube progenitors by node/notochord signals, or are specified earlier in the node and then intercalate into the neural tube during node regression, which also produces the notochord in its wake (Placzek and Briscoe, 2005). Although evidence in mice suggests that FP cells are not derived from the node but rather induced (Jeong and Epstein, 2003), their lineage relationship to other neural tube cell types appears to be distinct. For example, analysis of reporter expression from an *Olig2:Cre* line fails to label FP cells, showing that they do not share a common origin with pMN cells (Dessaud *et al.*, 2007). Thus, all cells within the three most ventral progenitor domains may not share a common lineage. Before reaching this conclusion, however, it will be important to determine whether FP antecedents express Nkx6.1, the earliest gene induced in the midline, or if there are axial differences in midline cell lineages. Thus, summarizing data in this and the preceding paragraph, it seems that the progressive fate acquisition model may not strictly apply to the FP domain.

13. What's It Gonna Be, Boy? GliA Versus GliR-Regulated Genes

As described in an earlier section, Shh activity inhibits GliR and promotes GliA formation. This fact raises the question of whether the expression of Shh–Gli target genes is controlled by derepression, direct activation, or some combination of both activities. Analysis of mouse mutants lacking specific Gli activities indicates that ventral progenitor target genes differ in their requirement. The requirement for GliR derepression can be inferred by identifying the ventral cell types that do not form in *Shh* mutants, which retain high levels of Gli3R activity, that are rescued in *Shh; Gli3*, *Smo;Gli3*, and *Gli2;3* double mutants that lack all Gli (GliA and GliR) activities (Bai *et al.*, 2004; Lei *et al.*, 2004; Litingtung and Chiang, 2000; Wijgerde *et al.*, 2002). One of the rescued genes is Olig2, a bHLH protein expressed in motoneuron progenitor (pMN) cells during neurogenesis and then later in oligodendrocyte progenitors (pOlig). Ectopic Olig2 expression can be induced by Shh–Gli signaling, and in the absence of Shh, Olig2 is lost. Gli-binding sites have been identified in a conserved region of the locus, consistent with the idea that Olig2 is a direct Shh–Gli target (Hallikas *et al.*, 2006). However, Olig2 expression is restored in *Shh;Gli3* and *Smo; Gli3* double mutants, and in embryos lacking both *Gli2* and *Gli3* (that are required to induce *Gli1* and whose combined loss therefore completely abrogates Shh signaling through Glis; Bai *et al.*, 2004; Lei *et al.*, 2004; Oh *et al.*, 2009; Wijgerde *et al.*, 2002). Thus, Olig2 appears to be a good example of a target gene whose expression is regulated by Gli3R derepression rather than GliA activation; although GliA can induce expression and may contribute to expression, the primary determinant is the presence of Gli3R which must be blocked to allow expression, presumably induced by Shh-independent factors such as retinoic acid (Novitch *et al.*, 2003; Pierani *et al.*, 1999). Similar genetic evidence supports the idea that Nkx6.1 and Nkx6.2, two of the earliest Shh target genes induced in the ventral neural tube, are primarily regulated by a derepression rather than direct activation mechanism (Vokes *et al.*, 2007).

Another interesting set of target genes that may be differentially regulated by Gli activities encode Shh pathway proteins. As mentioned above, *Gli1* is a positive target of Shh signaling, whose induction requires Gli2 and Gli3 transcriptional activation. In contrast, *Gli3* expression and activity are repressed by signaling (Marigo *et al.*, 1996). *Gli2* is not positively regulated by Shh signaling, and it remains unclear whether it might also be negatively regulated in the CNS like *Gli3*. *Ptc1* expression is also upregulated by Shh–Gli signaling but has more dynamic expression pattern suggesting a more complex mechanism involving both GliA and GliR input.

Why is derepression insufficient to induce all target genes? One explanation is that they differ in their requirement for GliA. If Shh pathway-independent activation is required in addition to Shh-dependent derepression, then as long as these additional signals are present, the conditions for activation can be met simply by removing GliR function. On the other hand, if the only (or predominant) positive inductive activity is from GliA, then both derepression and activation would be required; in this case, additional Shh input may be necessary to promote GliA formation. Thus, these genes would not be rescued by simultaneous loss of Shh and GliR. Two genes in this category are Nkx2.2, the p3 progenitor determinant, and FoxA2 (FP), both of which require the highest and most prolonged exposure to Shh–Gli activity for their expression (Dessaud *et al.*, 2007).

These data raise the question as to whether specific mechanisms are involved in shaping the patterned response of target genes regulated by either GliA or GliR activities. Prior studies of the regulation of *Nkx2.2* expression from our lab provide evidence in support of this possibility. In this work, we identified the *cis*-acting DNA regulatory module (CRM) that controls the p3-domain specific expression of *Nkx2.2* in the ventral neural tube (Lei *et al.*, 2006). A single Gli-binding site was identified that is required for Nkx2.2$^{p3\text{-}CRM}$ driven reporter expression in transgenic mouse lines. In addition, two conserved binding sites for Tcf proteins (van de Wetering *et al.*, 1991) were also identified in the same conserved region. Significantly, the Tcf sites were required to suppress enhancer expression outside the p3 domain; in their absence, reporter activity was induced in cells located at a greater distance from the ventral midline source of Shh in regions where the putative level of GliA would be predicted to be much lower than in the p3 region. This result indicates that *Nkx2.2* expression can be induced over a wide range of Shh–GliA levels *in vivo* but only in the absence of Tcf/Lef transcriptional repressor input. Thus, differences in GliA levels alone cannot account for its restricted expression pattern *in vivo*.

Integrating these data suggest a model whereby Shh–Gli signaling in the developing neural tube might be best though of as occurring in phases, with derepression of target genes preceding direct activation and requiring lower signaling levels. Initial activation of p2 genes (Nkx6.1) relies on GliR derepression and follows the Shh-dependent downregulation of dorsal "Class I" genes (such as *Pax6* and *Pax7*), which require the lowest levels of Shh (Ericson *et al.*, 1997a). This is consistent with the idea that lower levels are needed to block GliR than to activate GliA. Simultaneous regulation of cell cycle targets may be involved as well, perhaps controlling clonal progenitor expansion directly or indirectly via cross talk with Wnt-βcatenin signaling (Alvarez-Medina *et al.*, 2008, 2009; Ulloa and Briscoe, 2007; Ulloa *et al.*, 2007). Then, subsequent signaling begins to accumulate and become shaped by positive and negative feedback mechanisms. GliA form ventrally and act upon a subset of "Class II" targets, but differences in the Gli activity gradient are not sufficient on their own to restrict target gene responses for all domains: pMN and p3 progenitor segregation involves at least one additional input from Tcf repressors, while p3/FP segregation involves a temporal switch in competence (Lek *et al.*, 2010; Ribes *et al.*, 2010).

The role of Shh signaling during the next phase of spinal cord development is less well understood. It is unclear whether tissue expansion further restricts the diffusion of Shh closer to its FP source, or whether signaling can continue to extend progressively more dorsally over time, but in either case, this is likely to be accounted for by changes in the mechanisms that secrete and/transport Shh protein. At these stages, the range of gradient activity will be an important determinant of whether inhibition of GliR repressor or induction of GliA function controls target gene expression or other downstream effects.

14. New Game: Distinct Mechanisms for GliA-and GliR-Mediated Transcription

As described above and in other reviews, Shh signaling controls the development of early neuronal fates by orchestrating the expression of transcription factors that implement cell type-specific differentiation programs. One of the key issues in understanding this process is the identification of direct Shh–Gli target genes and characterizing their response to signaling input. Gli proteins recognize a consensus-binding sequence comprising nine core nucleotides (5′-GACCACCCA; Kinzler and Vogelstein, 1990). To date, most models of Hh-Gli signaling are based on the assumption that full-length Gli activators (GliA) and truncated GliR can bind the same consensus DNA motif and therefore function in an equivalent manner to repress or activate Hh target genes. However, the zinc-finger DNA-binding domains of the three Gli proteins are not perfectly conserved (Matise and Joyner, 1999), so it is conceivable that their binding affinities differ, although the functional significance of this has not been carefully evaluated *in vivo*. In addition, the selective association of GliA or GliR with distinct transcriptional cofactors might also modulate their binding specificity or affinity.

As described in an earlier section, the bifunctional transcriptional activities of the fly Ci protein are distributed unequally among the three vertebrate homologs, Gli1-3. Functional and genetic studies indicate that Gli1 can only serve as an activator in its full-length form, while Gli2 and Gli3 are capable of both activation and repression functions (although to different degrees; Bai and Joyner, 2001; Bai *et al.*, 2002, 2004; Lei *et al.*, 2004). Consistent with this, all three Gli proteins contain a highly conserved zinc-finger DNA-binding motif and a C-terminal transcriptional activation domain, while only Gli2 and Gli3 proteins have functional N-terminal repression domains (Sasaki *et al.*, 1999). It has been shown that C-terminal truncated forms of Gli2 and Gli3 can repress Shh–Gli target genes in both cell lines and embryos (Persson *et al.*, 2002), while N-terminal truncated forms of Gli2 and Gli3 have constitutive activation functions (Lei *et al.*, 2004; Sasaki *et al.*, 1999; Stamataki *et al.*, 2005). One question that remains is whether Gli proteins that share a common property (i.e., activation or repression) also share a common transcriptional mechanism to achieve this. The possibility that there are further differences among Gli proteins at this level would clearly extend the range of outcomes possible from their unique or combined activities.

Coactivators can be generally divided into three classes: the TRAP/DRIP/Mediator/ARC core transcriptional machinery recruitment class, the SWI/SNF ATP-dependent DNA unwinding class, and the HAT (histone acetylation transferase) histone-remodeling class (Spiegelman and Heinrich, 2004). The first class of coactivators is essential for the recruitment of general transcriptional polymerases, while the last two classes can facilitate transcriptional activation via a putative "open-chromatin" regulation. *In vitro* biochemical studies show that the C-terminal activation domain of human GLI1 (which is conserved in Gli2) can directly associate with $TAF_{II}31$ (TFIID TATA box-binding protein associated factor; Sasaki *et al.*, 1999; Yoon *et al.*, 1998). In separate studies, it was shown that the

Gli3 transactivation domain can interact with Med12, a subunit of the mediator-RNA polymerase II transcriptional initiation complex (Sasaki *et al.*, 1999; Yoon *et al.*, 1998; Zhou *et al.*, 2006). These results provide evidence that GliA can directly interact with the core transcriptional machinery. However, there are data from flies that suggest GliA could function via an epigenetic mechanism. It has been shown that Ci can directly recruit dCBP/p300, a HAT family member, to activate Hh target genes in *Drosophila* (Akimaru *et al.*, 1997). A conserved CBP-binding domain (CBD) has been identified in the C-termini of both Gli2 and Gli3, but not Gli1, raising the possibility that Gli proteins can potentially provide a target gene open-chromatin environment through the recruitment of CBP/p300 (Dai *et al.*, 1999; Sasaki *et al.*, 1999). However, to date no evidence supporting a functional or physical link between vertebrate Gli proteins and CBP/p300 has been provided.

There are also two general repression mechanisms that can inhibit gene expression. Repressors (e.g., GliR) could block expression through genetic competition with activator complexes (e.g., GliA) on the same/or proximal DNA-binding sites. Alternatively, repression can also occur through recruitment of histone-remodeling corepressors via an epigenetic mechanism. In the latter case, repression can be local or long range: repressors can associate with multiple corepressors to form "repressosomes" that could silence a broad chromosomal region by promoting a putative condensed chromatin environment (Courey and Jia, 2001). One of the identified classes of proteins in repressosomes is histone deacetylases (HDACs). In general, HATs and HDACs function in opposition to regulate chromatin remodeling in that hyperacetylation of histones is favorable for gene transcription, whereas hypoacetylation promotes transcriptional repression.

Several proteins have been proposed as potential cofactors to mediate GliR transcriptional repression, including Su(Fu), Ski, and SAP18 (Cheng and Bishop, 2002; Dai *et al.*, 2002; Paces-Fessy *et al.*, 2004). Genetic studies show that the N-terminal repression domain of Gli3 can bind directly to Ski and Su(Fu) and indirectly to SAP18 via Su(Fu). Interestingly, all these three proteins are components of the mSin3 complex that can recruit HDAC1 (Cheng and Bishop, 2002; Dai *et al.*, 2002). Consistent with this, Gli3 can be coimmunoprecipitated with HDAC1 probably through Ski (Dai *et al.*, 2002). Further, using an *in vitro* luciferase assay with HEK 293T cell lines, it was shown that SAP18 and Su(Fu) can function as corepressors to inhibit Gli-mediated transcription and that this effect can be antagonized in the presence of HDAC inhibitor TSA (Cheng and Bishop, 2002).

Together, these *in vitro* findings raise the possibility that GliR activity might involve HDAC1-mediated histone hypoacetylation at target loci. However, recent *in vitro* data call into question the idea that HDAC1 activity is required for Gli3R function (Tsanev *et al.*, 2009). Moreover, the N-terminal repression domains of Gli2 and Gli3, first identified in functional deletion studies (Sasaki *et al.*, 1999), do not overlap with the known Su(Fu)-binding sites in these proteins (Dunaeva *et al.*, 2003), consistent with recent genetic data showing the repressive activity of Gli3 in the CNS is independent of Su(Fu) (Wang *et al.*, 2010a). These data suggest that Su(Fu) is dispensable for Gli3R activity. In addition, luciferase assays with MEF cells suggest that neither Su(Fu) nor SAP18 can function to repress Gli mediated transcription (Chen *et al.*, 2009), in contrast to the work described above using HEK 293T cells, possibly because different cell lines were used in each assay. Together, these data call into question whether Gli3 repressor functions via a chromatin-condensing mechanism and, if so, whether it is Su(Fu) and/or HDAC dependent. Taken with the studies discussed above, these data highlight the need for further investigation into whether Gli proteins activate or repress target gene transcription via genetic and/or epigenetic mechanisms and whether they each have distinct mechanisms.

Finally, although only three bona fide Gli family proteins have been identified to date, there are additional related zinc-finger transcription factors comprising a broader Gli superfamily, including Zic proteins, Glis, and GliH (Nakashima *et al.*, 2002; Sakai-Kato *et al.*, 2008; Shimeld, 2008). Genetic evidence shows that Zic proteins can bind to similar DNA target motifs and can physically associate with Gli proteins via zinc-finger 3 through 5 (which also contact DNA; Koyabu *et al.*, 2001; Mizugishi *et al.*, 2001; Pavletich and Pabo, 1993). Thus, it is still unclear whether Gli proteins and other Gli superfamily members could function in a synergistic or antagonistic manner in certain developmental contexts.

15. Summary and Further Questions

Studies of Shh signaling in the CNS have given us a clearer picture of the mechanisms that are involved in controlling several aspects of its morphogenetic activity. In this review, we have focused primarily on the involvement of the "canonical" Shh–Gli pathway in controlling extracellular gradient formation, signal transduction, and target gene expression in the developing CNS. In all of these areas, there are significant gaps remaining in our understanding of the pathway and how it regulates cell fate specification, a process that ultimately plays itself out at the genetic and epigenetic level. A continued focus on elucidation of the genetic programs that are regulated by the signaling pathway should lead to new and important insights into its role in tissue patterning. However, there is also emerging evidence for "noncanonical" Shh pathway/s (Jenkins, 2009; Lauth and Toftgard, 2007; Yam *et al.*, 2009) that could be involved in diverse processes such as axon guidance and tissue growth. In addition, evidence for cross talk between the Shh pathway and others, including Wnts, indicates another level of regulation that has yet to be fully understood (Matise, 2007; Ulloa and Briscoe, 2007). Exploring these and other questions will ensure that research on the Shh pathway will continue to extend into the future.

REFERENCES

Aanstad, P., Santos, N., Corbit, K. C., Scherz, P. J., Trinh le, A., Salvenmoser, W., Huisken, J., Reiter, J. F., and Stainier, D. Y. (2009). The extracellular domain of Smoothened regulates ciliary localization and is required for high-level Hh signaling. *Curr. Biol.* **19,** 1034–1039.

Ahn, S., and Joyner, A. L. (2005). In vivo analysis of quiescent adult neural stem cells responding to Sonic hedgehog. *Nature* **437,** 894–897.

Akimaru, H., Chen, Y., Dai, P., Hou, D. X., Nonaka, M., Smolik, S. M., Armstrong, S., Goodman, R. H., and Ishii, S. (1997). Drosophila CBP is a co-activator of cubitus interruptus in hedgehog signalling. *Nature* **386,** 735–738.

Alcedo, J., Ayzenzon, M., Von Ohlen, T., Noll, M., and Hooper, J. E. (1996). The Drosophila *smoothened* gene encodes a seven-pass membrane protein, a putative receptor for the Hedgehog signal. *Cell* **86,** 221–232.

Allen, B. L., Tenzen, T., and McMahon, A. P. (2007). The Hedgehog-binding proteins Gas1 and Cdo cooperate to positively regulate Shh signaling during mouse development. *Genes Dev.* **21,** 1244–1257.

Alvarez-Medina, R., Cayuso, J., Okubo, T., Takada, S., and Marti, E. (2008). Wnt canonical pathway restricts graded Shh/Gli patterning activity through the regulation of Gli3 expression. *Development* **135,** 237–247.

Alvarez-Medina, R., Le Dreau, G., Ros, M., and Marti, E. (2009). Hedgehog activation is required upstream of Wnt signalling to control neural progenitor proliferation. *Development* **136,** 3301–3309.

Alves, G., Limbourg-Bouchon, B., Tricoire, H., Brissard-Zahraoui, J., Lamour-Isnard, C., and Busson, D. (1998). Modulation of Hedgehog target gene expression by the Fused serine-threonine kinase in wing imaginal discs. *Mech. Dev.* **78,** 17–31.

Ashe, H. L., and Briscoe, J. (2006). The interpretation of morphogen gradients. *Development* **133,** 385–394.

Ayers, K. L., and Therond, P. P. (2010). Evaluating Smoothened as a G-protein-coupled receptor for Hedgehog signalling. *Trends Cell Biol.* **20,** 287–298.

Bai, C. B., and Joyner, A. L. (2001). Gli1 can rescue the in vivo function of Gli2. *Development* **128,** 5161–5172.

Bai, C. B., Auerbach, W., Lee, J. S., Stephen, D., and Joyner, A. L. (2002). Gli2, but not Gli1, is required for initial Shh signaling and ectopic activation of the Shh pathway. *Development* **129,** 4753–4761.

Bai, C. B., Stephen, D., and Joyner, A. L. (2004). All mouse ventral spinal cord patterning by hedgehog is Gli dependent and involves an activator function of Gli3. *Dev. Cell* **6,** 103–115.

Balordi, F., and Fishell, G. (2007). Hedgehog signaling in the subventricular zone is required for both the maintenance of stem cells and the migration of newborn neurons. *J. Neurosci.* **27,** 5936–5947.

Barnfield, P. C., Zhang, X., Thanabalasingham, V., Yoshida, M., and Hui, C. C. (2005). Negative regulation of Gli1 and Gli2 activator function by Suppressor of fused through multiple mechanisms. *Differentiation* **73,** 397–405.

Barzi, M., Kostrz, D., Menendez, A., and Pons, S. (2011). Sonic Hedgehog-induced proliferation requires specific Gα inhibitory proteins. *J. Biol. Chem.* **286,** 8067–8074.

Beachy, P. A., Hymowitz, S. G., Lazarus, R. A., Leahy, D. J., and Siebold, C. (2011). Interactions between Hedgehog proteins and their binding partners come into view. *Genes Dev.* **24,** 2001–2012.

Belgacem, Y. H., and Borodinsky, L. N. (2011). Sonic hedgehog signaling is decoded by calcium spike activity in the developing spinal cord. *Proc. Natl. Acad. Sci. USA* **108**(11), 4482–4487.

Bijlsma, M. F., Peppelenbosch, M. P., and Spek, C. A. (2006). A dual role for 7-dehydrocholesterol reductase in regulating Hedgehog signalling? *Development* **133,** 3951, author reply 3952–3953.

Briscoe, J., and Ericson, J. (2001). Specification of neuronal fates in the ventral neural tube. *Curr. Opin. Neurobiol.* **11,** 43–49.

Briscoe, J., Pierani, A., Jessell, T. M., and Ericson, J. (2000). A homeodomain protein code specifies progenitor cell identity and neuronal fate in the ventral neural tube. *Cell* **101,** 435–445.

Buglino, J. A., and Resh, M. D. (2008). Hhat is a palmitoylacyltransferase with specificity for N-palmitoylation of Sonic Hedgehog. *J. Biol. Chem.* **283,** 22076–22088.

Bulgakov, O. V., Eggenschwiler, J. T., Hong, D. H., Anderson, K. V., and Li, T. (2004). FKBP8 is a negative regulator of mouse sonic hedgehog signaling in neural tissues. *Development* **131,** 2149–2159.

Buttitta, L., Mo, R., Hui, C. C., and Fan, C. M. (2003). Interplays of Gli2 and Gli3 and their requirement in mediating Shh-dependent sclerotome induction. *Development* **130,** 6233–6243.

Capurro, M. I., Xu, P., Shi, W., Li, F., Jia, A., and Filmus, J. (2008). Glypican-3 inhibits Hedgehog signaling during development by competing with patched for Hedgehog binding. *Dev. Cell* **14,** 700–711.

Carney, R. S., Mangin, J. M., Hayes, L., Mansfield, K., Sousa, V. H., Fishell, G., Machold, R. P., Ahn, S., Gallo, V., and Corbin, J. G. (2010). Sonic hedgehog expressing and responding cells generate neuronal diversity in the medial amygdala. *Neural Dev.* **5,** 14.

Carpenter, D., Stone, D. M., Brush, J., Ryan, A., Armanini, M., Frantz, G., Rosenthal, A., and de Sauvage, F. J. (1998). Characterization of two patched receptors for the vertebrate hedgehog protein family. *PNAS* **95,** 13630–13634.

Caspary, T., Garcia-Garcia, M. J., Huangfu, D., Eggenschwiler, J. T., Wyler, M. R., Rakeman, A. S., Alcorn, H. L., and Anderson, K. V. (2002). Mouse Dispatched homolog1 is required for long-range, but not juxtacrine Hh signaling. *Curr. Biol.* **12,** 1628–1632.

Caspary, T., Larkins, C. E., and Anderson, K. V. (2007). The graded response to Sonic Hedgehog depends on cilia architecture. *Dev. Cell* **12,** 767–778.

Chamberlain, C. E., Jeong, J., Guo, C., Allen, B. L., and McMahon, A. P. (2008). Notochord-derived Shh concentrates in close association with the apically positioned basal body in neural target cells and forms a dynamic gradient during neural patterning. *Development* **135,** 1097–1106.

Chan, J. A., Balasubramanian, S., Witt, R. M., Nazemi, K. J., Choi, Y., Pazyra-Murphy, M. F., Walsh, C. O., Thompson, M., and Segal, R. A. (2009). Proteoglycan interactions with Sonic Hedgehog specify mitogenic responses. *Nat. Neurosci.* **12,** 409–417.

Charron, F., Stein, E., Jeong, J., McMahon, A. P., and Tessier-Lavigne, M. (2003). The morphogen sonic hedgehog is an axonal chemoattractant that collaborates with netrin-1 in midline axon guidance. *Cell* **113,** 11–23.

Chaung, P.-T., and Kornberg, T. B. (2000). On the range of Hedgehog signaling. *Curr. Opin. Genet. Dev.* **10,** 515–522.

Chen, Y., Gallaher, N., Goodman, R. H., and Smolik, S. M. (1998). Protein kinase A directly regulates the activity and proteolysis of cubitus interruptus. *Proc. Natl. Acad. Sci. USA* **95,** 2349–2354.

Chen, J. K., Taipale, J., Cooper, M. K., and Beachy, P. A. (2002a). Inhibition of Hedgehog signaling by direct binding of cyclopamine to Smoothened. *Genes Dev.* **16,** 2743–2748.

Chen, J. K., Taipale, J., Young, K. E., Maiti, T., and Beachy, P. A. (2002b). Small molecule modulation of Smoothened activity. *Proc. Natl. Acad. Sci. USA* **99,** 14071–14076.

Chen, M. H., Li, Y. J., Kawakami, T., Xu, S. M., and Chuang, P. T. (2004a). Palmitoylation is required for the production of a soluble multimeric Hedgehog protein complex and long-range signaling in vertebrates. *Genes Dev.* **18,** 641–659.

Chen, W., Ren, X. R., Nelson, C. D., Barak, L. S., Chen, J. K., Beachy, P. A., de Sauvage, F., and Lefkowitz, R. J. (2004b). Activity-dependent internalization of smoothened mediated by beta-arrestin 2 and GRK2. *Science* **306,** 2257–2260.

Chen, M. H., Gao, N., Kawakami, T., and Chuang, P. T. (2005). Mice deficient in the fused homolog do not exhibit phenotypes indicative of perturbed hedgehog signaling during embryonic development. *Mol. Cell. Biol.* **25,** 7042–7053.

Chen, M. H., Wilson, C. W., Li, Y. J., Law, K. K., Lu, C. S., Gacayan, R., Zhang, X., Hui, C. C., and Chuang, P. T. (2009). Cilium-independent regulation of Gli protein function by Sufu in Hedgehog signaling is evolutionarily conserved. *Genes Dev.* **23,** 1910–1928.

Cheng, S. Y., and Bishop, J. M. (2002). Suppressor of Fused represses Gli-mediated transcription by recruiting the SAP18-mSin3 corepressor complex. *Proc. Natl. Acad. Sci. USA* **99,** 5442–5447.

Chiang, C., Litingtung, Y., Lee, E., Young, K. E., Corden, J. L., Westphal, H., and Beachy, P. A. (1996). Cyclopia and defective axial patterning in mice lacking *Sonic hedgehog* gene function. *Nature* **383,** 407–413.

Cho, A., Ko, H. W., and Eggenschwiler, J. T. (2008). FKBP8 cell-autonomously controls neural tube patterning through a Gli2- and Kif3a-dependent mechanism. *Dev. Biol.* **321,** 27–39.

Cocas, L. A., Miyoshi, G., Carney, R. S., Sousa, V. H., Hirata, T., Jones, K. R., Fishell, G., Huntsman, M. M., and Corbin, J. G. (2009). Emx1-lineage progenitors differentially contribute to neural diversity in the striatum and amygdala. *J. Neurosci.* **29,** 15933–15946.

Cooper, A. F., Yu, K. P., Brueckner, M., Brailey, L. L., Johnson, L., McGrath, J. M., and Bale, A. E. (2005). Cardiac and CNS defects in a mouse with targeted disruption of suppressor of fused. *Development* **132,** 4407–4417.

Corbit, K. C., Aanstad, P., Singla, V., Norman, A. R., Stainier, D. Y., and Reiter, J. F. (2005). Vertebrate Smoothened functions at the primary cilium. *Nature* **437,** 1018–1021.

Corcoran, R. B., and Scott, M. P. (2006). Oxysterols stimulate Sonic hedgehog signal transduction and proliferation of medulloblastoma cells. *Proc. Natl. Acad. Sci. USA* **103,** 8408–8413.

Courey, A. J., and Jia, S. (2001). Transcriptional repression: The long and the short of it. *Genes Dev.* **15,** 2786–2796.

Dai, P., Akimaru, H., Tanaka, Y., Maekawa, T., Nakafuku, M., and Ishii, S. (1999). Sonic Hedgehog-induced activation of the Gli1 promoter is mediated by GLI3. *J. Biol. Chem.* **274,** 8143–8152.

Dai, P., Shinagawa, T., Nomura, T., Harada, J., Kaul, S. C., Wadhwa, R., Khan, M. M., Akimaru, H., Sasaki, H., Colmenares, C., and Ishii, S. (2002). Ski is involved in transcriptional regulation by the repressor and full-length forms of Gli3. *Genes Dev.* **16,** 2843–2848.

Danesin, C., Agius, E., Escalas, N., Ai, X., Emerson, C., Cochard, P., and Soula, C. (2006). Ventral neural progenitors switch toward an oligodendroglial fate in response to increased

Sonic hedgehog (Shh) activity: Involvement of Sulfatase 1 in modulating Shh signaling in the ventral spinal cord. *J. Neurosci.* **26,** 5037–5048.
DeCamp, D. L., Thompson, T. M., de Sauvage, F. J., and Lerner, M. R. (2000). Smoothened activates Galphai-mediated signaling in frog melanophores. *J. Biol. Chem.* **275,** 26322–26327.
Dessaud, E., Yang, L. L., Hill, K., Cox, B., Ulloa, F., Ribeiro, A., Mynett, A., Novitch, B. G., and Briscoe, J. (2007). Interpretation of the sonic hedgehog morphogen gradient by a temporal adaptation mechanism. *Nature* **450,** 717–720.
Dessaud, E., McMahon, A. P., and Briscoe, J. (2008). Pattern formation in the vertebrate neural tube: A sonic hedgehog morphogen-regulated transcriptional network. *Development* **135,** 2489–2503.
Dessaud, E., Ribes, V., Balaskas, N., Yang, L. L., Pierani, A., Kicheva, A., Novitch, B. G., Briscoe, J., and Sasai, N. (2010). Dynamic assignment and maintenance of positional identity in the ventral neural tube by the morphogen sonic hedgehog. *PLoS Biol.* **8,** e1000382.
Ding, Q., Motoyama, J., Gasca, S., Mo, R., Sasaki, H., Rossant, J., and Hui, C. C. (1998). Diminished Sonic hedgehog signaling and lack of floor plate differentiation in *Gli2* mutant mice. *Development* **125,** 2533–2543.
Ding, Q., Fukami, S., Meng, X., Nishizaki, Y., Zhang, X., Sasaki, H., Dlugosz, A., Nakafuku, M., and Hui, C. (1999). Mouse suppressor of fused is a negative regulator of sonic hedgehog signaling and alters the subcellular distribution of Gli1. *Curr. Biol.* **9,** 1119–1122.
Dunaeva, M., Michelson, P., Kogerman, P., and Toftgard, R. (2003). Characterization of the physical interaction of Gli proteins with SUFU proteins. *J. Biol. Chem.* **278,** 5116–5122.
Dussillol-Godar, F., Brissard-Zahraoui, J., Limbourg-Bouchon, B., Boucher, D., Fouix, S., Lamour-Isnard, C., Plessis, A., and Busson, D. (2006). Modulation of the Suppressor of fused protein regulates the Hedgehog signaling pathway in Drosophila embryo and imaginal discs. *Dev. Biol.* **291,** 53–66.
Eaton, S. (2008). Multiple roles for lipids in the Hedgehog signalling pathway. *Nat. Rev. Mol. Cell Biol.* **9,** 437–445.
Echelard, Y., Epstein, D. J., St-Jacques, B., Shen, L., Mohler, J., McMahon, J. A., and McMahon, A. P. (1993). Sonic hedgehog, a member of a family of putative signaling molecules, is implicated in the regulation of CNS polarity. *Cell* **75,** 1417–1430.
Eggenschwiler, J. T., and Anderson, K. V. (2007). Cilia and developmental signaling. *Annu. Rev. Cell Dev. Biol.* **23,** 345–373.
Endoh-Yamagami, S., Evangelista, M., Wilson, D., Wen, X., Theunissen, J. W., Phamluong, K., Davis, M., Scales, S. J., Solloway, M. J., de Sauvage, F. J., and Peterson, A. S. (2009). The mammalian Cos2 homolog Kif7 plays an essential role in modulating Hh signal transduction during development. *Curr. Biol.* **19,** 1320–1326.
Engelhardt, S., and Rochais, F. (2007). G proteins: More than transducers of receptor-generated signals? *Circ. Res.* **100,** 1109–1111.
Ericson, J., Thor, S., Edlund, T., Jessell, T. M., and Yamada, T. (1992). Early stages of motor neuron differentiation revealed by expression of homeobox gene *Islet-1*. *Science* **256,** 1555–1560.
Ericson, J., Briscoe, J., Rashbass, P., van Heyningen, V., and Jessell, T. M. (1997a). Graded sonic hedgehog signaling and the specification of cell fate in the ventral neural tube. *Cold Spring Harb. Symp. Quant. Biol.* **62,** 451–466.
Ericson, J., Rashbass, P., Schedl, A., Brenner-Morton, S., Kawakami, A., van Heynigen, V., Jessell, T. M., and Briscoe, J. (1997b). Pax6 controls progenitor cell identity and neuronal fate in response to graded Shh signaling. *Cell* **90,** 169–180.

Etheridge, L. A., Crawford, T. Q., Zhang, S., and Roelink, H. (2010). Evidence for a role of vertebrate Disp1 in long-range Shh signaling. *Development* **137,** 133–140.

Gallet, A. (2011). Hedgehog morphogen: From secretion to reception. *Trends Cell Biol.* **21** (4), 238–246.

Goetz, J. A., Singh, S., Suber, L. M., Kull, F. J., and Robbins, D. J. (2006). A highly conserved amino-terminal region of sonic hedgehog is required for the formation of its freely diffusible multimeric form. *J. Biol. Chem.* **281,** 4087–4093.

Goodrich, L. V., Milenkovic, L., Higgins, K. M., and Scott, M. P. (1997). Altered neural cell fates and medulloblastoma in mouse *patched* mutants. *Science* **277,** 1109–1113.

Goulding, M. D., Lumsden, A., and Gruss, P. (1993). Signals from the notochord and floor plate regulate the region-specific expression of two Pax genes in the developing spinal cord. *Development* **117,** 1001–1016.

Gritli-Linde, A., Lewis, P., McMahon, A. P., and Linde, A. (2001). The whereabouts of a morphogen: Direct evidence for short- and graded long-range activity of hedgehog signaling peptides. *Dev. Biol.* **236,** 364–386.

Guerrero, I., and Chiang, C. (2007). A conserved mechanism of Hedgehog gradient formation by lipid modifications. *Trends Cell Biol.* **17,** 1–5.

Gurdon, J. B., and Bourillot, P. Y. (2001). Morphogen gradient interpretation. *Nature* **413,** 797–803.

Hallikas, O., Palin, K., Sinjushina, N., Rautiainen, R., Partanen, J., Ukkonen, E., and Taipale, J. (2006). Genome-wide prediction of mammalian enhancers based on analysis of transcription-factor binding affinity. *Cell* **124,** 47–59.

Han, Y. G., Spassky, N., Romaguera-Ros, M., Garcia-Verdugo, J. M., Aguilar, A., Schneider-Maunoury, S., and Alvarez-Buylla, A. (2008). Hedgehog signaling and primary cilia are required for the formation of adult neural stem cells. *Nat. Neurosci.* **11,** 277–284.

Hatten, M. E., and Roussel, M. F. (2011). Development and cancer of the cerebellum. *Trends Neurosci.* **34,** 134–142.

Hayhurst, M., Gore, B. B., Tessier-Lavigne, M., and McConnell, S. K. (2008). Ongoing sonic hedgehog signaling is required for dorsal midline formation in the developing forebrain. *Dev. Neurobiol.* **68,** 83–100.

Herron, B. J., Lu, W., Rao, C., Liu, S., Peters, H., Bronson, R. T., Justice, M. J., McDonald, J. D., and Beier, D. R. (2002). Efficient generation and mapping of recessive developmental mutations using ENU mutagenesis. *Nat. Genet.* **30,** 185–189.

Himmelstein, D. S., Bi, C., Clark, B. S., Bai, B., and Kohtz, J. D. (2010). Balanced Shh signaling is required for proper formation and maintenance of dorsal telencephalic midline structures. *BMC Dev. Biol.* **10,** 118.

Hooper, J. E., and Scott, M. P. (2005). Communicating with Hedgehogs. *Nat. Rev. Mol. Cell Biol.* **6,** 306–317.

Huang, X., Litingtung, Y., and Chiang, C. (2007a). Ectopic sonic hedgehog signaling impairs telencephalic dorsal midline development: Implication for human holoprosencephaly. *Hum. Mol. Genet.* **16,** 1454–1468.

Huang, X., Litingtung, Y., and Chiang, C. (2007b). Region-specific requirement for cholesterol modification of sonic hedgehog in patterning the telencephalon and spinal cord. *Development* **134,** 2095–2105.

Huangfu, D., and Anderson, K. V. (2006). Signaling from Smo to Ci/Gli: Conservation and divergence of Hedgehog pathways from Drosophila to vertebrates. *Development* **133,** 3–14.

Huangfu, D., Liu, A., Rakeman, A. S., Murcia, N. S., Niswander, L., and Anderson, K. V. (2003). Hedgehog signalling in the mouse requires intraflagellar transport proteins. *Nature* **426,** 83–87.

Huntzicker, E. G., Estay, I. S., Zhen, H., Lokteva, L. A., Jackson, P. K., and Oro, A. E. (2006). Dual degradation signals control Gli protein stability and tumor formation. *Genes Dev.* **20,** 276–281.

Iyer, G. H., Moore, M. J., and Taylor, S. S. (2005). Consequences of lysine 72 mutation on the phosphorylation and activation state of cAMP-dependent kinase. *J. Biol. Chem.* **280,** 8800–8807.

Jenkins, D. (2009). Hedgehog signalling: Emerging evidence for non-canonical pathways. *Cell. Signal.* **21,** 1023–1034.

Jeong, Y., and Epstein, D. J. (2003). Distinct regulators of Shh transcription in the floor plate and notochord indicate separate origins for these tissues in the mouse node. *Development* **130,** 3891–3902.

Jeong, J., and McMahon, A. P. (2005). Growth and pattern of the mammalian neural tube are governed by partially overlapping feedback activities of the hedgehog antagonists patched 1 and Hhip1. *Development* **132,** 143–154.

Jeong, Y., Dolson, D. K., Waclaw, R. R., Matise, M. P., Sussel, L., Campbell, K., Kaestner, K. H., and Epstein, D. J. (2011). Spatial and temporal requirements for sonic hedgehog in the regulation of thalamic interneuron identity. *Development* **138,** 531–541.

Jessell, T. M. (2000). Neuronal specification in the spinal cord: Inductive signals and transcriptional codes. *Nat. Rev. Genet.* **1,** 20–29.

Jia, J., Tong, C., and Jiang, J. (2003). Smoothened transduces Hedgehog signal by physically interacting with Costal2/Fused complex through its C-terminal tail. *Genes Dev.* **17,** 2709–2720.

Jia, J., Tong, C., Wang, B., Luo, L., and Jiang, J. (2004). Hedgehog signalling activity of Smoothened requires phosphorylation by protein kinase A and casein kinase I. *Nature* **432,** 1045–1050.

Kaesler, S., Luscher, B., and Ruther, U. (2000). Transcriptional activity of GLI1 is negatively regulated by protein kinase A. *Biol. Chem.* **381,** 545–551.

Kinzler, K. W., and Vogelstein, B. (1990). The GLI gene encodes a nuclear protein which binds specific sequences in the human genome. *Mol. Cell. Biol.* **10,** 634–642.

Kogerman, P., Grimm, T., Kogerman, L., Krause, D., Unden, A. B., Sandstedt, B., Toftgard, R., and Zaphiropoulos, P. G. (1999). Mammalian suppressor-of-fused modulates nuclear-cytoplasmic shuttling of Gli-1. *Nat. Cell Biol.* **1,** 312–319.

Kovacs, J. J., Whalen, E. J., Liu, R., Xiao, K., Kim, J., Chen, M., Wang, J., Chen, W., and Lefkowitz, R. J. (2008). Beta-arrestin-mediated localization of smoothened to the primary cilium. *Science* **320,** 1777–1781.

Koyabu, Y., Nakata, K., Mizugishi, K., Aruga, J., and Mikoshiba, K. (2001). Physical and functional interactions between Zic and Gli proteins. *J. Biol. Chem.* **276,** 6889–6892.

Krauss, S., Concordet, J. P., and Ingham, P. W. (1993). A functionally conserved homolog of the drosophila segment polarity gene *hh* is expressed in tissues with polarizing activity in Zebrafish embryos. *Cell* **75,** 1431–1444.

Kutejova, E., Briscoe, J., and Kicheva, A. (2009). Temporal dynamics of patterning by morphogen gradients. *Curr. Opin. Genet. Dev.* **19,** 315–322.

Lauth, M., and Toftgard, R. (2007). Non-canonical activation of GLI transcription factors: Implications for targeted anti-cancer therapy. *Cell Cycle* **6,** 2458–2463.

Lee, S.-K., and Pfaff, S. L. (2001). Transcriptional networks regulating neuronal identity in the developing spinal cord. *Nat. Neurosci.* **4,** 1183–1191.

Lee, C. S., Buttitta, L., and Fan, C. M. (2001). Evidence that the WNT-inducible growth arrest-specific gene 1 encodes an antagonist of sonic hedgehog signaling in the somite. *Proc. Natl. Acad. Sci. USA* **98,** 11347–11352.

Lee, Y., Miller, H. L., Russell, H. R., Boyd, K., Curran, T., and McKinnon, P. J. (2006). Patched2 modulates tumorigenesis in patched1 heterozygous mice. *Cancer Res.* **66,** 6964–6971.

Lei, Q., Zelman, A. K., Kuang, E., Li, S., and Matise, M. P. (2004). Transduction of graded Hedgehog signaling by a combination of Gli2 and Gli3 activator functions in the developing spinal cord. *Development* **131,** 3593–3604.

Lei, Q., Jeong, Y., Misra, K., Li, S., Zelman, A. K., Epstein, D. J., and Matise, M. P. (2006). Wnt signaling inhibitors regulate the transcriptional response to morphogenetic Shh-Gli signaling in the neural tube. *Dev. Cell* **11,** 325–337.

Lek, M., Dias, J. M., Marklund, U., Uhde, C. W., Kurdija, S., Lei, Q., Sussel, L., Rubenstein, J. L., Matise, M. P., Arnold, H. H., Jessell, T. M., and Ericson, J. (2010). A homeodomain feedback circuit underlies step-function interpretation of a Shh morphogen gradient during ventral neural patterning. *Development* **137,** 4051–4060.

Lewis, P. M., Dunn, M. P., McMahon, J. A., Logan, M., Martin, J. F., St-Jacques, B., and McMahon, A. P. (2001). Cholesterol modification of sonic hedgehog is required for long-range signaling activity and effective modulation of signaling by Ptc1. *Cell* **105,** 599–612.

Lewis, P. M., Gritli-Linde, A., Smeyne, R., Kottmann, A., and McMahon, A. P. (2004). Sonic hedgehog signaling is required for expansion of granule neuron precursors and patterning of the mouse cerebellum. *Dev. Biol.* **270,** 393–410.

Li, J., Wang, C., Pan, Y., Bai, Z., and Wang, B. (2011). Increased proteolytic processing of full-length Gli2 transcription factor reduces the hedgehog pathway activity in vivo. *Dev. Dyn.* **240**(4), 766–774.

Liem, K. F., Jr., He, M., Ocbina, P. J., and Anderson, K. V. (2009). Mouse Kif7/Costal2 is a cilia-associated protein that regulates Sonic hedgehog signaling. *Proc. Natl. Acad. Sci. USA* **106,** 13377–13382.

Litingtung, Y., and Chiang, C. (2000). Specification of ventral neuron types is mediated by an antagonistic interaction between Shh and Gli3. *Nat. Neurosci.* **3,** 979–985.

Liu, A., Wang, B., and Niswander, L. A. (2005). Mouse intraflagellar transport proteins regulate both the activator and repressor functions of Gli transcription factors. *Development* **132,** 3103–3111.

Low, W. C., Wang, C., Pan, Y., Huang, X. Y., Chen, J. K., and Wang, B. (2008). The decoupling of Smoothened from Galphai proteins has little effect on Gli3 protein processing and Hedgehog-regulated chick neural tube patterning. *Dev. Biol.* **321,** 188–196.

Lum, L., Zhang, C., Oh, S., Mann, R. K., von Kessler, D. P., Taipale, J., Weis-Garcia, F., Gong, R., Wang, B., and Beachy, P. A. (2003). Hedgehog signal transduction via Smoothened association with a cytoplasmic complex scaffolded by the atypical kinesin, Costal-2. *Mol. Cell* **12,** 1261–1274.

Ma, Y., Erkner, A., Gong, R., Yao, S., Taipale, J., Basler, K., and Beachy, P. A. (2002). Hedgehog-mediated patterning of the mammalian embryo requires transporter-like function of dispatched. *Cell* **111,** 63–75.

Maloverjan, A., Piirsoo, M., Michelson, P., Kogerman, P., and Osterlund, T. (2010). Identification of a novel serine/threonine kinase ULK3 as a positive regulator of Hedgehog pathway. *Exp. Cell Res.* **316,** 627–637.

Marigo, V., Johnson, R. L., Vortkamp, A., and Tabin, C. J. (1996). Sonic hedgehog differentially regulates expression of *GLI* and *GLI3* during limb development. *Dev. Biol.* **180,** 273–283.

Matise, M. P. (2007). Order in the classroom: Graded responses to instructive Hh signaling in the CNS. *Cell Cycle* **6,** 1194–1199.

Matise, M. P., and Joyner, A. L. (1997). Expression patterns of developmental control genes in normal and *Engrailed-1* mutant mouse spinal cord reveal early diversity in developing interneurons. *J. Neurosci.* **17,** 7805–7816.

Matise, M. P., and Joyner, A. L. (1999). Gli genes in development and cancer. *Oncogene* **18,** 7852–7859.

McCarthy, R. A., and Argraves, W. S. (2003). Megalin and the neurodevelopmental biology of sonic hedgehog and retinol. *J. Cell Sci.* **116,** 955–960.

McCarthy, R. A., Barth, J. L., Chintalapudi, M. R., Knaak, C., and Argraves, W. S. (2002). Megalin functions as an endocytic sonic hedgehog receptor. *J. Biol. Chem.* **277,** 25660–25667.

McDermott, A., Gustafsson, M., Elsam, T., Hui, C. C., Emerson, C. P., Jr., and Borycki, A. G. (2005). Gli2 and Gli3 have redundant and context-dependent function in skeletal muscle formation. *Development* **132,** 345–357.

Merchant, M., Vajdos, F. F., Ultsch, M., Maun, H. R., Wendt, U., Cannon, J., Desmarais, W., Lazarus, R. A., de Vos, A. M., and de Sauvage, F. J. (2004). Suppressor of fused regulates Gli activity through a dual binding mechanism. *Mol. Cell. Biol.* **24,** 8627–8641.

Merchant, M., Evangelista, M., Luoh, S. M., Frantz, G. D., Chalasani, S., Carano, R. A., van Hoy, M., Ramirez, J., Ogasawara, A. K., McFarland, L. M., Filvaroff, E. H., French, D. M., *et al.* (2005). Loss of the serine/threonine kinase fused results in postnatal growth defects and lethality due to progressive hydrocephalus. *Mol. Cell. Biol.* **25,** 7054–7068.

Milenkovic, L., Scott, M. P., and Rohatgi, R. (2009). Lateral transport of Smoothened from the plasma membrane to the membrane of the cilium. *J. Cell Biol.* **187,** 365–374.

Mizugishi, K., Aruga, J., Nakata, K., and Mikoshiba, K. (2001). Molecular properties of Zic proteins as transcriptional regulators and their relationship to GLI proteins. *J. Biol. Chem.* **276,** 2180–2188.

Motoyama, J., Heng, H., Crackower, M. A., Takabatake, T., Takeshima, K., Tsui, L. C., and Hui, C. (1998a). Overlapping and non-overlapping Ptch2 expression with Shh during mouse embryogenesis. *Mech. Dev.* **78,** 81–84.

Motoyama, J., Takabatake, T., Takeshima, K., and Hui, C. (1998b). Ptch2, a second mouse Patched gene is co-expressed with Sonic hedgehog. *Nat. Genet.* **18,** 104–106.

Nakashima, M., Tanese, N., Ito, M., Auerbach, W., Bai, C., Furukawa, T., Toyono, T., Akamine, A., and Joyner, A. L. (2002). A novel gene, GliH1, with homology to the Gli zinc finger domain not required for mouse development. *Mech. Dev.* **119,** 21–34.

Nishi, Y., Ji, H., Wong, W. H., McMahon, A. P., and Vokes, S. A. (2009). Modeling the spatio-temporal network that drives patterning in the vertebrate central nervous system. *Biochim. Biophys. Acta* **1789,** 299–305.

Novitch, B. G., Wichterle, H., Jessell, T. M., and Sockanathan, S. (2003). A requirement for retinoic acid-mediated transcriptional activation in ventral neural patterning and motor neuron specification. *Neuron* **40,** 81–95.

Nusslein-Volhard, C., and Wieschaus, E. (1980). Mutations affecting segment number and polarity in *Drosophila. Nature* **287,** 795–801.

Ogden, S. K., Ascano, M., Jr., Stegman, M. A., Suber, L. M., Hooper, J. E., and Robbins, D. J. (2003). Identification of a functional interaction between the transmembrane protein Smoothened and the kinesin-related protein Costal2. *Curr. Biol.* **13,** 1998–2003.

Oh, S., Huang, X., and Chiang, C. (2005). Specific requirements of sonic hedgehog signaling during oligodendrocyte development. *Dev. Dyn.* **234,** 489–496.

Oh, S., Huang, X., Liu, J., Litingtung, Y., and Chiang, C. (2009). Shh and Gli3 activities are required for timely generation of motor neuron progenitors. *Dev. Biol.* **331,** 261–269.

Okada, A., Charron, F., Morin, S., Shin, D. S., Wong, K., Fabre, P. J., Tessier-Lavigne, M., and McConnell, S. K. (2006). Boc is a receptor for sonic hedgehog in the guidance of commissural axons. *Nature* **444,** 369–373.

Paces-Fessy, M., Boucher, D., Petit, E., Paute-Briand, S., and Blanchet-Tournier, M. F. (2004). The negative regulator of Gli, Suppressor of fused (Sufu), interacts with SAP18, Galectin3 and other nuclear proteins. *Biochem. J.* **378,** 353–362.

Palma, V., Lim, D. A., Dahmane, N., Sanchez, P., Brionne, T. C., Herzberg, C. D., Gitton, Y., Carleton, A., Alvarez-Buylla, A., and Ruiz i Altaba, A. (2005). Sonic hedgehog controls stem cell behavior in the postnatal and adult brain. *Development* **132,** 335–344.

Pan, Y., and Wang, B. (2007). A novel protein-processing domain in Gli2 and Gli3 differentially blocks complete protein degradation by the proteasome. *J. Biol. Chem.* **282,** 10846–10852.

Pan, Y., Bai, C. B., Joyner, A. L., and Wang, B. (2006). Sonic hedgehog signaling regulates Gli2 transcriptional activity by suppressing its processing and degradation. *Mol. Cell. Biol.* **26,** 3365–3377.

Pan, Y., Wang, C., and Wang, B. (2009). Phosphorylation of Gli2 by protein kinase A is required for Gli2 processing and degradation and the Sonic Hedgehog-regulated mouse development. *Dev. Biol.* **326,** 177–189.

Pavletich, N. P., and Pabo, C. O. (1993). Crystal structure of a five-finger GLI-DNA complex: New perspectives on zinc fingers. *Science* **261,** 1701–1707.

Persson, M., Stamataki, D., te Welscher, P., Andersson, E., Bose, J., Ruther, U., Ericson, J., and Briscoe, J. (2002). Dorsal-ventral patterning of the spinal cord requires Gli3 transcriptional repressor activity. *Genes Dev.* **16,** 2865–2878.

Pham, A., Therond, P., Alves, G., Tournier, F. B., Busson, D., Lamour-Isnard, C., Bouchon, B. L., Preat, T., and Tricoire, H. (1995). The Suppressor of fused gene encodes a novel PEST protein involved in Drosophila segment polarity establishment. *Genetics* **140,** 587–598.

Philipp, M., Fralish, G. B., Meloni, A. R., Chen, W., MacInnes, A. W., Barak, L. S., and Caron, M. G. (2008). Smoothened signaling in vertebrates is facilitated by a G protein-coupled receptor kinase. *Mol. Biol. Cell* **19,** 5478–5489.

Pierani, A., Brenner-Morton, S., Chiang, C., and Jessell, T. (1999). A sonic hedgehog-independent, retinoid-activated pathway of neurogenesis in the ventral spinal cord. *Cell* **97,** 903–915.

Placzek, M., and Briscoe, J. (2005). The floor plate: Multiple cells, multiple signals. *Nat. Rev. Neurosci.* **6,** 230–240.

Porter, J. A., Ekker, S. C., Park, W.-J., von Kessler, D. P., Young, K. E., Chen, S.-H., Ma, Y., Woods, A. S., Cotter, R. J., Koonin, E. V., and Beachy, P. A. (1996). Hedgehog patterning activity: Role of a lipophilic modification mediated by the carboxy-terminal autoprocessing domain. *Cell* **86,** 21–34.

Preat, T. (1992). Characterization of Suppressor of fused, a complete suppressor of the fused segment polarity gene of *Drosophila melanogaster*. *Genetics* **132,** 725–736.

Price, M. A., and Kalderon, D. (1999). Proteolysis of Cubitus interruptus in *Drosophila* requires phosphorylation by Protein Kinase A. *Development* **126,** 4331–4339.

Qin, J., Lin, Y., Norman, R. X., Ko, H. W., and Eggenschwiler, J. T. (2011). Intraflagellar transport protein 122 antagonizes Sonic Hedgehog signaling and controls ciliary localization of pathway components. *Proc. Natl. Acad. Sci. USA* **108,** 1456–1461.

Rash, B. G., and Grove, E. A. (2007). Patterning the dorsal telencephalon: A role for sonic hedgehog? *J. Neurosci.* **27,** 11595–11603.

Reiter, J. F., and Skarnes, W. C. (2006). Tectonic, a novel regulator of the Hedgehog pathway required for both activation and inhibition. *Genes Dev.* **20,** 22–27.

Ribes, V., and Briscoe, J. (2009). Establishing and interpreting graded Sonic Hedgehog signaling during vertebrate neural tube patterning: The role of negative feedback. *Cold Spring Harb. Perspect. Biol.* **1,** a002014.

Ribes, V., Balaskas, N., Sasai, N., Cruz, C., Dessaud, E., Cayuso, J., Tozer, S., Yang, L. L., Novitch, B., Marti, E., and Briscoe, J. (2010). Distinct Sonic Hedgehog signaling dynamics specify floor plate and ventral neuronal progenitors in the vertebrate neural tube. *Genes Dev.* **24,** 1186–1200.

Riddle, R. D., Johnson, R. L., Laufer, E., and Tabin, C. (1993). *Sonic hedgehog* mediates the polarizing activity of the ZPA. *Cell* **75,** 1401–1416.

Riobo, N. A., Saucy, B., Dilizio, C., and Manning, D. R. (2006). Activation of heterotrimeric G proteins by Smoothened. *Proc. Natl. Acad. Sci. USA* **103,** 12607–12612.

Roelink, H., Augsburger, A., Heemskerk, J., Korzh, V., Norlin, S., Ruiz i Altaba, A., Tanabe, Y., Placzek, M., Edlund, T., Jessell, T. M., and Dodd, J. (1994). Floor plate and motor neuron induction by vhh-1, a vertebrate homolog of hedgehog expressed by the notochord. *Cell* **76,** 761–775.

Rohatgi, R., Milenkovic, L., and Scott, M. P. (2007). Patched1 regulates hedgehog signaling at the primary cilium. *Science* **317,** 372–376.

Rohatgi, R., Milenkovic, L., Corcoran, R. B., and Scott, M. P. (2009). Hedgehog signal transduction by Smoothened: Pharmacologic evidence for a 2-step activation process. *Proc. Natl. Acad. Sci. USA* **106,** 3196–3201.

Rubin, J. B., Choi, Y., and Segal, R. A. (2002). Cerebellar proteoglycans regulate sonic hedgehog responses during development. *Development* **129,** 2223–2232.

Ruel, L., and Therond, P. P. (2009). Variations in Hedgehog signaling: Divergence and perpetuation in Sufu regulation of Gli. *Genes Dev.* **23,** 1843–1848.

Ruel, L., Rodriguez, R., Gallet, A., Lavenant-Staccini, L., and Therond, P. P. (2003). Stability and association of Smoothened, Costal2 and Fused with Cubitus interruptus are regulated by Hedgehog. *Nat. Cell Biol.* **5,** 907–913.

Ruel, L., Gallet, A., Raisin, S., Truchi, A., Staccini-Lavenant, L., Cervantes, A., and Therond, P. P. (2007). Phosphorylation of the atypical kinesin Costal2 by the kinase Fused induces the partial disassembly of the Smoothened-Fused-Costal2-Cubitus interruptus complex in Hedgehog signalling. *Development* **134,** 3677–3689.

Sakai-Kato, K., Ishiguro, A., Mikoshiba, K., Aruga, J., and Utsunomiya-Tate, N. (2008). CD spectra show the relational style between Zic-, Gli-Glis-zinc finger protein and DNA. *Biochim. Biophys. Acta* **1784,** 1011–1019.

Sanchez-Camacho, C., and Bovolenta, P. (2009). Emerging mechanisms in morphogen-mediated axon guidance. *Bioessays* **31,** 1013–1025.

Sasaki, H., Nishizaki, Y., Hui, C., Nakafuku, M., and Kondoh, H. (1999). Regulation of Gli2 and Gli3 activities by an amino-terminal repression domain: Implication of Gli2 and Gli3 as primary mediators of Shh signaling. *Development* **126,** 3915–3924.

Shimeld, S. M. (2008). C2H2 zinc finger genes of the Gli, Zic, KLF, SP, Wilms' tumour, Huckebein, Snail, Ovo, Spalt, Odd, Blimp-1, Fez and related gene families from *Branchiostoma floridae*. *Dev. Genes Evol.* **218,** 639–649.

Smelkinson, M. G., Zhou, Q., and Kalderon, D. (2007). Regulation of Ci-SCFSlimb binding, Ci proteolysis, and hedgehog pathway activity by Ci phosphorylation. *Dev. Cell* **13,** 481–495.

Sousa, V. H., and Fishell, G. (2010). Sonic hedgehog functions through dynamic changes in temporal competence in the developing forebrain. *Curr. Opin. Genet. Dev.* **20,** 391–399.

Spiegelman, B. M., and Heinrich, R. (2004). Biological control through regulated transcriptional coactivators. *Cell* **119,** 157–167.

Spoelgen, R., Hammes, A., Anzenberger, U., Zechner, D., Andersen, O. M., Jerchow, B., and Willnow, T. E. (2005). LRP2/megalin is required for patterning of the ventral telencephalon. *Development* **132,** 405–414.

Stamataki, D., Ulloa, F., Tsoni, S. V., Mynett, A., and Briscoe, J. (2005). A gradient of Gli activity mediates graded Sonic Hedgehog signaling in the neural tube. *Genes Dev.* **19,** 626–641.

Svard, J., Heby-Henricson, K., Persson-Lek, M., Rozell, B., Lauth, M., Bergstrom, A., Ericson, J., Toftgard, R., and Teglund, S. (2006). Genetic elimination of Suppressor of fused reveals an essential repressor function in the mammalian Hedgehog signaling pathway. *Dev. Cell* **10,** 187–197.

Taipale, J., Cooper, M. K., Maiti, T., and Beachy, P. A. (2002). Patched acts catalytically to suppress the activity of Smoothened. *Nature* **418,** 892–897.

Tan, M., Hu, X., Qi, Y., Park, J., Cai, J., and Qiu, M. (2006). Gli3 mutation rescues the generation, but not the differentiation, of oligodendrocytes in Shh mutants. *Brain Res.* **1067,** 158–163.

Tempe, D., Casas, M., Karaz, S., Blanchet-Tournier, M. F., and Concordet, J. P. (2006). Multisite protein kinase A and glycogen synthase kinase 3beta phosphorylation leads to Gli3 ubiquitination by SCFbetaTrCP. *Mol. Cell. Biol.* **26,** 4316–4326.

Tenzen, T., Allen, B. L., Cole, F., Kang, J. S., Krauss, R. S., and McMahon, A. P. (2006). The cell surface membrane proteins Cdo and Boc are components and targets of the Hedgehog signaling pathway and feedback network in mice. *Dev. Cell* **10,** 647–656.

Therond, P., Alves, G., Limbourg-Bouchon, B., Tricoire, H., Guillemet, E., Brissard-Zahraoui, J., Lamour-Isnard, C., and Busson, D. (1996). Functional domains of fused, a serine-threonine kinase required for signaling in Drosophila. *Genetics* **142,** 1181–1198.

Thibert, C., Teillet, M. A., Lapointe, F., Mazelin, L., Le Douarin, N. M., and Mehlen, P. (2003). Inhibition of neuroepithelial patched-induced apoptosis by sonic hedgehog. *Science* **301,** 843–846.

Tian, H., Jeong, J., Harfe, B. D., Tabin, C. J., and McMahon, A. P. (2005). Mouse Disp1 is required in sonic hedgehog-expressing cells for paracrine activity of the cholesterol-modified ligand. *Development* **132,** 133–142.

Tokhunts, R., Singh, S., Chu, T., D'Angelo, G., Baubet, V., Goetz, J. A., Huang, Z., Yuan, Z., Ascano, M., Zavros, Y., Therond, P. P., Kunes, S., *et al.* (2010). The full-length unprocessed hedgehog protein is an active signaling molecule. *J. Biol. Chem.* **285,** 2562–2568.

Traiffort, E., Angot, E., and Ruat, M. (2010). Sonic Hedgehog signaling in the mammalian brain. *J. Neurochem.* **113,** 576–590.

Tsanev, R., Tiigimagi, P., Michelson, P., Metsis, M., Osterlund, T., and Kogerman, P. (2009). Identification of the gene transcription repressor domain of Gli3. *FEBS Lett.* **583,** 224–228.

Ulloa, F., and Briscoe, J. (2007). Morphogens and the control of cell proliferation and patterning in the spinal cord. *Cell Cycle* **6,** 2640–2649.

Ulloa, F., Itasaki, N., and Briscoe, J. (2007). Inhibitory Gli3 activity negatively regulates Wnt/beta-catenin signaling. *Curr. Biol.* **17,** 545–550.

Vaillant, C., and Monard, D. (2009). SHH pathway and cerebellar development. *Cerebellum* **8,** 291–301.

van de Wetering, M., Oosterwegel, M., Dooijes, D., and Clevers, H. (1991). Identification and cloning of TCF-1, a T lymphocyte-specific transcription factor containing a sequence-specific HMG box. *EMBO J.* **10,** 123–132.

Varjosalo, M., and Taipale, J. (2008). Hedgehog: Functions and mechanisms. *Genes Dev.* **22,** 2454–2472.

Varjosalo, M., Li, S. P., and Taipale, J. (2006). Divergence of hedgehog signal transduction mechanism between Drosophila and mammals. *Dev. Cell* **10,** 177–186.

Vokes, S. A., Ji, H., McCuine, S., Tenzen, T., Giles, S., Zhong, S., Longabaugh, W. J., Davidson, E. H., Wong, W. H., and McMahon, A. P. (2007). Genomic characterization of Gli-activator targets in sonic hedgehog-mediated neural patterning. *Development* **134,** 1977–1989.

Wang, B., and Li, Y. (2006). Evidence for the direct involvement of {beta}TrCP in Gli3 protein processing. *Proc. Natl. Acad. Sci. USA* **103,** 33–38.

Wang, B., Fallon, J. F., and Beachy, P. A. (2000a). Hedgehog-regulated processing of Gli3 produces an anterior/posterior repressor gradient in the developing vertebrate limb. *Cell* **100,** 423–434.

Wang, G., Amanai, K., Wang, B., and Jiang, J. (2000b). Interactions with Costal2 and suppressor of fused regulate nuclear translocation and activity of cubitus interruptus. *Genes Dev.* **14,** 2893–2905.

Wang, Y., Zhou, Z., Walsh, C. T., and McMahon, A. P. (2009). Selective translocation of intracellular Smoothened to the primary cilium in response to Hedgehog pathway modulation. *Proc. Natl. Acad. Sci. USA* **106,** 2623–2628.

Wang, C., Pan, Y., and Wang, B. (2010a). Suppressor of fused and Spop regulate the stability, processing and function of Gli2 and Gli3 full-length activators but not their repressors. *Development* **137,** 2001–2009.

Wang, Y., Martin, J. F., and Bai, C. B. (2010b). Direct and indirect requirements of Shh/Gli signaling in early pituitary development. *Dev. Biol.* **348,** 199–209.

Wechsler-Reya, R. J., and Scott, M. P. (1999). Control of neuronal precursor proliferation in the cerebellum by Sonic Hedgehog [see comments]. *Neuron* **22,** 103–114.

Wen, X., Lai, C. K., Evangelista, M., Hongo, J. A., de Sauvage, F. J., and Scales, S. J. (2010). Kinetics of hedgehog-dependent full-length Gli3 accumulation in primary cilia and subsequent degradation. *Mol. Cell. Biol.* **30,** 1910–1922.

Wicher, G., and Aldskogius, H. (2008). Megalin deficiency induces critical changes in mouse spinal cord development. *Neuroreport* **19,** 559–563.

Wicher, G., Larsson, M., Rask, L., and Aldskogius, H. (2005). Low-density lipoprotein receptor-related protein (LRP)-2/megalin is transiently expressed in a subpopulation of neural progenitors in the embryonic mouse spinal cord. *J. Comp. Neurol.* **492,** 123–131.

Wijgerde, M., McMahon, J. A., Rule, M., and McMahon, A. P. (2002). A direct requirement for Hedgehog signaling for normal specification of all ventral progenitor domains in the presumptive mammalian spinal cord. *Genes Dev.* **16,** 2849–2864.

Wilson, C. W., and Chuang, P. T. (2010). Mechanism and evolution of cytosolic Hedgehog signal transduction. *Development* **137,** 2079–2094.

Wilson, C. W., Chen, M. H., and Chuang, P. T. (2009a). Smoothened adopts multiple active and inactive conformations capable of trafficking to the primary cilium. *PLoS One* **4,** e5182.

Wilson, C. W., Nguyen, C. T., Chen, M. H., Yang, J. H., Gacayan, R., Huang, J., Chen, J. N., and Chuang, P. T. (2009b). Fused has evolved divergent roles in vertebrate Hedgehog signalling and motile ciliogenesis. *Nature* **459,** 98–102.

Wolff, C., Roy, S., Lewis, K. E., Schauerte, H., Joerg-Rauch, G., Kirn, A., Weiler, C., Geisler, R., Haffter, P., and Ingham, P. W. (2004). iguana encodes a novel zinc-finger protein with coiled-coil domains essential for Hedgehog signal transduction in the zebrafish embryo. *Genes Dev.* **18,** 1565–1576.

Wong, S. Y., and Reiter, J. F. (2008). The primary cilium at the crossroads of mammalian hedgehog signaling. *Curr. Top. Dev. Biol.* **85,** 225–260.

Xie, J., Murone, M., Luoh, S. M., Ryan, A., Gu, Q., Zhang, C., Bonifas, J. M., Lam, C. W., Hynes, M., Goddard, A., Rosenthal, A., Epstein, E. H., Jr., *et al.* (1998). Activating Smoothened mutations in sporadic basal-cell carcinoma. *Nature* **391,** 90–92.

Xu, Q., Guo, L., Moore, H., Waclaw, R. R., Campbell, K., and Anderson, S. A. (2010). Sonic hedgehog signaling confers ventral telencephalic progenitors with distinct cortical interneuron fates. *Neuron* **65,** 328–340.

Yam, P. T., Langlois, S. D., Morin, S., and Charron, F. (2009). Sonic hedgehog guides axons through a noncanonical, Src-family-kinase-dependent signaling pathway. *Neuron* **62,** 349–362.

Yoon, J. W., Liu, C. Z., Yang, J. T., Swart, R., Iannaccone, P., and Walterhouse, D. (1998). GLI activates transcription through a Herpes Simplex viral protein 16-like activation domain. *J. Biol. Chem.* **273,** 3496–3501.

Yu, W., Wang, Y., McDonnell, K., Stephen, D., and Bai, C. B. (2009). Patterning of ventral telencephalon requires positive function of Gli transcription factors. *Dev. Biol.* **334,** 264–275.

Zhang, W., Zhao, Y., Tong, C., Wang, G., Wang, B., Jia, J., and Jiang, J. (2005). Hedgehog-regulated Costal2-kinase complexes control phosphorylation and proteolytic processing of Cubitus interruptus. *Dev. Cell* **8,** 267–278.

Zhang, W., Kang, J. S., Cole, F., Yi, M. J., and Krauss, R. S. (2006). Cdo functions at multiple points in the Sonic Hedgehog pathway, and Cdo-deficient mice accurately model human holoprosencephaly. *Dev. Cell* **10,** 657–665.

Zheng, X., Mann, R. K., Sever, N., and Beachy, P. A. (2010). Genetic and biochemical definition of the Hedgehog receptor. *Genes Dev.* **24,** 57–71.

Zhou, H., Kim, S., Ishii, S., and Boyer, T. G. (2006). Mediator modulates Gli3-dependent Sonic hedgehog signaling. *Mol. Cell. Biol.* **26,** 8667–8682.

Zhu, A. J., and Scott, M. P. (2004). Incredible journey: How do developmental signals travel through tissue? *Genes Dev.* **18,** 2985–2997.

CHAPTER FIVE

Activity-Mediated Synapse Formation: A Role for Wnt-Fz Signaling

Macarena Sahores *and* Patricia C. Salinas

Contents

Abstract

Synapse formation is a critical step in the assembly of neuronal circuits. Both secreted and membrane-associated proteins contribute to the assembly and maturation of synapses. In addition, neuronal activity regulates the formation of neuronal circuits through the stimulation of growth factor secretion and the localization of receptors such as NMDA and AMPA receptors (NMDAR and AMPAR, respectively). Little is known, however, about the role of activity in the localization and function of receptors for synaptogenic molecules. Wnts are secreted proteins that play a role in synapse formation by regulating pre- and postsynaptic assembly at central and peripheral synapses. Wnts can signal through different receptors including Frizzleds (Fzs), the LRP5/6 coreceptors, Ror and Ryk. Fz receptors have been shown to mediate Wnt function during synapse formation. At the cell surface, Fz receptors are located at synaptic and extrasynaptic sites. Importantly, synaptic localization of Fzs is regulated by neuronal activity in a Wnt-dependent manner. In this review, we discuss the

Department of Cell and Developmental Biology, University College London, London, United Kingdom

Current Topics in Developmental Biology, Volume 97

ISSN 0070-2153, DOI: 10.1016/B978-0-12-385975-4.00011-5

function of Wnt-Fz signaling in the assembly of central and peripheral synapses and the evidence supporting a role for Wnt ligands and their Fz receptors in activity-mediated synapse formation.

1. Introduction

Neuronal activity plays a key role in the development and maintenance of neuronal circuits. Activity can control synaptic stability and the rate of synapse formation through changes in axonal and dendritic filopodia dynamics and by controlling gene expression (Hua and Smith, 2004; McAllister, 2007). Neuronal activity also regulates the localization and trafficking of synaptic vesicles, scaffolding proteins, cytoplasmic signaling molecules, and neurotransmitter receptors (McAllister, 2007; Newpher and Ehlers, 2008; Saneyoshi *et al.*, 2008; Waites *et al.*, 2005). In addition, neuronal activity modulates the release of synaptogenic proteins such as brain-derived neurotrophic factor (BDNF) and neuronal activity-regulated pentraxin (Lever *et al.*, 2001; O'Brien *et al.*, 1999; Tsui *et al.*, 1996; Vicario-Abejon *et al.*, 1998). Interestingly, the surface levels of the BDNF receptor TrkB increase following high-frequency stimulation (HFS) (Du *et al.*, 2000). Thus, neuronal activity regulates synapse formation through multiple mechanisms.

Wnt proteins are well-established synaptogenic factors that modulate the formation of synaptic connections in several developmental systems (Cerpa *et al.*, 2008; Hall *et al.*, 2000; Lucas and Salinas, 1997; Packard *et al.*, 2002). New studies demonstrate that neuronal activity regulates the release or expression of Wnts (Ataman *et al.*, 2008; Chiang *et al.*, 2009; Gogolla *et al.*, 2009; Sahores *et al.*, 2010; Singh *et al.*, 2010; Wayman *et al.*, 2006; Yu and Malenka, 2003). Moreover, different patterns of neuronal activity differentially regulate the synaptic localization of Wnt receptors (Sahores *et al.*, 2010). In this review, we focus our attention on the emerging links between Wnt signaling, neuronal circuit formation, and neuronal activity.

2. Wnts and Their Receptors

Wnts are secreted glycoproteins that play key roles in many aspects of development such as embryonic patterning, cell fate determination, and cell proliferation (Angers and Moon, 2009; Clevers, 2006; Davidson and Niehrs, 2010; Petersen and Reddien, 2009). Mutations in these factors or in components of the Wnt signaling pathway have been implicated in pathological conditions such as cancer, late-onset Alzheimer's disease, bone density defects, Familial Exudative Vitreoretinopathy, and Type II diabetes (Clevers, 2006; De Ferrari *et al.*, 2007; Milat and Ng, 2009; Nikopoulos *et al.*, 2010; Welters and Kulkarni, 2008). In the nervous system, Wnts regulate several developmental processes, such as neural tube patterning, neuronal differentiation, axon guidance, dendritic development, and synapse formation (Fradkin *et al.*, 2010; Hur and Zhou, 2010; Korkut and Budnik, 2009; Salinas and Zou, 2008; Ulloa and Marti, 2010).

Wnts, belonging to a family of 19 members (van Amerongen and Nusse, 2009), bind to a variety of receptors at the cell surface. The most studied receptor is the seven-pass transmembrane protein Frizzled (Fz), which can act alone or in combination with the single-pass transmembrane receptor LDL receptor-related protein (LRP5 or LRP6) (Angers and Moon, 2009; MacDonald *et al.*, 2009; van Amerongen and Nusse, 2009). In addition, Wnts can signal through the tyrosine kinase receptors Derailed (Drl)/Ryk and Ror (van Amerongen and Nusse, 2009). Through these different receptors, Wnts can activate distinct signaling cascades (Angers and Moon, 2009; van Amerongen and Nusse, 2009).

The best-characterized pathways activated by Wnts are the canonical (or β-catenin), the planar cell polarity (PCP), and the calcium pathways. In the canonical signaling pathway, Fz associates with LRP5 or LRP6 upon Wnt binding (Angers and Moon, 2009; MacDonald *et al.*, 2009; van Amerongen and Nusse, 2009). This association triggers a signaling pathway that requires the scaffold protein Dishevelled (Dvl) and inhibits the serine/threonine kinase Gsk3β. This inhibition leads to the accumulation of β-catenin which otherwise is phosphorylated by Gsk3β and earmarked for degradation. The accumulation of β-catenin and its translocation to the nucleus leads to the transcriptional activation of specific target genes through the transcription factors TCF or LEF. However, activation of this pathway does not always lead to transcription, since a divergent pathway downstream of Gsk3β or β-catenin regulates cytoskeleton dynamics (Salinas, 2007). The PCP pathway is also activated by Wnt binding to their Fz receptors. However, LRP5/6 receptors are not required (Simons and Mlodzik, 2008; van Amerongen and Nusse, 2009). Downstream of Fz, Dvl activates the small GTPases Rac1 and RhoA resulting in the subsequently activation of c-Jun-N-terminal kinase (JNK) and Rock, respectively. Activation of this pathway results in changes in cytoskeleton dynamics (Wang and Nathans, 2007). The third most-studied pathway is the calcium pathway, where Wnt-Fz signaling through Dvl activates phospholipase C (PLC) (Kohn and Moon, 2005). This is followed by a rise in inositol triphosphate levels, activation of calcium channels, and increase in intracellular Ca^{2+}, resulting in the activation of protein kinase C, calcium–calmodulin kinase (CaMKII), and calmodulin (Kohn and Moon, 2005). This calcium pathway regulates both the cytoskeleton and transcriptional events (Kohn and Moon, 2005). Thus, Wnts can trigger different signaling cascades. Importantly, the same Wnt can activate different signaling pathways even in the same cells

(Davis *et al.*, 2008; Rosso *et al.*, 2005). Further studies are needed to establish the mechanisms that determine the specificity by which these different signaling pathways are activated.

3. The Role of Wnt Signaling in Synapse Formation

3.1. Central synapses

Wnt proteins promote the initial stages of synaptic assembly (Cerpa *et al.*, 2008; Hall *et al.*, 2000; Lucas and Salinas, 1997; Packard *et al.*, 2002). The first evidence came from studies using cultured cerebellar neurons, where Wnt7a, secreted by granule cells, induces the clustering of the presynaptic marker Synapsin-1, a hallmark of presynaptic assembly (Lucas and Salinas, 1997). Subsequently, it was demonstrated that Wnt7a acts as a retrograde signal that regulates the formation of the mossy fiber-granule cell synapse in the cerebellum (Hall *et al.*, 2000). *In vivo* analyses of Wnt7a mutant mice revealed a defect in the accumulation of presynaptic proteins at cerebellar glomerular rosettes (multisynaptic structures formed by a mossy fiber axon and several granule cell dendrites) (Hall *et al.*, 2000). The prosynaptogenic effect of Wnts is not exclusive to the cerebellum. For example, Wnt3 stimulates presynaptic differentiation in dorsal root ganglia (Krylova *et al.*, 2002), whereas Wnt7a, Wnt7b, and Wnt3a promote synaptogenesis in cultured hippocampal neurons (Cerpa *et al.*, 2008; Davis *et al.*, 2008; Sahores *et al.*, 2010; Varela-Nallar *et al.*, 2009) (Fig. 5.1). Wnt5a also regulates synapse formation but its role is more controversial. One study showed that 36-h exposure to Wnt5a decreases the number of presynaptic puncta in cultured hippocampal neurons, suggesting that Wnt5a acts as an antisynaptogenic factor (Davis *et al.*, 2008). However, a different study showed that 5-day exposure to Wnt5a-expressing astrocytes increases presynaptic clustering in cultured neurons (Paganoni *et al.*, 2010). Interestingly, other studies showed that short exposure to Wnt5a does not stimulate presynaptic differentiation (Farias *et al.*, 2009) but rather induces the clustering of PSD-95, a postsynaptic protein, and regulates the recycling of $GABA_A$ receptors (Cuitino *et al.*, 2010; Farias *et al.*, 2009; Varela-Nallar *et al.*, 2010) (Fig. 5.1). The reason for these very opposite effects observed with Wnt5a is currently unclear. Loss-of-function studies using the Wnt5a mutant mice might clarify the role of Wnt5a in synapse formation. Collectively, studies on different Wnts demonstrate that these signaling molecules act as pro- or antisynaptogenic factors at central synapses.

As secreted synaptogenic factors, Wnts bind to their receptors at the surface of target cells. In the hippocampus, surface Fz5 localizes to synaptic sites at the time when synapses form, and its surface expression is

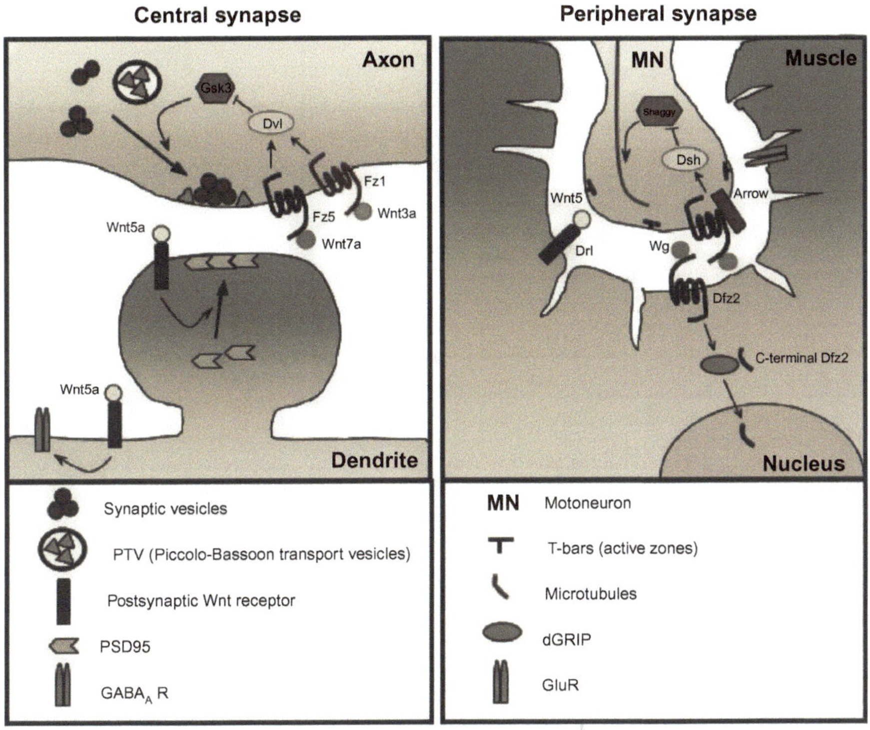

Figure 5.1 Regulation of synaptic development by Wnt signaling. At central synapses, the binding of Wnt7a to presynaptic Fz5 leads to Dvl activation resulting in the inhibition of GSK3β. This inhibition leads to the recruitment of synaptic vesicles and cytomatrix proteins such as Bassoon to the zone of nascent synapses. Wnt3a has been shown to induce vesicle clustering through Fz1. Postsynaptically, Wnt5a triggers the clustering of PSD-95 and regulates GABA$_A$R clustering through an unidentified receptor. At peripheral synapses, Wg binds to Dfz2, which is located at pre- and postsynaptic sites. Presynaptically, Dsh inhibits Shaggy by a divergent canonical pathway that requires the coreceptor Arrow and leads to changes in microtubule dynamics. Postsynaptically, Wg triggers the endocytosis and cleavage of the C-terminal of Dfz2, which is transported to the nucleus by dGRIP. In addition to Wg, Wnt5 also regulates presynaptic development. Wnt5 binds to its postsynaptic Drl receptor to regulate bouton development. (See Color Insert.)

developmentally regulated (Sahores *et al.*, 2010). Interestingly, Wnt7a binds to Fz5, and axonal Fz5 expression mimics Wnt7a activity by increasing the clustering of Synapsin-1 and Bassoon puncta, a hallmark of presynaptic assembly (Fig. 5.1). Conversely, shRNA knockdown or blockade of Fz5 with a secreted form of the extracellular CRD domain of the receptor (Fz5-CRD) decreases the number of presynaptic sites. Importantly, both approaches abolish the synaptogenic effect of Wnt7a, indicating that Fz5 is required for Wnt7a-dependent synaptogenesis in hippocampal neurons

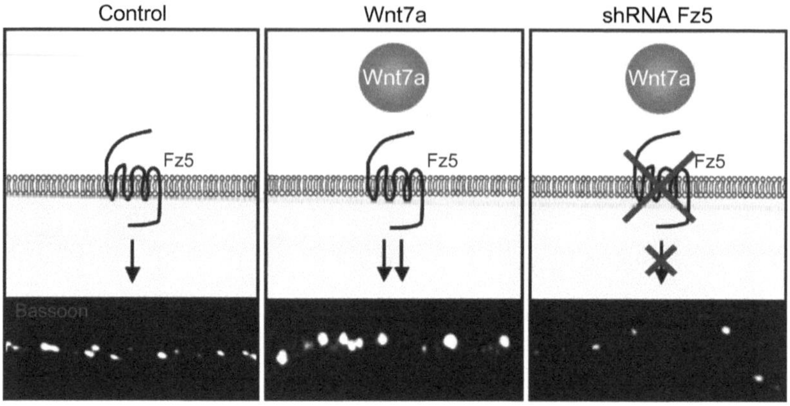

Figure 5.2 Fz5 as a receptor for Wnt7a during synapse formation. On cultured hippocampal neurons, Wnt7a or Fz5 expression induces the clustering of presynaptic markers such as Bassoon. ShRNAs to Fz5 completely abolish Wnt7a-induced presynaptic clustering, suggesting that Fz5 is the main receptor for Wnt7a to promote presynaptic assembly. (For color version of this figure, the reader is referred to the web version of this chapter.)

(Fig. 5.2). Another Fz expressed in the developing hippocampus is Fz1, which localizes to the axons of cultured neurons (Varela-Nallar *et al.*, 2009). The CRD domain of Fz1 blocks the clustering of presynaptic markers induced by Wnt3a, suggesting that Fz1 mediates Wnt3a-induced presynaptic differentiation (Fig. 5.1). These findings suggest that different Wnts might use different Fz receptors to regulate synaptogenesis.

The tyrosine kinase receptors Ror1 and Ror2 are also involved in the formation of synapses. Both receptors are present in axons and dendrites of cultured hippocampal neurons and their expression is developmentally regulated (Paganoni and Ferreira, 2003). Ror1 and/or Ror2 knockdown induces a significant decrease in the clustering of the presynaptic marker Synaptophysin (Paganoni *et al.*, 2010). In addition, RNA knockdown of Ror1 and/or Ror2 abolishes the reported presynaptic clustering activity of Wnt5a, a protein that binds to Ror receptors. These findings demonstrate that several receptors mediate Wnt-induced synapse formation.

What is the signaling pathway downstream of Wnts that regulates synapse formation? Dvl1 is required presynaptically, since *in vivo Dvl1* loss-of-function mimics the phenotype of the *Wnt7a* mutant by exhibiting defects in the accumulation of presynatic proteins in cerebellar glomerular rosettes. Moreover, double *Wnt7a/Dvl1* mutant mice exhibit a stronger synaptogenic defect in the cerebellum (Ahmad-Annuar *et al.*, 2006). Importantly, *Dvl1* mutant axons fail to fully respond to exogenous Wnt7a. Conversely, expression of Dvl1 in cerebellar granule cell axons increases presynaptic puncta and the number of recycling sites (Ahmad-Annuar *et al.*,

2006), mimicking the activity of Wnt7a. Downstream of Dvl, activation of the canonical Wnt pathway promotes presynaptic differentiation as inhibition of GSK3β (Hall *et al.*, 2000) or expression of β-catenin increases the number of presynaptic sites (Davis *et al.*, 2008) (Fig. 5.1). However, it remains to be determined whether transcription through TCF factors is required for Wnt-mediated synaptic assembly.

Postsynaptically, Wnts seem to activate a non-canonical pathway. Wnt5a stimulates PSD-95 clustering, and JNK inhibitors block this effect (Farias *et al.*, 2009) (Fig. 5.1). In addition, Wnt5a regulates the clustering of $GABA_AR$ through a mechanism that requires calcium signaling (Cuitino *et al.*, 2010). However, it is unclear whether Wnt5a directly signals to the dendrite or whether this is an indirect effect due to presynaptic action of Wnt5a. Collectively, these results demonstrate that Wnts activate different signaling mechanisms in axons and dendrites to stimulate pre- or postsynaptic assembly.

3.2. Peripheral synapses

Wnt proteins also play a key role in the development of peripheral synapses (Speese and Budnik, 2007; Wu *et al.*, 2010). At the *Drosophila* neuromuscular junction (NMJ), Wingless (Wg) is expressed by motoneurons and released at the synaptic cleft through a mechanism that involves exosome-like vesicles (Korkut *et al.*, 2009; Packard *et al.*, 2002). *wg*-deficient flies exhibit fewer synaptic boutons and defects in the structure of the active zones and postsynaptic apparatus. Importantly, these defects are associated with changes in the organization of microtubules at presynaptic boutons. *wg*-deficient flies also present a diffused distribution of glutamate receptors in the muscle. Thus, Wg is essential for synapse formation at the *Drosophila* NMJ.

Wg signals through Dfz2, an Fz receptor, that is present at both sides of the NMJ (Bhanot *et al.*, 1996; Packard *et al.*, 2002) (Fig. 5.1). Wg binds to Dfz2 at pre- and postsynaptic sites, and is internalized by the postsynaptic muscle cells (Packard *et al.*, 2002). Interestingly, muscle cells overexpressing Dfz2 or a dominant negative Dfz2 exhibit similar phenotypes to *wg* mutants, with a decreased number of synaptic boutons and boutons with aberrant morphology. These results suggest that Wg might regulate bouton formation by binding to postsynaptic Dfz2. The effects observed with overexpression of wild type or mutant Dfz2 on muscle cells might be due to Dfz2 acting as a sink for Wg resulting in presynaptic defects. Dfz2 regulates NMJ formation through a process that involves endocytosis of the receptor from the muscle plasma membrane (Mathew *et al.*, 2005) (Fig. 5.1). Internalization is followed by the cleavage of Dfz2 and translocation of its C-terminus domain to the nucleus via a pathway that requires dGRIP and Importins-β11 and α2 (Ataman *et al.*, 2006; Mathew *et al.*, 2005; Mosca and Schwarz, 2010).

Another Wnt receptor, Arrow/LRP6, is localized to pre- and postsynaptic sides of the *Drosophila* NMJ (Miech *et al.*, 2008) (Fig. 5.1). *arrow* mutants exhibit a phenotype similar to *wg* mutants, with fewer synaptic boutons that display abnormal size and morphology and disrupted presynaptic microtubules (Miech *et al.*, 2008; Packard *et al.*, 2002). Interestingly, expression of Arrow in both neurons and muscles partially rescues the phenotype of *arrow* mutants suggesting that this receptor might act at both sides of the NMJ. Dsh/Dvl and Shaggy/GSKβ act downstream of Arrow/LRP6 to regulate the number and structure of synaptic boutons, as well as microtubule looping (Miech *et al.*, 2008). This activity seems to be independent of β-catenin and transcriptional regulation (Miech *et al.*, 2008), suggesting that a divergent canonical pathway regulates the initial stages of presynaptic assembly. Thus, Wg, which is released anterogradely by motoneurons, signals bidirectionally to both pre- and postsynaptic sites to activate distinct signaling pathways that regulate the assembly of the NMJ.

Wg is not the only Wnt protein shown to regulate NMJ development in *Drosophila*. Wnt5 and its receptor Drl are both expressed at the *Drosophila* NMJ (Liebl *et al.*, 2008; Yoshikawa *et al.*, 2003) (Fig. 5.1). *wnt5* mutants have fewer boutons and active zones (Liebl *et al.*, 2008). Although *drl* mutants also exhibit a reduced bouton number, they have no defect in the number of active zones, suggesting that Wnt5 might act through another receptor. Altogether, these results suggest that several Wnts regulate NMJ development by acting in parallel.

Interestingly, Wnts do not always promote the formation of synapses. In the DA9 motoneuron of *Caenorhabditis elegans,* the Wnt protein lin-44 inhibits presynaptic assembly, as *lin-44* mutants exhibit ectopic synapses at the proximal segment of the axon, close to the source of lin-44 in wild-type animals (Klassen and Shen, 2007). Lin-17/Fz is normally recruited to a subregion of the axon that remains devoid of synaptic sites by lin-44. Indeed, ectopic expression of lin-44/Wnt induces the expansion of lin-17/Fz expression in the axon resulting in the lack of presynaptic terminals where lin-17 is present. These findings suggest that lin-44/Wnt and lin-17/Fz inhibit presynaptic assembly. Dsh acts downstream of lin-44/Wnt and lin-17/Fz, while members of the Wnt canonical pathway, such as β-catenins, Axin, or Arrow/LRP, appear not to be involved in the inhibition of synapse formation (Klassen and Shen, 2007). Therefore, Wnt controls the site of synaptic assembly of peripheral synapses by regulating the localization of its receptor.

In vertebrates, Wnts regulate two aspects of NMJ development: the prepatterning of postsynaptic receptors, which precedes the arrival of motoneurons, and the assembly of the synapses. Studies in zebrafish have demonstrated that Wnt11r is required for the formation of acetylcholine receptor (AChR) clusters in the central region of the muscle before the arrival of motor axons (Jing *et al.*, 2009), a process called prepatterning. Prepatterning depends on the expression of MuSK, the receptor for the well-known synaptic organizer Agrin, but not on Agrin itself or the presence of motoneurons (Lin *et al.*, 2001). Silencing of Wnt11r with morpholinos gives a similar phenotype to the MuSK/*unplugged* zebrafish mutant, presenting defects in the formation of the AChR clusters (Jing *et al.*, 2009; Lin *et al.*, 2001). Indeed, Wnt11r binds to MuSK, a tyrosine receptor that contains an extracellular domain with high homology to the CRD domain of Fzs (Saldanha *et al.*, 1998). These findings indicate that Wnts can signal through other receptors such as MuSK to regulate the clustering of postsynaptic receptors.

Wnts can also promote the clustering of muscle AChRs during NMJ assembly in vertebrates. Gain-of-function studies using cell implantation experiments demonstrate that Wnt3, normally expressed by motoneurons (Krylova *et al.*, 2002), enhances the clustering of AChRs in chick wings (Henriquez *et al.*, 2008). Conversely, blockade of endogenous Wnts with the secreted Wnt antagonist Sfrp1 decreases AChR cluster formation. The *Dvl1* mutant mouse also exhibits defects in the distribution of AChRs in the diaphragm (Henriquez *et al.*, 2008). A similar phenotype, albeit stronger, has been observed in other mutants such as *Agrin, HB9*, *ChAT*, and *cdk5* (Fu *et al.*, 2005; Gautam *et al.*, 1996; Lin *et al.*, 2001; Misgeld *et al.*, 2005). Interestingly, Dvl1 interacts with MuSK, and disruption of the Dvl/MuSK interaction also inhibits AChR clustering (Luo *et al.*, 2002). These results demonstrate that Wnts, possibly Wnt3 from motoneurons, promote postsynaptic assembly at the vertebrate NMJ by inducing AChR clustering through a pathway that requires Dvl1 and MuSK.

How do Wnts promote postsynaptic receptor clustering? In cultured myotubes, Wnt3 induces the rapid formation of AChR microclusters through activation of Rac1, independently of the canonical pathway (Henriquez *et al.*, 2008). These microclusters coalesce into large and stable clusters only in the presence of Agrin. Wnt3 activates Rac1 but not RhoA, whereas Agrin activates both proteins to promote clustering. Thus, Wnt3 collaborates with Agrin to regulate postsynaptic assembly. However, further studies are required to determine the precise mechanism by which Wnt collaborates with Agrin.

Wnt3a, a highly related protein to Wnt3, exhibits the opposite effect to Wnt3 as it inhibits Agrin-induced AChR clustering in both cultured myotubes and *in vivo* (Wang *et al.*, 2008). In cultured myotubes, Wnt3a, which is expressed by muscles, decreases the levels of Rapsyn, a protein essential for cluster formation (Wang *et al.*, 2008). Moreover, β-catenin expression mimics Wnt3a by inducing the dispersal of AChR clusters, and Dkk-1, a secreted protein that blocks canonical signaling, suppresses the anticlustering activity of Wnt3a (Wang *et al.*, 2008). Collectively, the studies on Wnt3 and Wnt3a suggest that Wnts can promote or inhibit postsynaptic assembly. These findings raise the interesting possibility that motoneuron-derived Wnts promote NMJ assembly in collaboration with Agrin by activating a non-canonical pathway, whereas muscle-derived Wnts, through the activation of the canonical Wnt pathway, induce AChR dispersal. Consistent with this view, conditional mutant mice deficient in muscle β-catenin exhibit larger and more disperse AChR clusters than wild-type mice (Li *et al.*, 2008). These opposite functions of Wnts would ensure the assembly of the NMJ just in direct apposition to the presynaptic motoneuron bouton.

Although the receptor for Wnt3 at the NMJ has not been identified, recent studies demonstrate that LRP4 is crucial for NMJ formation. *Lrp4* mutants exhibit the same phenotype to MuSK mutants, in which AChRs fail to cluster (Kim *et al.*, 2008; Weatherbee *et al.*, 2006; Zhang *et al.*, 2008). Biochemical studies demonstrate that Agrin binds to LRP4 in heterologous cells. Previous studies have shown that Agrin binds to MuSK in muscle cells but not in heterologous cells, suggesting that Agrin binds to MuSK using a coreceptor that is only expressed in muscle cells. Indeed, Agrin forms a complex with MuSK and LRP4 (Kim *et al.*, 2008). These new experiments strongly suggest that LRP4 acts as a coreceptor when Agrin binds to MuSK (Kim *et al.*, 2008; Zhang *et al.*, 2008). Given that LRP4 and LRP5/6 share a high level of homology on their extracellular domain (Nykjaer and Willnow, 2002), these results raise the possibility that Wnts such as Wnt3 might collaborate with Agrin by binding to LRP4. Further studies will elucidate whether the interplay of these receptors and coreceptors modulates Wnt and Agrin function during NMJ differentiation.

4. Role of Neuronal Activity in Wnt-Mediated Dendrite Morphogenesis

Wnt signaling plays a key role in dendrite morphogenesis. In hippocampal cultures, loss- and gain-of-function studies demonstrate that Wnt7b and Dvl1 promote the formation of long, branched dendrites through a non-canonical pathway that involves Rac1 and JNK (Rosso *et al.*, 2005). In addition, the secreted Wnt antagonist Sfrp1 not only blocks Wnt7b activity on dendrites but also impairs dendritic development on its own, suggesting that endogenous Wnts regulate this process.

Wnt signaling, together with neuronal activity, regulates the arborization of dendrites. Neuronal activity has been well established as a key regulator of dendritic development (Heiman and Shaham, 2010; Hua and Smith, 2004). Depolarization with high K^+ increases the dendritic arborization of cultured hippocampal neurons (Redmond *et al.*, 2002; Yu and Malenka, 2003). Interestingly, the effect of neuronal activity on dendrites is blocked by Dkk1, suggesting that depolarization increases the release and/

or the expression of endogenous Wnts to regulate dendritic arborization (Yu and Malenka, 2003). Further studies led to the discovery that depolarization increases dendritic growth through a mechanism that involves CaMKI, ERK, and the transcription factor CREB (Wayman *et al.*, 2006). By performing a SACO (serial analysis of chromatin occupancy) screen, Wnt-2 was identified as an activity-dependent CREB-responsive gene. Interestingly, neuronal activity increases CREB-dependent transcription of Wnt2, and neurons expressing Wnt2 display enhanced dendritic length and branching, similar to that observed following depolarization (Wayman *et al.*, 2006). Thus, neuronal activity regulates dendritic growth by modulating the expression and/or secretion of Wnts.

Neuronal activity through Wnt signaling also regulates dendrite refinement. In *Drosophila*, wide-field serotonergic neurons (CSDn) were used to study dendritic growth and refinement, since these neurons undergo remodeling during metamorphosis and present very few dendrites in the adult. To test the role of activity, the effect of expression of the K^+ channel Kir2.1, which leads to hyperpolarization, was studied (Singh *et al.*, 2010). Adult flies expressing Kir2.1 exhibit an enlarged dendritic field, suggesting that neuronal activity is necessary for dendritic refinement during metamorphosis. Interestingly, CSDn expresses both Wg and Dfz2. Moreover, neuropil Wg expression increases following K^+ depolarization, and Wg heretozygotes present mild but significant dendritic defects. Mutants for *Dfz2* or *Dsh*, as well as overexpression of the Wnt antagonists Axin and Shaggy, exhibit increased dendritic fields. Importantly, in the absence of sensory input, dendritic arbors do not retract, an effect that can be rescued by activation of the Wnt signaling (Singh *et al.*, 2010). Interestingly, Wnt5 and Drl act synergistically with Wg/Dfz2 to regulate dendritic refinement (Singh *et al.*, 2010). These results suggest that the sensory experience through the canonical Wnt signaling pathway promotes the retraction of dendritic arbors and therefore contributes to the refinement of neuronal circuits.

5. Role of Neuronal Activity in Wnt-Mediated Synapse Formation

5.1. Central synapses

In addition to promoting Wnt expression and/or secretion, neuronal activity regulates the localization of Wnt receptors. In developing hippocampal cultures, neuronal activity regulates not only the levels of Fz5 at the cell surface but also its synaptic insertion (Sahores *et al.*, 2010) (Fig. 5.3). Interestingly, diverse patterns of stimulation elicit different responses. HFS increases the amount of Fz5 at the surface and synaptic sites, whereas

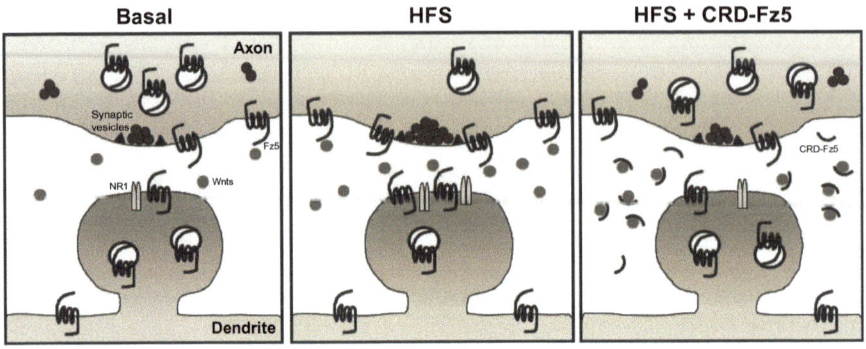

Figure 5.3 Activity-mediated Fz5 insertion at synaptic sites. A fraction of surface Fz5 is localized to synaptic sites in cultured hippocampal neurons under basal conditions. Changes in neuronal activity elicited by high-frequency stimulation (HFS) induce the mobilization of Fz5 receptor to the cell surface and to synapses. The soluble extracellular domain of Fz5 (CRD-Fz5), which sequesters endogenous Wnts, blocks the effect of HFS on Fz5 localization to the plasma membrane and to synapses. In addition, CRD-Fz5 completely abolishes activity-induced synapse formation. (For color version of this figure, the reader is referred to the web version of this chapter.)

low-frequency stimulation decreases both surface and synaptic Fz5, without affecting the total levels of the protein. Blockade of Wnts with Sfrps or CRD-Fz5 suppresses the mobilization of the receptor to synapses that is induced by HFS (Fig. 5.3). These results suggest that HFS elicits the release of endogenous Wnts, which then in turn regulate the localization of Fz receptor at synaptic sites.

Blockade of Fz5 function during HFS suppresses synapse formation. It is well established that HFS, a paradigm that induces long-term potentiation, increases the number of synapses (Bolshakov *et al.*, 1997; Bozdagi *et al.*, 2000). Interestingly, Sfrps or the soluble CRD-Fz5 block the synaptogenic function of HFS, as the number of colocalized vGlut1 and NR1 puncta significantly decreases (Sahores *et al.*, 2010). These experiments suggest that neuronal activity regulates the expression, release, or activity of Wnts, which then in turn promote synapse formation. These results support the view that Wnt signaling cooperates with neuronal activity to regulate synapse formation, partly by regulating the localization of its receptors, and that Wnt signaling through Fz5 significantly contributes to activity-mediated synapse formation.

Further evidence for a role of Wnt signaling in activity-mediated synaptic connectivity comes from studies at the mossy fiber-CA3 synapse in the adult hippocampus. Increasing activity by raising animals in an enriched environment (EE) induces an increase in the number and remodeling of mossy fiber terminals, spine densities, and synapses (Gogolla *et al.*, 2009). Exposure to an EE enhances Wnt7a/b expression in CA3 region.

Importantly, local infusion of Sfrp1 blocks the EE-induced increase of synapse number, whereas Wnt7a application mimics the effects of EE on the number of synaptic sites and mossy fiber terminal remodeling. Collectively, the results from HFS and EE experiments demonstrate that Wnt signaling mediates some of the effects of neuronal activity on the regulation of synapse formation.

5.2. Peripheral synapses

Wnts are involved in activity-induced changes also at peripheral synapses. At the *Drosophila* NMJ, patterned stimulation using K^+ induces the formation of undifferentiated "ghost" boutons that develop into mature boutons and enhances spontaneous neurotransmitter release (Ataman *et al.*, 2008). Neuronal activity also induces the secretion of Wg at synaptic sites. Interestingly, the formation of ghost boutons induced by K^+ depolarization is absent in *wg* heterozygous- or *wg* temperature-sensitive mutants, suggesting that Wg is required for the structural changes induced by neuronal activity. The same study shows that presynaptic expression of Shaggy/GSK3β, known to antagonize Wnt signaling, markedly decreases the formation of new boutons. In addition, patterned stimulation increases the postsynaptic nuclear import of Dfz2 C-terminal, whereas blockade of neurotransmitter release has the opposite effect. Therefore, patterned stimulation induces the release of Wg, which then signals to both sides of the synapse to regulate synaptic structure and function.

Through Wg, neuronal activity also maintains neuronal connections in the *Drosophila* adult olfactory system. Olfactory sensory neurons (OSNs) project to glomeruli in the antennal lobe where they make synaptic contacts. Such glomeruli express low levels of Wg, but patterned depolarization increases the levels of endogenous Wg protein in the antennal lobes. Activity blockade using a *shibire* mutant that impairs synaptic vesicle recycling, or a mutant for the olfactory coreceptor Orb83, involved in spontaneous neuronal activity, leads to axonal degeneration in OSNs (Chiang *et al.*, 2009). Interestingly, inhibition of Wnt signaling by expression of a truncated version of Fz8, a truncated version of Arrow, a DN mutant of Dsh, or overexpression of Axin also results in neurodegeneration. Thus, neuronal activity regulates the maintenance of axon stability through Wnt signaling.

An intriguing observation that emerges from studies on Wnt signaling is that canonical Wnt signaling seems to have opposite effect on axons and dendrites. In axons, canonical Wnt signaling protects against degeneration induced by blockade of neuronal activity (Chiang *et al.*, 2009). In dendrites, in contrast, canonical Wnt signaling promotes retraction when dendrites begin to receive sensory inputs (Singh *et al.*, 2010). How these two opposing effects are achieved in these two different neuronal compartments remain to be investigated.

6. Concluding Remarks

In recent years, great progress has been made on the role of Wnt signaling in synapse formation. The growing evidence reveals a link between Wnt signaling and activity-dependent synaptogenesis. A model emerges where neuronal activity induces the release of Wnts, which bind to their receptors to regulate different aspects of neuronal connectivity at the central and peripheral nervous systems. By increasing or decreasing the levels of Wnts, which also affect the level and localization of their receptors, neuronal activity would ensure that Wnts act preferentially on active neurons, providing the means to promote the formation and perhaps the maintenance of synapses within active parts of the circuit.

Given the number of possible combinations of Wnts and their receptors, specific Wnt-receptor pairs could regulate distinct functions. However, the binding between Wnts and their receptors does not seem to be exclusive, as the same Wnt can bind to many receptors (Bhanot *et al.*, 1996; Hsieh *et al.*, 1999). Thus, the specificity of Wnt signaling is achieved by a different, as of yet unknown, mechanism. For example, the density and localization of a Wnt receptor, together with the appropriate signaling machinery and the site of release of Wnt, could provide the specificity to regulate pre- and postsynaptic processes. In addition, neuronal activity could provide a further level of control by regulating the levels and/or localization of Wnts and their receptors.

Given that signaling molecules that collaborate with Wnts (such as Agrin) are also regulated by neuronal activity (O'Connor *et al.*, 1995), Wnts might also collaborate with other activity-regulated secreted molecules to regulate synapse formation at central synapses. Future studies will elucidate the potential interactions of Wnts and their receptors with other signaling pathways that contribute to activity-mediated synapse formation.

ACKNOWLEDGMENTS

We would like to thank the members of our laboratory for useful discussions and comments. Our work is supported by grants from the MRC, The Wellcome Trust, BBSRC, and the European Union (MOLPARK).

REFERENCES

Ahmad-Annuar, A., *et al.* (2006). Signaling across the synapse: A role for Wnt and Dishevelled in presynaptic assembly and neurotransmitter release. *J. Cell Biol.* **174,** 127–139.

Angers, S., and Moon, R. T. (2009). Proximal events in Wnt signal transduction. *Nat. Rev. Mol. Cell Biol.* **10,** 468–477.

Ataman, B., *et al.* (2006). Nuclear trafficking of Drosophila Frizzled-2 during synapse development requires the PDZ protein dGRIP. *Proc. Natl. Acad. Sci. USA* **103,** 7841–7846.

Ataman, B., *et al.* (2008). Rapid activity-dependent modifications in synaptic structure and function require bidirectional Wnt signaling. *Neuron* **57,** 705–718.

Bhanot, P., *et al.* (1996). A new member of the frizzled family from Drosophila functions as a Wingless receptor. *Nature* **382,** 225–230.

Bolshakov, V. Y., *et al.* (1997). Recruitment of new sites of synaptic transmission during the cAMP-dependent late phase of LTP at CA3-CA1 synapses in the hippocampus. *Neuron* **19,** 635–651.

Bozdagi, O., *et al.* (2000). Increasing numbers of synaptic puncta during late-phase LTP: N-cadherin is synthesized, recruited to synaptic sites, and required for potentiation. *Neuron* **28,** 245–259.

Cerpa, W., *et al.* (2008). Wnt-7a modulates the synaptic vesicle cycle and synaptic transmission in hippocampal neurons. *J. Biol. Chem.* **283,** 5918–5927.

Chiang, A., *et al.* (2009). Neuronal activity and Wnt signaling act through Gsk3-{beta} to regulate axonal integrity in mature Drosophila olfactory sensory neurons. *Development* **136,** 1273–1282.

Clevers, H. (2006). Wnt/beta-catenin signaling in development and disease. *Cell* **127,** 469–480.

Cuitino, L., *et al.* (2010). Wnt-5a modulates recycling of functional GABAA receptors on hippocampal neurons. *J. Neurosci.* **30,** 8411–8420.

Davidson, G., and Niehrs, C. (2010). Emerging links between CDK cell cycle regulators and Wnt signaling. *Trends Cell Biol.* **20,** 453–460.

Davis, E. K., *et al.* (2008). Wnts acting through canonical and noncanonical signaling pathways exert opposite effects on hippocampal synapse formation. *Neural Dev.* **3,** 32–36.

De Ferrari, G. V., *et al.* (2007). Common genetic variation within the low-density lipoprotein receptor-related protein 6 and late-onset Alzheimer's disease. *Proc. Natl. Acad. Sci. USA* **104,** 9434–9439.

Du, J., *et al.* (2000). Activity- and Ca(2+)-dependent modulation of surface expression of brain-derived neurotrophic factor receptors in hippocampal neurons. *J. Cell Biol.* **150,** 1423–1434.

Farias, G. G., *et al.* (2009). WNT-5a/JNK signaling promotes the clustering of PSD-95 in hippocampal neurons. *J. Biol. Chem.* **284,** 15857–15866.

Fradkin, L. G., *et al.* (2010). Ryks: New partners for Wnts in the developing and regenerating nervous system. *Trends Neurosci.* **33,** 84–92.

Fu, A. K., *et al.* (2005). Aberrant motor axon projection, acetylcholine receptor clustering, and neurotransmission in cyclin-dependent kinase 5 null mice. *Proc. Natl. Acad. Sci. USA* **102,** 15224–15229.

Gautam, M., *et al.* (1996). Defective neuromuscular synaptogenesis in agrin-deficient mutant mice. *Cell* **85,** 525–535.

Gogolla, N., *et al.* (2009). Wnt signaling mediates experience-related regulation of synapse numbers and mossy fiber connectivities in the adult hippocampus. *Neuron* **62,** 510–525.

Hall, A. C., *et al.* (2000). Axonal remodeling and synaptic differentiation in the cerebellum is regulated by WNT-7a signaling. *Cell* **100,** 525–535.

Heiman, M. G., and Shaham, S. (2010). Twigs into branches: How a filopodium becomes a dendrite. *Curr. Opin. Neurobiol.* **20,** 86–91.

Henriquez, J. P., *et al.* (2008). Wnt signaling promotes AChR aggregation at the neuromuscular synapse in collaboration with agrin. *Proc. Natl. Acad. Sci. USA* **105,** 18812–18817.

Hsieh, J. C., *et al.* (1999). Biochemical characterization of Wnt-frizzled interactions using a soluble, biologically active vertebrate Wnt protein. *Proc. Natl. Acad. Sci. USA* **96,** 3546–3551.

Hua, J. Y., and Smith, S. J. (2004). Neural activity and the dynamics of central nervous system development. *Nat. Neurosci.* **7,** 327–332.

Hur, E. M., and Zhou, F. Q. (2010). GSK3 signalling in neural development. *Nat. Rev. Neurosci.* **11,** 539–551.

Jing, L., *et al.* (2009). Wnt signals organize synaptic prepattern and axon guidance through the zebrafish unplugged/MuSK receptor. *Neuron* **61,** 721–733.

Kim, N., *et al.* (2008). Lrp4 is a receptor for Agrin and forms a complex with MuSK. *Cell* **135,** 334–342.

Klassen, M. P., and Shen, K. (2007). Wnt signaling positions neuromuscular connectivity by inhibiting synapse formation in C. elegans. *Cell* **130,** 704–716.

Kohn, A. D., and Moon, R. T. (2005). Wnt and calcium signaling: Beta-catenin-independent pathways. *Cell Calcium* **38,** 439–446.

Korkut, C., and Budnik, V. (2009). WNTs tune up the neuromuscular junction. *Nat. Rev. Neurosci.* **10,** 627–634.

Korkut, C., *et al.* (2009). Trans-synaptic transmission of vesicular Wnt signals through Evi/Wntless. *Cell* **139,** 393–404.

Krylova, O., *et al.* (2002). WNT-3, expressed by motoneurons, regulates terminal arborization of neurotrophin-3-responsive spinal sensory neurons. *Neuron* **35,** 1043–1056.

Lever, I. J., *et al.* (2001). Brain-derived neurotrophic factor is released in the dorsal horn by distinctive patterns of afferent fiber stimulation. *J. Neurosci.* **21,** 4469–4477.

Li, X. M., *et al.* (2008). Retrograde regulation of motoneuron differentiation by muscle beta-catenin. *Nat. Neurosci.* **11,** 262–268.

Liebl, F. L., *et al.* (2008). Derailed regulates development of the Drosophila neuromuscular junction. *Dev. Neurobiol.* **68,** 152–165.

Lin, W., *et al.* (2001). Distinct roles of nerve and muscle in postsynaptic differentiation of the neuromuscular synapse. *Nature* **410,** 1057–1064.

Lucas, F. R., and Salinas, P. C. (1997). WNT-7a induces axonal remodeling and increases synapsin I levels in cerebellar neurons. *Dev. Biol.* **193,** 31–44.

Luo, Z. G., *et al.* (2002). Regulation of AChR clustering by Dishevelled interacting with MuSK and PAK1. *Neuron* **35,** 489–505.

MacDonald, B. T., *et al.* (2009). Wnt/beta-catenin signaling: Components, mechanisms, and diseases. *Dev. Cell* **17,** 9–26.

Mathew, D., *et al.* (2005). Wingless signaling at synapses is through cleavage and nuclear import of receptor DFrizzled2. *Science* **310,** 1344–1347.

McAllister, A. K. (2007). Dynamic aspects of CNS synapse formation. *Annu. Rev. Neurosci.* **30,** 425–450.

Miech, C., *et al.* (2008). Presynaptic local signaling by a canonical wingless pathway regulates development of the Drosophila neuromuscular junction. *J. Neurosci.* **28,** 10875–10884.

Milat, F., and Ng, K. W. (2009). Is Wnt signalling the final common pathway leading to bone formation? *Mol. Cell. Endocrinol.* **310,** 52–62.

Misgeld, T., *et al.* (2005). Agrin promotes synaptic differentiation by counteracting an inhibitory effect of neurotransmitter. *Proc. Natl. Acad. Sci. USA* **102,** 11088–11093.

Mosca, T. J., and Schwarz, T. L. (2010). The nuclear import of Frizzled2-C by Importins-beta11 and alpha2 promotes postsynaptic development. *Nat. Neurosci.* **13,** 935–943.

Newpher, T. M., and Ehlers, M. D. (2008). Glutamate receptor dynamics in dendritic microdomains. *Neuron* **58,** 472–497.

Nikopoulos, K., *et al.* (2010). Overview of the mutation spectrum in familial exudative vitreoretinopathy and Norrie disease with identification of 21 novel variants in FZD4, LRP5, and NDP. *Hum. Mutat.* **31,** 656–666.

Nykjaer, A., and Willnow, T. E. (2002). The low-density lipoprotein receptor gene family: A cellular Swiss army knife? *Trends Cell Biol.* **12,** 273–280.

O'Brien, R. J., *et al.* (1999). Synaptic clustering of AMPA receptors by the extracellular immediate-early gene product Narp. *Neuron* **23,** 309–323.
O'Connor, L. T., *et al.* (1995). Expression of agrin mRNA is altered following seizures in adult rat brain. *Brain Res. Mol. Brain Res.* **33,** 277–287.
Packard, M., *et al.* (2002). The Drosophila Wnt, wingless, provides an essential signal for pre- and postsynaptic differentiation. *Cell* **111,** 319–330.
Paganoni, S., and Ferreira, A. (2003). Expression and subcellular localization of Ror tyrosine kinase receptors are developmentally regulated in cultured hippocampal neurons. *J. Neurosci. Res.* **73,** 429–440.
Paganoni, S., *et al.* (2010). Ror1-Ror2 complexes modulate synapse formation in hippocampal neurons. *Neuroscience* **165,** 1261–1274.
Petersen, C. P., and Reddien, P. W. (2009). Wnt signaling and the polarity of the primary body axis. *Cell* **139,** 1056–1068.
Redmond, L., *et al.* (2002). Calcium regulation of dendritic growth via CaM kinase IV and CREB-mediated transcription. *Neuron* **34,** 999–1010.
Rosso, S. B., *et al.* (2005). Wnt signaling through Dishevelled Rac and JNK regulates dendritic development. *Nat. Neurosci.* **8,** 34–42.
Sahores, M., *et al.* (2010). Frizzled-5, a receptor for the synaptic organizer Wnt7a, regulates activity-mediated synaptogenesis. *Development* **137,** 2215–2225.
Saldanha, J., *et al.* (1998). Identification of a Frizzled-like cysteine rich domain in the extracellular region of developmental receptor tyrosine kinases. *Protein Sci.* **7,** 1632–1635.
Salinas, P. C. (2007). Modulation of the microtubule cytoskeleton: A role for a divergent canonical Wnt pathway. *Trends Cell Biol.* **17,** 333–342.
Salinas, P. C., and Zou, Y. (2008). Wnt signaling in neural circuit assembly. *Annu. Rev. Neurosci.* **31,** 339–358.
Saneyoshi, T., *et al.* (2008). Activity-dependent synaptogenesis: Regulation by a CaM-kinase kinase/CaM-kinase I/betaPIX signaling complex. *Neuron* **57,** 94–107.
Simons, M., and Mlodzik, M. (2008). Planar cell polarity signaling: From fly development to human disease. *Annu. Rev. Genet.* **42,** 517–540.
Singh, A. P., *et al.* (2010). Dendritic refinement of an identified neuron in the Drosophila CNS is regulated by neuronal activity and Wnt signaling. *Development* **137,** 1351–1360.
Speese, S. D., and Budnik, V. (2007). Wnts: Up-and-coming at the synapse. *Trends Neurosci.* **30,** 268–275.
Tsui, C. C., *et al.* (1996). Narp, a novel member of the pentraxin family, promotes neurite outgrowth and is dynamically regulated by neuronal activity. *J. Neurosci.* **16,** 2463–2478.
Ulloa, F., and Marti, E. (2010). Wnt won the war: Antagonistic role of Wnt over Shh controls dorso-ventral patterning of the vertebrate neural tube. *Dev. Dyn.* **239,** 69–76.
van Amerongen, R., and Nusse, R. (2009). Towards an integrated view of Wnt signaling in development. *Development* **136,** 3205–3214.
Varela-Nallar, L., *et al.* (2009). Role of the Wnt receptor Frizzled-1 in presynaptic differentiation and function. *Neural Dev.* **4,** 41.
Varela-Nallar, L., *et al.* (2010). Wingless-type family member 5A (Wnt-5a) stimulates synaptic differentiation and function of glutamatergic synapses. *Proc. Natl. Acad. Sci. USA.* **107,** 21164–21169.
Vicario-Abejon, C., *et al.* (1998). Neurotrophins induce formation of functional excitatory and inhibitory synapses between cultured hippocampal neurons. *J. Neurosci.* **18,** 7256–7271.
Waites, C. L., *et al.* (2005). Mechanisms of vertebrate synaptogenesis. *Annu. Rev. Neurosci.* **28,** 251–274.
Wang, Y., and Nathans, J. (2007). Tissue/planar cell polarity in vertebrates: New insights and new questions. *Development* **134,** 647–658.

Wang, J., *et al.* (2008). Wnt/beta-catenin signaling suppresses Rapsyn expression and inhibits acetylcholine receptor clustering at the neuromuscular junction. *J. Biol. Chem.* **283,** 21668–21675.

Wayman, G. A., *et al.* (2006). Activity-dependent dendritic arborization mediated by CaM-kinase I activation and enhanced CREB-dependent transcription of Wnt-2. *Neuron* **50,** 897–909.

Weatherbee, S. D., *et al.* (2006). LDL-receptor-related protein 4 is crucial for formation of the neuromuscular junction. *Development* **133,** 4993–5000.

Welters, H. J., and Kulkarni, R. N. (2008). Wnt signaling: Relevance to beta-cell biology and diabetes. *Trends Endocrinol. Metab.* **19,** 349–355.

Wu, H., *et al.* (2010). To build a synapse: Signaling pathways in neuromuscular junction assembly. *Development* **137,** 1017–1033.

Yoshikawa, S., *et al.* (2003). Wnt-mediated axon guidance via the Drosophila Derailed receptor. *Nature* **422,** 583–588.

Yu, X., and Malenka, R. C. (2003). Beta-catenin is critical for dendritic morphogenesis. *Nat. Neurosci.* **6,** 1169–1177.

Zhang, B., *et al.* (2008). LRP4 serves as a coreceptor of agrin. *Neuron* **60,** 285–297.

CHAPTER SIX

Signaling Pathways and Axis Formation in the Lower Metazoa

Thomas W. Holstein, Hiroshi Watanabe, *and* Suat Özbek

Contents

Department of Molecular Evolution and Genomics, Centre for Organismal Studies, Heidelberg University, Heidelberg, Germany

Current Topics in Developmental Biology, Volume 97
ISSN 0070-2153, DOI: 10.1016/B978-0-12-385975-4.00012-7

Abstract

The determination of the body axis in the last common ancestor of bilaterian animals is still a matter of debate. While *Hox* genes pattern the formation of the primary, anteroposterior body axis in bilaterians, there is growing evidence from lower metazoans that the Wnt/β-catenin pathway acts as the primordial signaling system in this process. This review summarizes molecular data from recent genomic analyses of basal model organisms with a focus on the evolution of signaling pathways involved in the establishment of the primary and successive body axes during early metazoan evolution.

1. Introduction

The evolution of multicellular organisms from single cells is a major evolutionary transition (Szathmary and Smith, 1995). For eukaryotes, all recent molecular phylogenies indicate that animals (metazoans) constitute a monophyletic group that has a common origin in protozoans (Ruiz-Trillo *et al.*, 2008; Schierwater *et al.*, 2009) (Fig. 6.1). Within the highly diverse

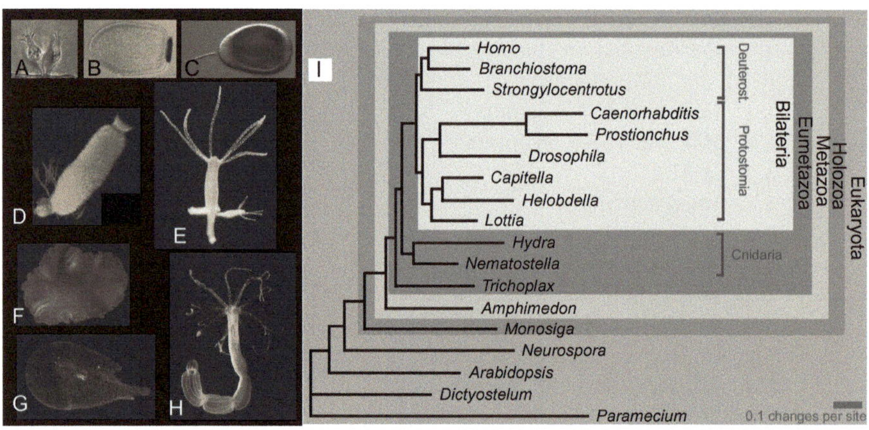

Figure 6.1 Basal metazoan model systems. (A) Colony of the choanoflagellate Salpingoeca, (B) larval stage of the sponge *Amphimedon queenslandica*, (C) planula larva of *Nematostella vectensis*, (D) *Sycon*, (E) freshwater polyp *Hydra vulgaris* with buds, (F) placozoan *Trichoplax adhaerens*, (G) acoel flatworm *Convoluta triloba*, (H) polyp of the starlet sea anemone *N. vectensis*, (I) molecular phylogeny of metazoans. Reproduced with permission from (A) © Whiley from Nielsen (2008) (Michael Plewka Plingfactory.de); (B) © Nature from Srivastava *et al.* (2010b); (C) © The Company of Biologists Ltd. from Technau and Steele (2011); (D) Nielsen (2008), Whiley; Fredrik Pleijel, Tjaå Nrnoån, Marine Biological Laboratory; (E) © Holstein and Nüchter; (F) © Nature from Srivastava *et al.* (2008); (G) © Elsevier from Sikes and Bely (2010); (H) © Nature from Kusserow *et al.* (2005) (Holstein and Nüchter); (I) ©Nature from Srivastava *et al.* (2010b). (For color version of this figure, the reader is referred to the Web version of this chapter.)

protists, the small group of choanoflagellates has been generally considered to represent the common root in animal evolution (King *et al.*, 2003, 2008; von Salvini-Plawen, 1978). However, metazoans do also share a number of common molecular traits with fungi, and they exhibit a poor resolution of their metazoan relationships that may result from the rapidity with which metazoan phyla diversified (Rokas *et al.*, 2005). Due to the poor resolution of early metazoan molecular phylogenies (Rokas *et al.*, 2005), it is therefore still a matter of debate how the earliest metazoans were morphologically and molecularly organized. The fossil record from 580-Myr-old Ediacaran assemblages indicates that the first metazoans were similar to modern sponges and cnidarians that exhibit a diploblastic organization with an outer ectoderm and an inner endoderm (Xiao and Laflamme, 2009). Within a relatively short period of time, a major diversification of body plans occurred, which is documented in the Burgess Shale assemblages of the lower Cambium (Conway Morris, 1998, 2000). In these assemblages, all major body plans including vertebrate-like organisms can be identified. Since the fully sequenced genomes of major representatives from sponges and cnidarians demonstrate a high complexity of the cnidarian genome, it is an intriguing question how the signaling pathways of basal metazoans and, in particular, of cnidarians contributed to bilaterian animals.

A unifying feature of all animals is a common set of major evolutionary innovations including an embryonic development with a blastula and gastrula stage. During gastrulation, at least one body axis and two germ layers are formed that have contractile and sensory properties. The basal metazoan groups including sponges, ctenophores, and cnidarians are lacking a third germ layer and a defined second body axis. These basal groups are therefore separated (Radiata) from the further evolved metazoans with a bilateral symmetry (bilaterians) and most of the bilaterian animals possess a through gut with a separate mouth and anus. According to the fate of the initial blastopore opening (mouth or anus) in the embryo, bilaterians are further split into protostomes and deuterostomes (Brusca and Brusca, 2002; Grobben, 1908; Hyman, 1940). The protostomes, which are characterized by a ventral nerve chord and a dorsal brain ganglion, comprise two major groups of invertebrate phyla designated as lophotrochozoans and ecdysozoans, the latter including two prominent model organisms, *Drosophila melanogaster* and *Caenorhabditis elegans*. The deuterostomes, which develop a dorsal central nervous system (CNS), are split into chordates, including vertebrates and cephalochordates (*Amphioxus*), hemichordates, and echinoderms.

How the transition from radial to bilateral symmetry occurred is still an enigma in metazoan evolution. One line of hypotheses proposes that the mouth and anus evolved from a slit-like blastopore by closure of the tissue in between (Arendt *et al.*, 2001). The Nielsen's trochaea (Nielsen, 2001), Remane's enterocoely (Remane, 1950), and Jägersten's bilatero-gastraea

(Jägersten, 1955) are all hypotheses that are in line with this basic assumption (Hejnol and Martindale, 2009; Martindale and Hejnol, 2009). In an alternative hypothesis, a radially symmetric planula-like organism evolved into a bilateral symmetric organism (Hyman, 1940; von Salvini-Plawen, 1978). In this scenario, in a bilateral organism, the posterior blastopore was shifted on the ventral side anteriorly to give rise to one ventral opening (mouth and anus) that evolved independently in the protostome and deuterostome lineages (acoela–planula hypothesis; Hejnol and Martindale, 2009; Hyman, 1940; Martindale and Hejnol, 2009; von Salvini-Plawen, 1978). In both of these major evolutionary scenarios, all bilaterians evolved from a common ur-bilaterian ancestor (De Robertis, 2010; De Robertis and Sasai, 1996). This was questioned by Hans Meinhardt who proposed a third major scenario according to which protostomes and deuterostome lineages evolved independently from a radial symmetric cnidarian-like organism. It was proposed that the body of a gastrula-like organism gave rise to the head of more evolved organisms by intercalating the trunk in the sub-blastoporal region giving rise to two fundamentally different modes of midline formation (Meinhardt, 2002, 2004, 2006).

All animals having a bilateral symmetry are patterned along two major body axes (Niehrs, 2010): the primary anterior–posterior (AP) axis, which is oriented parallel to the gut and, in a perpendicular orientation, the secondary dorsal–ventral (DV) axis. In most bilaterians, a controlled expression of *Hox* cluster genes directs the formation of the AP axis. Recent data suggest that a localized *wnt/β-catenin* expression determines the orientation of the primary body axis in both, bilaterians and nonbilaterians, indicating that the conserved Wnt signaling pathway is at the base of polarized development in all metazoans (Petersen and Reddien, 2009). The secondary DV axis, which is patterned by the antagonistic Chordin–BMP (bone morphogenetic protein) network, has also been found to be conserved between vertebrate and invertebrate phyla (De Robertis, 2010). It is striking that the molecular vectors of both body axes are already present in cnidarians and in part also in other prebilaterian animals (Chapman *et al.*, 2010; Putnam *et al.*, 2007; Rentzsch *et al.*, 2007; Srivastava *et al.*, 2008, 2010b; Steele *et al.*, 2011; Technau and Steele, 2011). Nodal signaling controls the third body axis, defined by left-right symmetry, and recent reports on conserved nodal/*pitx* genes in lophotrochozoa suggest a prebilaterian origin also for this signaling system (Grande and Patel, 2009b).

Thus, the genetic toolkits for body axis formation seem to be conserved throughout most bilaterian and nonbilaterian animals. The prebilaterian ancestor therefore is believed to be an unexpectedly complex organism. Does this notion also hold true for the Urmetazoa? The genomic data made available recently for choanoflagellates and basal metazoans like sponges and cnidarians have confirmed the emerging view of highly conserved common themes in the development of multicellular animals.

2. Animal Relationships and the Origin of the Basal Metazoa

2.1. Choanoflagellates

Choanoflagellates are small unicellular protists comprising both marine and freshwater species (Fig. 6.1A). According to current molecular phylogenies, choanoflagellates are the closest unicellular relative of metazoans (King *et al.*, 2008). The genome of the recently sequenced choanoflagellate *Monosiga brevicollis* contains approximately 9200 genes, including a number of genes that encode domains of metazoan-specific cell adhesion and signaling proteins (King *et al.*, 2008). Choanoflagellates are morphologically similar to the choanocytes of sponges and were therefore proposed to represent the closest living relatives of metazoans (King *et al.*, 2008; von Salvini-Plawen, 1978).

2.2. Placozoa

Placozoans are disk-shaped organisms without any gastric cavity (Fig. 6.1F). They are composed of one outer epithelial layer with ciliated epithelial cells that exhibits different functions on the upper and lower side of the flattened animal. The bottom layer has a nutritive function with gland cells that release enzymes and epithelial cells that phagocytose the digested food; the upper epithelium seems to have a protective function. Between both layers are contractile, multinucleate fiber cells; neuronal and muscle cells are apparently absent (Srivastava *et al.*, 2008). *Trichoplax adherens* is the only described species of this phylum. Its only body axis is the top versus bottom axis (Srivastava *et al.*, 2008). *Trichoplax* reproduces mainly asexually by fission, and its embryology is almost unknown although population genetics suggest genetic recombination (Signorovitch *et al.*, 2005; Srivastava *et al.*, 2008). The *Trichoplax* genome was estimated to contain 11,514 protein-coding genes (Srivastava *et al.*, 2008). These genes are highly conserved compared to those in other animals (87%), in particular, in their exon–intron structure (Srivastava *et al.*, 2008). Eighty-three percent of all genes that are conserved between sea anemones and bilaterians have homologues in *Trichoplax* (Srivastava *et al.*, 2008).

The phylogenetic relationship of placozoans was since their discovery a matter of debate. On the morphological level, they were considered to represent a neotenic sponge or cnidarian larvae or an early branching metazoan phylum possibly related to the early transitional state from blastula to gastrula-like organisms as proposed by Bütschli (1884), Grell and Benwitz (1971), Ivanov (1973) and reviewed by von Salvini-Plawen (1978). On the molecular level, small ribosomal RNA and mitochondrial DNA analyses suggested that placozoans are either a sister-group to bilaterians (Collins,

1998; Ruiz-Trillo *et al.*, 2008) or the earliest eumetazoan branch (Dellaporta *et al.*, 2006; Signorovitch *et al.*, 2007).

2.3. Porifera

Sponges are among the oldest metazoan phyla and they occupy the ecological niche of sessile filter feeding. Since they are lacking any nervous system, they were classified as "Parazoa" in distinction to the "Eumetazoa" exhibiting a proper nervous system (i.e., cnidarians, ctenophores, and bilaterians; Hyman, 1940). Despite the fact that sponge morphology is highly diverse and the phylogeny is only poorly resolved (Dunn *et al.*, 2008; Pick *et al.*, 2010; Schierwater *et al.*, 2009; Srivastava *et al.*, 2010a), the basic body plan of a sponge is simple (Fig. 6.1D) and can be derived from a diploblastic tube that upon growth gets invaginated giving rise to a branched flagellated channel system with filtering chambers formed by choanocytes (Brusca and Brusca, 2002). The choanocytes are similar in morphology to the choanoflagellates (King *et al.*, 2008), and they probably represent the ancestral metazoan cell type (King *et al.*, 2008; Srivastava *et al.*, 2010a; von Salvini-Plawen, 1978).

Sponge development encompasses the formation of a blastula that develops into a diploblastic free-living ciliated embryo (Fig. 6.1B), which settles and undergoes metamorphosis to form the adult sponge (Leys, 2004). The formation of the diploblastic embryo occurs typically by delamination, ingression, or egression (Leys, 2004) and is highly reminiscent to the gastrulation in eumetazoans (Borojevic, 1967; Borojevic and Levi, 1965; Brien, 1967; Leys, 2004; von Salvini-Plawen, 1978). The sponge larvae have several characters in common with that of cnidarians (Brien, 1967), such as a body axis with locomotory flagellation at one (anterior) end and a poorly flagellated posterior end. During metamorphosis, the germ layers are inverted so that the cell layers of the adult sponge are reversed compared to those of higher animals (Amano and Hori, 2001; Brien, 1967; Leys, 2004; von Salvini-Plawen, 1978). As a result of this inversion, the food-collecting flagellated cells of the sponge larva are transformed to the choanocyte-lined inner cavity of the adult sponge. Therefore, the poriferan choanosome corresponds to the eumetazoan ectoderm (von Salvini-Plawen, 1978). A diverging view assumes that the metamorphosis represents a gastrulation step in sponge development, although there is no formation of an archenteron equivalent to the gastrulation in other metazoa. The sequenced genome of the desmosponge *Amphimedon queenslandica* reveals a high degree of conservation of developmental genes in sponges (Srivastava *et al.*, 2010a). There are no molecular data showing the homology of the germ layers between sponges and eumetazoans so far.

2.4. Cnidaria

The Cnidaria are diploblastic, aquatic animals with a simple body plan that is reminiscent of a bilaterian gastrula. The two body layers, an outer ectoderm and an inner endoderm, are separated by an acellular mesoglea, a composite connective tissue and basal lamina. The fossil record reveals that cnidarians are >500-Myr-old and molecular phylogenies identified the Cnidaria as a sister group to the Bilateria (Chen *et al.*, 2002; Conway Morris, 2000).

Today, there are two major cnidarian genetic models: the first one is the freshwater polyp *Hydra* (Fig. 6.1E), which is well known for a long time for its high regeneration capacity, and which has been studied in detail at the cellular level (Steele, 2006). *Hydra* has three independent cell lines: ectodermal and endodermal epitheliomuscular cells and interstitial stem cells that give rise to nerve cells (including nematocytes), gland cells, and gametes (Steele, 2002, 2006). Classical transplantation experiments have revealed gradients of morphogenetic signals that govern the patterning of the oral–aboral body axis (Bode and Bode, 1980, 1984; MacWilliams, 1983a,b; Meinhardt, 2002). During regeneration, these morphogenetic signals can be activated or enhanced at the site of wounding so that a new head or foot forms within 36 h (Bode, 2003; Holstein *et al.*, 2003). *Hydra* can be also dissociated into a suspension of single cells and regenerate a new organism after reaggregation and *de novo* pattern formation (Gierer *et al.*, 1972; Steele, 2006; Technau *et al.*, 2000).

The other model system is the starlet sea anemone *Nematostella vectensis*, (Fig. 6.1H) which was introduced by the pioneering work of Cadet Hand (Darling *et al.*, 2005; Holland, 2004). The great advantage of *Nematostella* is the availability of gametes and embryos that can be harvested in the lab at any time (Fritzenwanker and Technau, 2002). *Nematostella* embryos form a coeloblastula at the stage of 32–64 cells. Afterward, blastulae start to invaginate and form a gastrula, which develops into a planula larva that metamorphoses into a primary polyp (Fritzenwanker *et al.*, 2007; Kraus and Technau, 2006; Lee *et al.*, 2007; Fig. 6.1).

The genomes of both cnidarian models have been sequenced. The *Nematostella* genome project revealed that the cnidarian genome is complex, with a gene repertoire, exon–intron structure, and large-scale gene linkage more similar to vertebrates than to flies or nematodes, implying that the genome of the eumetazoan ancestor was similarly complex (Putnam *et al.*, 2007; see also Extavour, 2005; Finnerty *et al.*, 2004; Fritzenwanker *et al.*, 2004; Hayward *et al.*, 2004; Magie *et al.*, 2005; Martindale, 2004; Martindale *et al.*, 2004; Matus *et al.*, 2006b; Miller *et al.*, 2005; Scholz and Technau, 2003; Spring *et al.*, 2000, 2002; Technau and Bode, 1999; Technau and Scholz, 2003; Technau *et al.*, 2005; Torras *et al.*, 2004). We therefore presume that the genetic repertoire responsible for the formation of a specific bilaterian body plan already existed in the common ancestor of cnidarians and bilaterians (Guder *et al.*, 2006a).

In both model systems, molecular and genomic tools have been developed to study both organisms at a functional level (Guder *et al.*, 2006a). Besides ISH/IHC protocols and the full set of the genomic tools (BAC, EST, and cDNA libraries), RNAi, morpholinos, transient, and stable transfection are available or under optimization procedure. The transgenic approach is successfully working in all hydra labs and was introduced by the Bosch lab (Wittlieb *et al.*, 2006). Stable transgenic *Hydra* can be produced by injecting reporter constructs driven by a *Hydra* actin gene promoter (or other promoters) into blastomeres of two- to eight-cell *Hydra* embryos. In *Nematostella*, genomic integration of transgenes was established by using the I-SceI meganuclease system (Renfer *et al.*, 2010). Thus, due to their morphological simplicity and remarkable regeneration capacity, cnidarians can serve as an important model to understand basic functions of Wnt signaling in gastrulation, axis formation, germ-layer specification, regeneration, and cell differentiation (Guder *et al.*, 2006a).

2.5. Ctenophora

Based on the diploblastic body plan with a primitive gastric cavity, Ctenophores have been previously placed together with cnidarians in the group Coelenterata. Although the main body axis of ctenophores is also the oral–aboral axis, with a mouth at one end and a sensory organ at the opposite end, they are quite different from cnidarians. Ctenophores have rows of cilia arrays (comb plates) that are used for locomotion and they are lacking nematocytes (Brusca and Brusca, 2002; Pang and Martindale, 2008). The development of ctenophores also differs in that they are direct developing forms with a larval stage, the cydippid larva that is similar to the adult animal (Pang and Martindale, 2008). Based on recent molecular studies, ctenophores were even placed as the earliest branching extant group, diverging before sponges (Dunn *et al.*, 2008), which is, however, in debate again (Pick *et al.*, 2010). Whole genome data are essential to get a better resolution of these deep evolutionary nodes (Pang and Martindale, 2008).

2.6. Acoela and Turbellaria

Flatworms are among the first bilaterian animals and represent a diverse group of ciliated protostomes with an elaborated but primitive gut that frequently has only one pharyngeal opening. This is different from most bilaterian animals that possess a gut with a separate mouth and anus. Morphologically, in particular, the acoel flatworms are similar to planula larvae of cnidarians. Based on new molecular phylogenies, the evolutionary position of the acoel flatworms has been recently reinvestigated and they are considered to represent the sister group to the remaining bilaterians (Arendt and Nubler-Jung, 1994, 1997; Baguna and Riutort, 2004; Ruiz-Trillo *et al.*, 1999). In the acoel *Convolutriloba longifissura*, it was found that brachyury and goosecoid are expressed in association with the acoel mouth, which is similar to other bilaterians (Arendt *et al.*, 2001).

Planarians are freshwater flatworms and members of the Lophotrochozoa; they are famous for their ability to regenerate a head or a tail following transection (Reddien and Sanchez Alvarado, 2004). There is a large set of data addressing the function of Wnt and BMP signaling in planarian regeneration.

3. Wnt Signaling

The *wnt* genes are highly specific for metazoans, and so far, no *wnt* genes have been described from any unicellular eukaryotes, neither from cellular slime molds (*Dictyostelium discoideum*) nor from choanoflagellate Protozoa (*Monosiga*) that are closely related to the Metazoa (Guder *et al.*, 2006a; King *et al.*, 2003). A *β-catenin*-related gene (*Aardvark*) and a *Fzd*-related gene have been described from *D. discoideum*. However, Aardvark shares structural similarities to certain plant proteins that also exhibit armadillo repeat domains (Coates, 2003; Grimson *et al.*, 2000; Srivastava *et al.*, 2010b). Since no other *Fzd*-like genes have been found in any other nonmetazoan organisms, the ancestry of this protein family is also unclear (Srivastava *et al.*, 2010b). So far, the only eukaryotic member of the Wnt pathway that can be found outside the Metazoa is the kinase GSK-3 (Srivastava *et al.*, 2010b).

3.1. Wnt ligands

The number of *wnt* genes in deuterostomes ranges from 11 to 19 (Logan and Nusse, 2004; Nelson and Nusse, 2004), with the number increasing from basal deuterostomes such as sea urchins to humans among the vertebrates. These different *wnt* genes cluster in 12 subfamilies. When did these genes arise during metazoan evolution? Since the genome of flies (*D. melanogaster*) and nematode worms (*C. elegans*) contain only a small number (five) of *wnt* gene subfamilies, it was postulated that there was a gradual increase in complexity during metazoan evolution. However, a comprehensive analysis of cnidarian *wnt* genes in *Nematostella* and *Hydra* revealed the surprising result that all bilaterian *Wnt* gene subfamilies are already present in cnidarian genomes (Guder *et al.*, 2006a; Kusserow *et al.*, 2005; Lee *et al.*, 2006a; Lengfeld *et al.*, 2009). This complete set of *Wnt* gene subfamilies indicates that already the common ancestor of cnidarians and bilaterians must have invented the complex repertoire of *wnt* gene ligands (Kusserow *et al.*, 2005).

A clustering of the *wnt1/6/10/9/3* subfamilies in the phylogenetic analyses that was supported by the syntenic *wnt* organization between *Drosophila* and vertebrates (Guder *et al.*, 2006a; Kusserow *et al.*, 2005) suggested an ancestral cluster of *wnt genes* in the evolution of the common ancestor of cnidarians and bilaterians. Recent analysis of the genomic organization revealed that the *wnt6–wnt10* cluster observed in *D. melanogaster* is conserved in the cnidarian lineage (Sullivan *et al.*, 2007). Also a novel cluster comprising *wnt5–wnt7/wnt7b* was observed in *Nematostella* (Sullivan *et al.*, 2007). Despite the intriguing expression patterns of *wnt* genes along the oral–aboral axis (see below), *wnt* genes do not exhibit a Hox-like colinearity on the genomic level (Sullivan *et al.*, 2007). Future work must clarify to what extent *Wnt* genes are expressed under the influence of common enhancers or coordinately regulated by higher-order chromatin organization (Guder *et al.*, 2006a; Spitz and Duboule, 2005; Spitz *et al.*, 2003; Sullivan *et al.*, 2007).

New data indicate that cnidarian Wnt ligands are involved in all major Wnt-related pathways, that is, canonical, noncanonical planar cell polarity (PCP), and Wnt-Ca^{2+} signaling (see below) similar to bilaterian animals (Price, 2006; Strutt *et al.*, 2006) (Fig. 6.2). On the genomic level, the orthologs of Strabismus/Van Gogh, the GTPase RhoA, the Rho-associated coiled-coil-containing protein kinase 2 (Rock-2), the small GTPase RAC1 and the stress-activated Jun N-terminal kinase (JNK) are present. Although the function of these proteins in cytoskeleton remodeling has not been shown for cnidarian cell adhesion and motility so far, it is likely that they fulfill similar functions. Also the genes encoding for Wnt-mediated Ca^{2+} signaling are present, suggesting that a Wnt-induced activation of phospholipase C (PLC) induces a transient increase in cytoplasmic-free calcium that in turn activates the protein kinase C (PKC), calcium calmodulin-mediated kinase II (CaM-KII), or phosphatase calcineurin (CaN). Experiments using transgenic *Hydra* expressing a *Hydra* codon optimized version of the recombinant Ca^{2+} indicator GCaMP revealed significant changes in the Ca^{2+} patterns in response to pattern formation stimuli (Jana Schlüter *et al.*, unpublished).

The invention of Wnt ligands is tightly coupled with the Wnt secretion machinery. The Wnt secretion factor Wntless was reported to be conserved in the ctenophore *Mnemiopsis* (Pang *et al.*, 2010) and it is present in *Hydra* and *Nematostella* as well (Fig. 6.2). In sponges, it has so far not been identified. As Ching *et al.* (2008) have shown that *Drosophila* WntD is independent of Wntless and Porcupine, it might be speculated that sponge Wnts could follow a similar way of secretion.

The *wnt* gene repertoire of sponges, placozoans, and ctenophores is markedly less complex. The *A. queenslandia* genome contains three Wnt ligand genes, which are difficult to assign to any known Wnt family

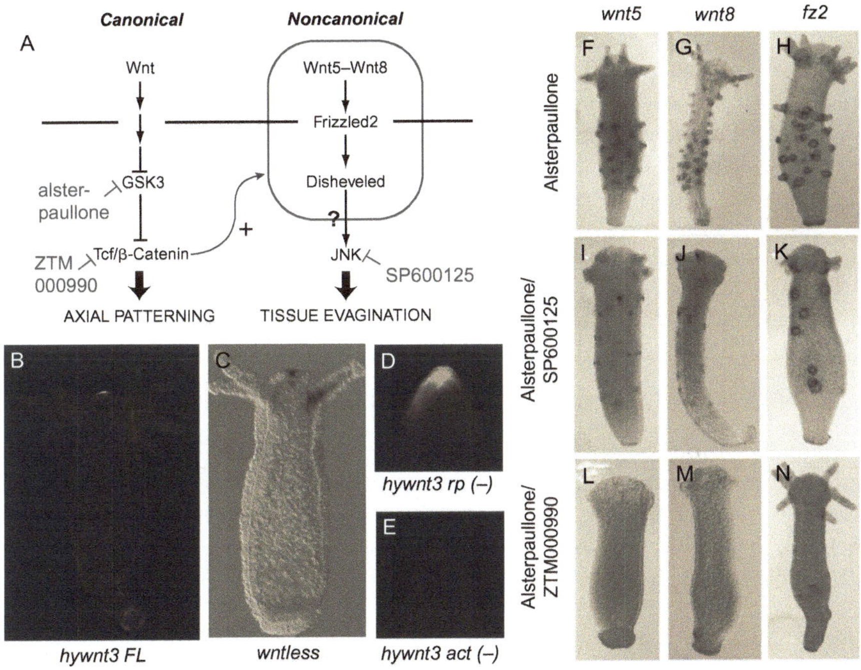

Figure 6.2 Wnt signaling in the cnidarian model *Hydra*. (A) Model of the interaction between canonical and noncanonical Wnt signaling in *Hydra* showing that noncanonical Wnt signaling acts downstream of canonical Wnt signaling in the formation of signaling centers. (B) *Hydra vulgaris* carrying the *hyactin-RFP::hywnt3-EGFP* construct where *hyactin-RFP* is expressed ubiquitously (pink), while *hywnt3FL-EGFP* (green) is localized to the hypostome, which is derived from the blastopore of the gastrula stage and represents the polyp's major signaling center, that is, the "head organizer." (C) *wntless* expression in the hypostome and at the sites of tentacle formation. (D and E) Autoregulatory and repressive elements restrict *wnt3* expression to the organizer. (F–N) Polyps treated with 5 M alsterpaullone show multiple ectopic *hywnt5, hywnt8, hyfz2* expressing cell clusters and ectopic tentacle formation. (E–G) Cotreatment with 5 M alsterpaullone/25 M SP600125 results in formation of multiple ectopic cell clusters expressing noncanonical *wnt* genes, but not in ectopic tentacle formation. (L–N) Cotreatment with 5 M alsterpaullone/25 M ZTM000990 blocks ectopic activation of noncanonical *wnt* genes and tentacle formation. © PNAS from Philipp *et al.* (2009) (A, F–N) and Nakamura *et al.* (2011). (See Color Insert.)

(Adamska *et al.*, 2010; Srivastava *et al.*, 2010b). *Oscarella* has the lowest number of Wnt ligands in any metazoan genome with two reported genes (Lapebie *et al.*, 2009). The genome of the placozoan *T. adherens* contains three so far unclassified *wnt* genes. Also in ctenophores, only a comparatively minimal set of four Wnt ligands (*wntA, wnt6, wnt9, wntx*) was isolated from *Mnemiopsis leidyi* (Pang *et al.*, 2010). Members of the protostome lineages exhibit a much lower number of *wnt* gene subfamilies, indicating that a significant loss of *wnt* genes must have occurred during early lophotrochozoan and/or ecdysozoan evolution (Kusserow *et al.*, 2005).

When and to what extent did this gene loss start? Since no complete genomes are available from lower protostomes, this question cannot be clearly answered yet. In the turbellarian *Schmidtea mediterranea*, nine *wnt* genes were found that cluster into four subfamilies (i.e., *wnt1, -2, -5, and -11*). The lower complexity of the protostome *wnt* gene repertoire compared to deuterostomes and cnidarians suggests, however, that a significant loss of family members took place during the early evolution of protostomes (Kusserow *et al.*, 2005; Lengfeld *et al.*, 2009).

3.2. Wnt receptors and target genes

Wnt ligands bind to Frizzled (Fzd) receptors in a complex with low-density lipoprotein receptor-related proteins (LRP5/6) and recruit Disheveled (Dsh) to the membrane leading to the phosphorylation of the LRP5/6 receptor by CK1, GSK3, and other kinases (Heuberger and Birchmeier, 2010). The phosphorylated LRP5/6 in turn recruits Axin to the membrane resulting in the stabilization of β-catenin. In sponges, Fzd receptors comprise a conserved motif believed to mediate the canonical Wnt pathway and have been shown to represent the minimal metazoan repertoire of two *Fzd* genes belonging to the Fzd1/2/7 (and chordate Fzd3/6) and Fzd3 families, respectively (Adamska *et al.*, 2010). One Lrp5/6/arrow-encoding gene was recovered in the same study. The genome analysis of the placozoan *Trichoplax* revealed a similar repertoire of two frizzled receptors and one *LRP5/6* gene (Srivastava *et al.*, 2008). Also in the ctenophore, *Mnemiopsis* two Fzd-7-related receptors and one *LRP5/6* gene were found (Pang *et al.*, 2010).

In cnidarians, several Fzd receptor genes have been identified, four in *Hydra* and six in *Nematostella* (Guder *et al.*, 2006a; Lee *et al.*, 2006b; Minobe *et al.*, 2000). Therefore, the general picture is emerging that the radiation of *wnt* genes was accompanied by a diversification of Wnt receptors, although the number of receptors is significantly lower than that of the ligands. *LRP5/6* is present in *Nematostella* and can be also identified in the *Hydra* genome. Interestingly, in *Hydra* a related protein was identified as the head activator-binding protein (Hampe *et al.*, 1999).

In the absence of nuclear β-catenin, TCF/LEF proteins form a transcriptional repression complex by recruiting the corepressor Groucho and histone deacetylases. When Wnt signaling stimulates the nuclear accumulation of β-catenin, Groucho is displaced and a transcriptional activation complex of β-catenin, Tcf/Lef, and the histone acetylase CBP is formed instead. There is no evidence for TCF/LEF and Groucho outside the metazoans indicating that these transcriptions factors also arose within the metazoan stem group.

3.3. Wnt destruction complex

LRP5/6 binds to Axin, which is important for the release of the cytosolic destruction complex that phosphorylates and finally degrades cytosolic β-catenin. Core members of this destruction complex are Axin, APC, and GSK, which are all present in sponges, placozoans, ctenophores, and cnidarians. It was pointed out that sponge and cnidarian APC and Axin are lacking β-catenin-binding motifs that are required for the correct formation of the destruction complex in their bilaterian counterparts (Srivastava *et al.*, 2010b). However, GSK-3 has been shown to phosphorylate β-catenin by its own (Yost *et al.*, 1996), and upon inactivation by alsterpaullone, it can therefore activate canonical Wnt signaling in *Hydra* (Broun *et al.*, 2005). Since GSK-3 can be also found outside the Metazoa, for example, in *Monosiga* and *Dictyostelium*, it is considered to be a pan-eucaryotic kinase (Srivastava *et al.*, 2008).

Previous cloning work and genomic data from *Nematostella* and *Hydra* databases also revealed orthologs of CamKII, casein kinase 1a, 1d, 1g2, 1g3, and 1e, Dsh, Flamingo, JNK, PKC, Tcf/Lef, Van Gogh, and Wntless (Hobmayer *et al.*, 1996, 2000a; Lee *et al.*, 2006b; Lengfeld, 2009; Philipp *et al.*, 2005; Rentzsch *et al.*, 2005; Technau *et al.*, 2005). These data indicate that in addition to the canonical Wnt pathway also noncanonical Wnt pathways, for example, the Wnt/PCP and the Wnt/Ca^{2+} pathways known from bilaterian animals (Price, 2006; Strutt *et al.*, 2006) are present in cnidarians. By applying specific activators and inhibitors Hobmayer *et al.* have demonstrated that canonical Wnt signaling in *Hydra* acts upstream of the PCP pathway and thus controls patterning by both signaling centers (Fig. 6.2).

4. TGF-β and BMP Signaling

The transforming growth factor-β (TGF-β) signaling pathway (see reviews by Massagué; Heldin *et al.*, 2009; Massague, 1998; Massague and Wotton, 2000; Moustakas and Heldin, 2009; Wu and Hill, 2009) is metazoan-specific and neither ligand nor receptor molecules have been identified outside the animal kingdom so far (Srivastava *et al.*, 2010b). The TGF-β superfamily consists of two major groups. The first are the TGF-β members, which include the name-giving group TGF-β and the related activins and GDFs (growth differentiation factors). The second group comprises the bone morphogenetic proteins (BMPs) including nodal proteins. TGF-β family members are involved in different basic cell functions, for example, cell proliferation and apoptosis, but also cell differentiation and migration. TGF-β ligands bind to the type II receptor tyrosine kinase, which upon binding recruit type I receptors that are phosphorylated and activated by the type II receptor. Activated type I receptors phosphorylate receptor-mediated Smads (R-Smads). The phosphorylated R-Smads (Smad-1/5/8,

2/3) bind to co-Smads (Smad-4) and are translocated into the nucleus. This heteromeric complex can activate specific downstream target genes. Inhibitory Smads (Smad-6, -7) can bind to R-Smads and have a competing function (Moustakas and Heldin, 2009; Wu and Hill, 2009).

Comparative genomics indicated that the TGF-β superfamily and its corresponding signaling pathways are less complete at the base of metazoan evolution than the Wnt pathway. The first complete BMP/TGF-β pathways are present in the *Trichoplax* and cnidarian genomes including type I and type II receptors, ligands (Dpp/BMP orthologs), and antagonists (Noggin), receptor-, inhibitory-, and co-SMAD transcription factors.

4.1. BMPs and TGF-β ligands

Any TGF-β-related ligands are absent from *Monosiga*, *Dictyostelium*, and other nonmetazoan species (see above). The phylogenetic analysis by Srivastava *et al.* (2010a,b) indicates that the metazoan TGF-β superfamily, which can be subdivided into two major clades of ligands, the TGF-β *sensu stricto* and TGF-β related (e.g., Activins, Leftys, GDF8s), and the BMP related (e.g., BMPs, Nodals; see also Matus *et al.*, 2006a), is already present in sponges. The *Amphimedon* genome analysis revealed a cluster of five *Amphimedon*-specific genes at the base of the metazoan *TGF-β* superfamily suggesting an independent expansion of this ancestral gene in sponges. Two genes have been identified in *Amphimedon* that cluster within the *TGF-β* (s.s.) clade (Srivastava *et al.*, 2010b). However, since TGF-β (s.s) ligands have been identified so far only in deuterostomes, this placement of *Amphimedon* ligands warrants further investigation (Srivastava *et al.*, 2010b). Also the occurrence of other *TGF-β* members in basal metazoans remains uncertain, since only activin was identified as a member of this group in the *Nematostella* genome (Srivastava *et al.*, 2010b).

No BMP-related ligands were found in sponges, suggesting that these genes diversified only later in Eumetazoan evolution (Srivastava *et al.*, 2010b). By comparison, cnidarians evolved several *BMP*-related genes, for example, *BMP2/4*, *BMP6/7*, and *GDF2/5* (Putnam *et al.*, 2007; Srivastava *et al.*, 2010b) that share strong sequence homologies with their bilaterian counterparts.

4.2. BMP and TGF-β receptors

TGF-β pathway receptors are serine–threonine kinases (STKRs) which are absent in unicellular organisms, but already present in sponges. Tyrosine kinases have been identified in *Monosiga* and in *Dictyostelium*, but they are considered to be distinct from the TGF- β-related STKR family (Manning *et al.*, 2008; Srivastava *et al.*, 2010b). Among the *Amphimedon*, STKR, one protein model clusters together with the eumetazoan activin type

2 receptors, while the others cannot be assigned to a specific eumetazoan subfamily (Srivastava *et al.*, 2010b). A complete set of type 1 and type 2 activin receptors could be first identified in *Trichoplax* and in cnidarians. Also the type 1 and type 2 BMP receptors are present in cnidarians, while *Trichoplax* exhibits only BMP type 1 receptor (See Table 6.1).

4.3. SMADs

The Smad proteins that are required for the transmission of the TGF-β signal to the nucleus are also a metazoan invention. As recently pointed out by Adamska *et al.* (2010) and Srivastava *et al.* (2010a,b), proteins with a Smad-like MH2 domain can be found in *Monosiga*, but this domain is coupled with a C2H2 zinc finger domain as opposed to the metazoan Smads that are characterized by a MH1 and MH2 domain only. Both R-Smads and Co-Smads are found in *Amphimedon*, *Trichoplax*, and the cnidarian genomes indicating that the common ancestor of sponges and eumetazoans already invented the complete signal transduction cascade. Inhibitory Smads (I-Smads, Smad6/7) that interfere with the phosphorylation of R-Smads and/or the formation of the Rsmad/Co-Smad complexes are missing in sponges suggesting that they only evolved after the divergence of Porifera and eumetazoans. In the nucleus, Smad complexes do recruit a number of transcription factor proteins including the coactivators CBFβ and CBP as well as the corepressor Ski/Sno, which are also specific for the metazoans (Srivastava *et al.*, 2010b).

A further negative regulator of TGF-β signaling is the E3 ubiquitin ligase Smurf, which can target R-Smads and receptors for degradation. Smurfs have been only identified in bilaterian and cnidarian genomes so far.

4.4. Nodal signaling

The *nodal* gene was first identified as a mesoderm-inducing factor in the mouse (Conlon *et al.*, 1991) and encodes a secreted factor belonging to the TGF-β superfamily. In vertebrates, nodal-related genes are expressed in the organizer and are essential for mesoderm and endoderm induction and the establishment of the DV axis (Duboc and Lepage, 2008; Duboc *et al.*, 2008). In addition, during later embryonic development, Nodal signaling, in concert with its extracellular antagonist Lefty, is responsible for the establishment of left–right asymmetry. This function of Nodal is conserved in deuterostomes including invertebrate groups like ascidians and sea urchins (Chea *et al.*, 2005). Recently, Grande and Patel have demonstrated a functional Nodal/Pitx (a downstream target of Nodal signaling) cassette in the determination of chirality in snails, a bilaterian nondeuterostome group. In chordates, nodal is expressed on the left side of the embryo but on the right side in echinoderms, while in lophotrochozoans expression is related to body chirality (Grande and Patel, 2009a,b).

Table 6.1 Wnt, TGF-b, and Hedgehog signaling pathways in metazoan evolution (Dd *Dictyostelium discoideum*, Mb *Monosiga brevicolis*, Aq *Amphimedon queenslandica*, Ta *Trichoplax adherens*, Hm *Hydra magnipapillata*, Nv *Nematostella vectensis*, Dm *Drosophila melanogaster*, Ci *Ciona intestinalis*, Hs *Homo sapiens*)

			Dd	Mb	Aq	Ta	Hm	Nv	Dm	Ci	Hs
Wnt pathway	Ligand	Wnt			■	■	■	■	■	■	■
Canonical	Receptors	SFRPs	▨		■	■	■	■	■	■	■
		LRP5/6			■	■	■	■	■	■	■
	Associated	Dally							■	■	■
		Disheveled			■	■	■	■	■	■	■
		CK1				■	■	■	■	■	■
		CK2				■	■	■	■	■	■
		GSK-3			■	■	■	■	■	■	■
		GBP							■	■	■
		Axin			■	■	■	■	■	■	■
		PP2a				■	■	■	■	■	■
		APC			▨	▨	▨	▨	■	■	■
	Nuclear factor	β-Catenin			■	■	■	■	■	■	■
		Tcf/Lef			■	■	■	■	■	■	■
		CBP	▨	▨	■	■	■	■	■	■	■
		Pygopus							■	■	■
		Groucho			■	■	■	■	■	■	■
	Antagonist	SFRPs			■	■	■	■	■	■	■
		Dkk					■	■		■	■
PCP	Receptor	Strabismus				■	■	■	■	■	■
	Associated	Knypek								■	■
		RhoA				■	■	■	■	■	■
		Rock2				■	■	■	■	■	■
		Rac				■	■	■	■	■	■
		JNK				■	■	■	■	■	■
		PLC				■	■	■	■	■	■
Ca^{2+}	Associated	CaMKII				■	■	■	■	■	■
		PKC				■	■	■	■	■	■
		CaN	■	■	■	■	■	■	■	■	■

Table 6.1 *(continued)*

			Dd	Mb	Aq	Ta	Hm	Nv	Dm	Ci	Hs
Wnt pathway	Ligand	Wnt									
TGF-β/ BMP	Ligand	BMP/Dpp									
	Receptor	BMP R1									
		BMP R2									
	Associated	Smad-1/5/8									
		Smad-6/7									
	Nuclear factor	bHLH inhibitor ID									
		Smad-4									
	Antagonist	Noggin									
		Chordin									
TGF-β	Ligand	TGF-β									
	Receptor	TGF-β R1									
		TGF-β R2									
	Associated	Smad-2/3									
		Smurf1									
		RhoA									
		Rock1									
		P70S6K									
	Antagonist	Decorin									
Activin	Ligand	Activin									
	Receptor	Activin R1									
		Activin R2									
	Antagonist	FST									
Nodal	Ligand	Nodal									
	Nuclear factor	Pitx									
	Antagonist	Lefty									

(continued)

Table 6.1 (*continued*)

			Dd	Mb	Aq	Ta	Hm	Nv	Dm	Ci	Hs
Wnt pathway	Ligand	Wnt			■	■	■	■	■	■	■
Hedgehog	Ligand	Hh		░	░		■	■	■	■	■
	Receptor	Ptc		░	░		■	■	■	■	■
		Smo					■	■	■	■	■
	Associated	Megalin						■	■	■	■
		Rab23			■	■	■	■	■	■	■
		Cos			■	■		■	■	■	■
		Ci/Gli			■		■	■	■	■	■
		Fu	■	■	■	■	■	■	■	■	■
		Su(fu)			■		■	■	■	■	■
		Slim b			■	■	■		■	■	■
		PKA					■	■	■	■	■
		GSK-3			■	■	■	■	■	■	■
		Ck1			■	■	■	■	■	■	■
	Nuclear factor	Zic2					■	■	■	■	■

Nodal-related factors have so far not been identified in ecdysozoans and any nonbilaterian phyla suggesting an early bilaterian origin of left–right axis determination. However, based on the resolution of current TGF-β molecular phylogenies, it cannot be excluded that nodal-like proteins already exist in basal metazoans. Nodal signaling might be an alternative evolutionary path to break the radial symmetry, for example, in branching cnidarians. A functional nodal-like TGF-β ligand would require at least conserved downstream target genes of the Nodal-signaling *nodal/pitx* cassette (see below).

5. Negative Regulators of BMP and Wnt Signaling

Secreted extracellular factors can exhibit a strong antagonizing or modulating effect on BMPs and Wnts by "ligand trapping" (Davidson *et al.*, 2002; Jones and Jomary, 2002; Kawano and Kypta, 2003; Mao and Niehrs, 2003). Wnt antagonists like the family of secreted Frizzled-related proteins (sFRPs), the Wnt inhibitory factor (WIF), Cerberus, and the Dickkopf (Dkk) proteins have been now identified from *Hydra* and *Nematostella* (Guder *et al.*, 2006b; Lee *et al.*, 2006b; Technau *et al.*, 2005), some of them (Dkk and Cerberus) are absent from the *Drosophila* and *C. elegans* genome databases. In *Nematostella*, eight sFRPs exist, and in *Hydra* also several sFRPs have been identified in the *Hydra* genome. From *Hydra* and *Nematostella*, two *dickkopf* orthologs have been cloned. One is an ortholog to *dkk-3* and the other is an ortholog to the *dkk1/2/4* subfamily found in vertebrates (Fedders *et al.*, 2004; Guder *et al.*, 2006b). In *Hydra*, *dkk-3* has been suggested to be involved in neuronal differentiation (Fedders *et al.*, 2004). *dkk1/2/4* is proposed to be a putative precursor to the *dkk1*, *dkk2*, and *dkk4* subfamilies found in vertebrates (Augustin *et al.*, 2006; Guder *et al.*, 2006a,b). Expression analysis in *Hydra* and *Nematostella* reveals that *dkk1/2/4* is expressed opposite to the *wnt* expression domains (Guder *et al.*, 2006a; Lee *et al.*, 2006b) suggesting a Wnt-antagonizing function. Overexpression studies of *hydkk1/2/4A* in *Xenopus* embryos indicate that it is a functional ortholog of vertebrate *Dkk1* and *Dkk4* that it is capable of inhibiting Wnt signaling (Guder *et al.*, 2006a,b). An activation of canonical Wnt signaling by alsterpaullone treatment causes the complete, but reversible downregulation of *hydkk1/2/4*, showing that *hydkk1/2/4 is* negatively regulated by Wnt signaling.

In sponges, only *Oscarella* has been shown to possess a *dkk* gene (Nichols *et al.*, 2006), while *Amphimedon* lacks dkk orthologs (Adamska *et al.*, 2010). Ctenophores also lack Wnt antagonists related to *dkk* or other gene families (Pang *et al.*, 2010). There is a general paucity in sponges and ctenophores concerning soluble Wnt inhibitory ligands. A new aspect of Wnt regulation acting on the transcriptional level has been recently demonstrated in *Hydra.* Autocatalytic and repressive inputs restrict *Wnt* expression to the site of the signaling center (Nakamura *et al.*, 2011). A *HyWnt3rep* repressor element is necessary and sufficient to localize *HyWnt3* expression to the head organizer independent of Dkk1/2/4 (Nakamura *et al.*, 2011). When the HyWnt3rep element was removed an expansion of gene expression dependent on Tcf-binding sites was observed (Fig. 6.2). This suggests that *HyWnt3* has the potential to be activated in a Wnt/β-catenin–dependent manner and in a broad domain even in the context of intact HyDkk1/2/4 function (Nakamura *et al.*, 2011). It was proposed that this repressive activity of *HyWnt3rep* gets locally abolished at the onset of regeneration and budding, thereby permitting local *HyWnt3* transcription and new organizer formation (Nakamura *et al.*, 2011). It will be interesting to identify similar cis-and transregulatory elements in bilaterians to understand the evolution of the blastoporal organizer.

In BMP/TGF-β signaling, there are three structurally different antagonizing protein families: Noggin/Chordin, Follistatin, and the large CAN family (for review, see Zakin and De Robertis, 2010). On the level of Chordin, interaction with Cross-veinless-2 (CV2), Twisted Gastrulation (Tsg), and the Tolloid metalloproteinases can further regulate Chordin diffusion by cleavage and/or trapping (Zakin and De Robertis, 2010). Also Sizzled, which is a frizzled-related diffusible protein (sFRP), acts as a competitive inhibitor of Tolloid metalloproteinases. Follistatin is an inhibitor of Activin, and the members of the CAN family (Cerberus/Gremlin/Coco/Dan) are multivalent BMP inhibitors that also affect nodal and Wnt signaling. Finally, also the extracellular matrix itself, for example, Collagen IV might bind BMPs and thereby modify their diffusion range.

Chordin proteins are characterized by well-defined CHRD domains and do only appear on the level of cnidarians. However, *chordin*-like genes with CR domains are already present in sponges and *Trichoplax* (Srivastava *et al.*, 2010b). The CR domain is a common motif in many secreted proteins (e.g., Kielin, CRIM, and Cross-veinless) and members of the ECM (e.g., thrombospondins, von Willebrand factor, cysteine knots, and insulin growth factor-binding protein) that all share the capacity to bind BMP and TGF-β ligands and thereby alter the diffusibility of these growth factors (Garcia Abreu *et al.*, 2002).

6. Axis Formation and Regeneration in Basal Metazoans

The Wnt, BMP, and Nodal signaling pathways represent three molecular coordinates defining the anterior-posterior, dorso-ventral and left-right axis in bilaterians. Their appearance in metazoan evolution was tightly coupled to major evolutionary transitions and the formation of the body axes outlined in Figure 6.3.

Wnt signaling has a role in body axis formation already in sponges, cnidarians (Broun *et al.*, 2005; Hobmayer *et al.*, 2000b; Kusserow *et al.*, 2005) (Fig. 6.3), and planarians (Gurley *et al.*, 2008; Iglesias *et al.*, 2008; Petersen and Reddien, 2008). A Wnt gradient also patterns the AP axis of the CNS in frogs (Kiecker and Niehrs, 2001; Niehrs, 2001, 2004) and it was proposed that patterning by Wnt gradients has been conserved throughout the evolution of the AP body axis (De Robertis, 2010; Niehrs, 2010). We propose that prebilaterian animals already evolved BMP and Wnt signaling gradients that specify the primary body axis in the common ancestor of bilaterian animals. This clearly indicates that posteriorizing Wnt gradients are not a novelty restricted to chordates or secondarily co-opted in various organisms (De Robertis, 2010; Hejnol and Martindale, 2009; Martindale and Hejnol, 2009), but that they probably represent an ancient feature (Guder *et al.*, 2006a; Niehrs, 2010; Rentzsch *et al.*, 2006, 2007).

The evolution of the second body axis in bilaterian animals is less clear. According to the current view, the common ancestor of bilaterians evolved a second axis that is perpendicular to the primary Wnt body axis and based

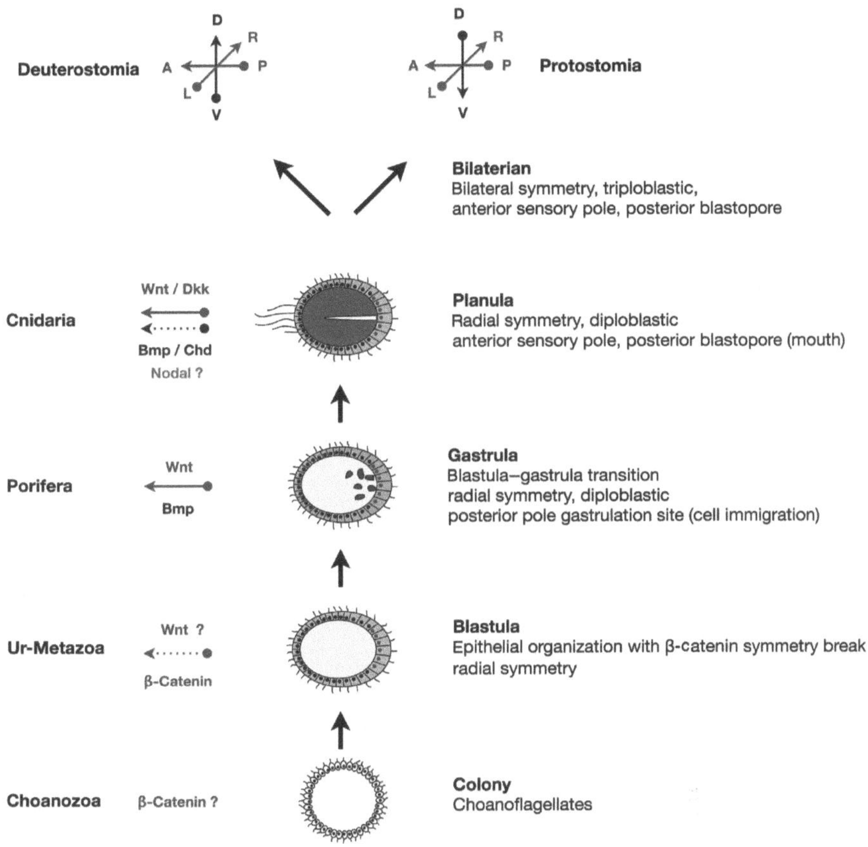

Figure 6.3 Signaling pathways and axis formation in the basal metazoans. Major evolutionary transitions show the function of β-catenin and Wnt signaling in the formation of the primary body axis (red). In cnidarians, a functional BMP pathway (blue) is acting in parallel to the Wnt pathway and gives rise to the second (DV) body axis in bilaterians. In bilaterians, the nodal pathway (green) determines the third spatial coordinate, the left–right axis. It is yet unclear whether this pathway is already acting in cnidarians. (See Color Insert.)

on the BMP gradient. This BMP gradient constitutes the DV axis in bilaterian animals (De Robertis, 2010; Niehrs, 2010; Rentzsch *et al.*, 2007).

In such a Cartesian coordinate system (De Robertis, 2010; Niehrs, 2010), not only the body (polarity) axes are important but also the symmetry axes and the spatial coordinates, which all are not identical (Manuel, 2009). In a bilateral animal, three spatial coordinates do exist, the AP, the DV, and the left–right coordinate which are all defined by distinct signaling pathways, that is, the Wnt/Dpp, BMP/Chd, and the Nodal/Lefty pathway. Cnidarians are radial symmetrical organisms that exhibit striking branching patterns and some groups exhibit even traces of bilateral symmetry. In terms of pattern formation, these asymmetries correspond to a break in the radial symmetry of the polyp and could principally explain the origin of a second body axis. In *Nematostella*, these asymmetries are correlated with a transient expression of *BMP/Chd* that is perpendicular to the primary body axis (Rentzsch *et al.*, 2006). But Manuel (2009) pointed out that the bilateral symmetries in anthozoans and bilaterians may not be comparable. Accordingly, the bilaterality of anthozoans would be superimposed on a fundamentally radial anatomy. In *Nematostella*, the asymmetric expression of *chordin* starts indeed from a radial symmetry around the blastopore similar to *Hydra* (Rentzsch *et al.*, 2006, 2007). But future work must show whether the anthozoan bilateral structures are evolutionary acquisitions (synapomorphies) of anthozoans within the cnidarian phylum, or features shared with bilaterians (plesiomorphies).

6.1. AP axis in prebilaterians

One major function of Wnt signaling is in gastrulation and axis formation. This is obvious from the striking localization of nearly all *wnt* genes at the site of the blastopore during and after gastrulation of the *Nematostella* embryo. The expression domains for *Nematostella wnt* genes reveal a characteristic pattern in staggered domains that span the oral–aboral axis except for the aboral pole itself. These distinct regional and germ-layer-specific expression patterns of the *wnts* suggest that the ancestral role of these genes was in specifying position along the main body axis (Kusserow *et al.*, 2005). Although an analysis of the genomic organization of *wnt* genes in *Nematostella* indicates only a limited syntenic organization of *wnt* genes, it is tempting to speculate that there is a regulative hierarchy of *wnt* gene expression during *Nematostella* embryogenesis (Guder *et al.*, 2006a; Spitz and Duboule, 2005; Spitz *et al.*, 2003; Sullivan *et al.*, 2007; see above).

Functional data from *Nematostella* support a function of canonical Wnt signaling in the gastrulation process. β-catenin becomes translocated into nuclei in cells at the site of the blastopore during gastrulation (Wikramanayake *et al.*, 2003), and lithium chloride treatment, which blocks GSK-3-mediated degradation of β-catenin, results in extended gastrulation movements. Lee *et al.* (2007) have recently done an important step forward by showing that NvDsh is necessary for the localization of nuclear β-catenin (Lee *et al.*, 2007). The mechanism of Dsh enrichment at the site of gastrulation is unclear so far but might be explained by degradation of Dsh at the opposite pole and/or binding to the cytoskeleton (Lee *et al.*, 2007). The fact that cnidarians can start gastrulation at the animal pole was surprising since it is different from deuterostomes, which gastrulate at the vegetal pole. These findings stress the importance of understanding the Dsh/β-catenin interaction with the cytoskeleton at the site of gastrulation (Lee *et al.*, 2007; Momose and Houliston, 2007; Momose and Schmid, 2006).

The blastoporal signaling center in cnidarian embryos is reminiscent of the other signaling center, that is, the *Hydra* head organizer, which is located at the apical tip of the polyp's hypostome. Actually, the hypostome of a cnidarian polyp corresponds to the blastopore of a cnidarian gastrula (Fritzenwanker *et al.*, 2007; Lee *et al.*, 2007; Momose and Schmid, 2006). When transplanted to an ectopic site, the head organizer induces a secondary body axis by recruiting host tissue (Broun and Bode, 2002; Broun *et al.*, 2005; Browne, 1909; MacWilliams, 1983b). *wnt* genes and β-catenin-dependent signaling play a major role in head induction (Cramer von Laue, 2003; Hobmayer *et al.*, 2000a; Rentzsch *et al.*, 2005). Treatment with the GSK-3β inhibitor alsterpaullone increases the accumulation of β-catenin in nuclei and can stimulate head formation (Broun *et al.*, 2005). Alsterpaullone treatment also causes an expansion of the normal expression domains of *hywnt*, *hytcf*, and *hybra1*, the *Hydra brachyury* ortholog and induces numerous spots of *hywnt3a* expression seen along the body column (Broun *et al.*, 2005).

Wnt/β-catenin signaling acts also during *Hydra* head regeneration (Hobmayer *et al.*, 2000a; see also Muller *et al.*, 2007; Plickert *et al.*, 2006) and during *de novo* pattern formation in reaggregates (Technau *et al.*, 2000). The characteristic expression of various *wnt* genes and pharmacological intervention with β-catenin and JNK inhibitors indicate a hierarchy of Wnt signaling in head regeneration (Fig. 6.2 A, F-N) (Philipp *et al.*, 2009).

In summary, one can conclude that the similar activation of overlapping *wnt* expression domains during gastrulation, early embryogenesis, and regeneration indicates various Wnt pathways, that is, the Wnt/PCP, the Wnt/Ca^{2+}, and the canonical Wnt pathway act concomitantly during gastrulation, axis formation, and regeneration. We are far away from having a mechanistic and complete view, how these various Wnt pathways may interact, and which downstream target genes are activated to induce cell differentiation, for example, neuronal differentiation (Lie *et al.*, 2005; Onai *et al.*, 2004; Teo *et al.*, 2006). However, the genomic and molecular tools are on hand to unravel these fundamental processes.

6.2. AP axis in basal deuterostomes

The function and localization of β-catenin signaling in deuterostomes AP-axis formation has been recently reviewed in detail (De Robertis, 2010; Niehrs, 2010). The general picture emerging from many studies in deuterostomes and protostomes is that β-catenin has an important function in the early polarization of the animal–vegetal axis of the embryo (Martin and

Kimelman, 2009). Similar to prebilaterian animals, β-catenin is vegetally localized in many basal deuterostomes and protostomes.

In sea urchins (echinoderms), development starts with one clear polarity from the animal to the vegetal pole that resembles the oral–aboral body axis in cnidarians. β-Catenin signaling marks the vegetal pole of the sea urchin embryo, and an overactivation of β-catenin with LiCl results in vegetalized embryos with ectopic endoderm (Logan *et al.*, 1999; Wikramanayake *et al.*, 1998, 2004).

Also in many species of the hemichordates and chordates, there is a clear polarization along the AP axis with a peak expression of β-catenin at the vegetal and blastoporal side. In ascidians (urochordates) and in amphioxus (cephalochordates), the nuclear-localized β-catenin marks the vegetal and blastopore pole, which marks the future posterior end of the embryo. (Imai *et al.*, 2000; Miyawaki, 2003). Lithium chloride treatment causes an activation of β-catenin signaling and a posteriorization of the amphioxus embryo, which is accompanied by a loss of the neural plate (Onai *et al.*, 2009; Yu *et al.*, 2007). A gradient of nuclear β-catenin from posterior to anterior can be also found in vertebrates. Similar to the posterior growth in amphioxus tail buds (Schubert *et al.*, 2001), Wnt signaling is involved in tail development in zebrafish (Agathon *et al.*, 2003; Shimizu *et al.*, 2005; Thorpe *et al.*, 2005) where Wnt3a determines the AP position of the somite determination front (Aulehla *et al.*, 2003; De Robertis, 2010; Dunty *et al.*, 2008; Niehrs, 2010). In summary, the data from basal deuterostomes and chordates indicate that canonical Wnt/β-catenin signaling acts to polarize early embryos and to pattern the posterior–anterior body axis (De Robertis, 2010; Holland, 2002; Niehrs, 2010).

6.3. AP axis in basal protostomes

There is a deep split between protostomes and deuterostomes, since both groups diverged almost over 500 million years ago, which is almost as old as the split between the major cnidarian groups. It is generally assumed that protostomes and deuterostomes evolved from a common prebilaterian ancestor. According to modern molecular phylogenies, protostome can be subdivided into two major groups, Lophotrochozoa and Ecdysozoa.

Planarians exhibit nine *wnt* genes: five of them are expressed in *S. mediterranea* along the AP-axis, predominantly at the posterior end, and a Wnt inhibitor-like gene is expressed at the anterior pole (Petersen and Reddien, 2008). Downregulation of β-catenin in regenerating planarians results in an enhancement of anterior head structures suggesting that β-catenin controls the posterior–anterior (tail-versus-head) axis (Gurley *et al.*, 2008; Iglesias *et al.*, 2008; Petersen and Reddien, 2008). Inhibition of the β-catenin-inhibitory APC-like gene results in regeneration of tails from anterior-facing wounds (Gurley *et al.*, 2008). *wnt* genes are required for this polarity process (Adell *et al.*, 2009; Petersen and Reddien, 2009).

These data indicate that a posterior source of Wnt signaling together with an anterior localization of Wnt was already present in the common ancestor of bilaterians and constitutes a conserved and basic feature of axial patterning in protostome and deuterostome bilaterians.

6.4. DV axis in basal bilaterians

The DV axis is the second body axis required for defining a bilaterian animal, that is, the position of the nerve chord, heart, and intestine in bilaterian animals and it is perpendicular to the AP-axis (De Robertis, 2010; Niehrs, 2010). The position of the organs along the DV axis is inversely related in chordates and in protostomes. In chordates, the nervous chord is located at the dorsal side and at the ventral side in invertebrates (Brusca and Brusca, 2002). On the molecular level, however, the DV axis is uniformly determined by BMP/Chordin signaling. BMP/Dpp and its antagonist Chordin are expressed at opposite ends of the DV axis (De Robertis, 2010; Niehrs, 2010). This BMP–Chordin antagonism is highly conserved in bilaterian evolution, and it is used to specify the different cell types of the mesoderm and nervous system in a concentration-dependent manner (Niehrs, 2010). High concentrations of BMP/Dpp induce various structures in the ecto-, endo-, and mesoderm in all bilaterians (Table 6.2). By comparison, low concentrations of BMP/Dpp together with high Chordin concentrations lead to neuronal structures (Bier, 1997; Mizutani *et al.*, 2006; Sutherland *et al.*, 2003).

The conserved BMP/Chordin antagonism in vertebrates and insects has led to the revival of Geoffroy Saint-Hilaire's hypothesis (Geoffroy St-Hilaire, 1822) postulating that the DV axis is homologous among all bilaterian animals but was inverted during evolution (Arendt and Nubler-Jung, 1994; De Robertis and Sasai, 1996). Since all bilaterians are generally assumed to be monophyletic, this raises questions on the molecular and morphological identity of the last common bilaterian ancestor. It also addresses the question whether modern bilaterians exhibit features of an axis inversion as proposed (Arendt and Nubler-Jung, 1994; De Robertis and Sasai, 1996) axis inversion. The most basal protostomes are acoel and turbellarian flatworms, and among the deuterostomes, the hemichordates belong to the most ancient forms. In both groups, there is clear evidence for the involvement of the BMP–Chordin system in patterning the DV axis (Lowe *et al.*, 2006; Molina *et al.*, 2007). In planarians, the CNS develops on the ventral side. It can be duplicated at the dorsal side after RNAi knockdown of BMP or Smad-1 (Molina *et al.*, 2007; Adell *et al.*, 2010).

Table 6.2 Molecular vectors defining metazoan body axes[a]

	First axis vector	β-Catenin/Wnt	References	Second axis vector	BMP	References	Third axis vector	Nodal	References
Diploblasts									
Porifera	X	Wnt expressed at the posterior pole of Amphimedon larvae	Adamska *et al.* (2007, 2010), Windsor and Leys (2010)						
Placozoa		*Trichoplax adherens* n.d.							
Ctenophora	X	Oral–aboral expression in *Mnemiopsis*	Pang *et al.* (2010)						
Cnidaria/ Medusozoa	X	Regulates oral–aboral patterning *Clytia* planulae larvae and *Hydra polyps* and regeneration	Broun *et al.* (2005), Guder *et al.* (2006a), Hobmayer *et al.* (2000a), Momose *et al.* (2008), Momose and Houliston (2007)		BMP5/6 expressed along the oral–aboral axis			Unknown	
Cnidaria/ Anthozoa	X	Regulates oral–aboral patterning *Nematostella* embryos, planula larvae, and polyps	Kusserow *et al.* (2005), Wikramanayake *et al.* (2003)	X	BMP2/4 expressed asymmetrically in the gastrulating embryo	Finnerty *et al.* (2004), Hayward *et al.* (2002)		Unknown	
Triploblasts/deuterostomes									
Echinodermata	X	Regulates animal-vegetal embryonic patterning in the sea urchin *Lytechinus* and *Paracentrotus*	Logan *et al.* (1999), Wikramanayake *et al.* (1998)	X	Regulates oral–aboral axis	Duboc *et al.* (2004)	X	Expression restricted to the presumptive oral ectoderm at late gastrulation	Duboc *et al.* (2004), Flowers *et al.* (2004)
Hemichordata	X	Defines posterior organizer and specifies endomesodermin Saccoglossus	Darras *et al.* (2011)	X	Regulates Saccoglossus DV-axis	Lowe *et al.* (2006)	X	Unknown	

Cephalochordata	X	Posterior Wnt, anterior Wnt-antagonists expression; LiCl posteriorizes Amphioxus embryos	Yu *et al.* (2007)	X	Regulates Amphioxus DV-axis	Yu *et al.* (2007)	X	Asymmetric (left) expression of AmphiNodal in all three germ layers	Yu *et al.* (2002)
Urochordata	X	Regulates endodermal (vegetal) cell fate in the sea squirt *Ciona*	Imai *et al.* (2000)	X	Regulates DV neural tube patterning	Darras *et al.* (2011)	X	Asymmetric (left) nodal expression after gastrulation (*Halocynthia*)	Morokuma *et al.* (2002)
Osteichthyes	X	Regulates AP CNS and mesodermal patterning	Kim *et al.* (2000), Agathon *et al.* (2001), Erter *et al.* (2001)	X	Gradient regulates DV mesodermal and ectodermal patterning	Neave *et al.* (1997)	X	Spaw expressed in the left lateral plate mesoderm (LPM)	Long *et al.* (2003)
Amphibia	X	Regulates early AP CNS patterning	Kiecker and Niehrs (2001)	X	gradient regulates DV mesodermal and ectodermal patterning	Dosch *et al.* (1997), Jones and Smith (1998), Tribulo *et al.* (2003), Knecht and Harland (1997)	X	Xnr-1 in the left side of the LPM	Lustig *et al.* (1996)
Aves	X	Regulates CNS patterning and anterior mesoderm (heart) specification	Marvin *et al.* (2001)	X	Regulates DV mesodermal patterning	Streit *et al.* (1998), Joubin and Stern (1999)	X	Asymmetric (left) nodal expression and function	Levin (1998), Levin *et al.* (1995)
Mammalia	X	Regulates AP-axis before and during gastrulation	Kelly *et al.* (2000), Huelsken *et al.* (2000), Liu *et al.* (1999)	X	Expression of BMP antagonists expressed in node	Klingensmith *et al.* (1999)	X	Node expression and restricted to the left LPM	Collignon *et al.* (1996), Lowe *et al.* (2006)
Triploblasts/protostomes									
Turbellarians	X	Requirement for β-catenin in AP-axis formation of intact and regenerating planarians (*Schmidtea*)	Gurley *et al.* (2008), Iglesias *et al.* (2008), Petersen and Reddien (2008)	X	Requirement for dorsally expressed dpp/BMP2–4 for DV patterning in intact and regenerating planarians	Reddien *et al.* (2007)	X	Unknown	
Nemerteans	X	Requirement for β-catenin in animal-vegetal embryonic patterning in *Cerebratulus*	Henry *et al.* (2008)	X	Unknown		X	Unknown	

(*continued*)

Table 6.2 *(continued)*

	First axis vector	β-Catenin/Wnt	References	Second axis vector	BMP	References	Third axis vector	Nodal	References
Nematodes	X	Specifies global embryonic AP cell fates in *C. elegans*	Kaletta *et al.* (1997), Lin *et al.* (1998)	X	Unknown		X[b]	Not present	
Annelids	X	β-Catenin asymmetries regulate global animal-vegetal development in *Platynereis*	Schneider and Bowerman (2007)	X	Expressed dorsally in Platynereis and regulates mediolateral genes in neuroectoderm	Denes *et al.* (2007)	X	Unknown	
Molluscs	X	LiCl affects animal vegetal axis in *Loligo*	Crawford (2003)	X	Unknown		X	Expressed on the right side of the embryo in the dextral *Lottia* and on the left side in the sinistral *Biomphalaria*	Grande and Patel (2009b)
Arthropods	X	Regulates AP segment polarity in *Drosophila* and embryonic posterior development in *Gryllus* and *Achaearanea*	Nusslein-Volhard and Wieschaus (1980), Miyawaki *et al.* (2004), McGregor *et al.* (2008)	X	Regulates embryonic DV patterning in *Drosophila* and *Achaearanea*	Akiyama-Oda and Oda (2006), Holley *et al.* (1995)	X	Not present	

[a] Adopted from Niehrs, 2010.
[b] Evidence for noncanonical wnt signalling in *C. elegans* (Pohl and Bao, 2010).

7. Conclusions and Outlook

The broad variety in animal forms appears to be founded on astonishingly few signaling pathways that are shared by all major metazoan phyla. Although the molecular components of these signaling networks can in part be traced down to premetazoan origins, the major wave of innovation is evidently linked to the emergence of multicellularity and the metazoan stem (Fig. 6.3). Cnidarians are distinctive in representing the earliest phylum, in which a functional repertoire of core signaling molecules is present. While in some few cases the ligand and receptor varieties were reduced during evolution, there is a general trend toward an expansion of the existing set of membrane-associated molecule families. We anticipate three lines for future research addressing major unsolved questions:

(i) The origin of protostome and deuterostomes clades is still unclear. This question is tightly coupled to the problem of the origin of the second body axis (DV axis) including the molecular identity of the signaling pathways specifying the additional spatial coordinates. Is there a common bilaterian ancestor, for example, acoel flatworms or is there a deeper split of bilaterian animals with an independent origin of the BMP/Chd-directed DV body axis?

(ii) The fact that the Wnt- and BMP-signaling pathways act along the same body axis in prebilaterian animals suggests that the molecular mechanisms integrating positional information of both pathways are very old. We propose that Wnt signaling and BMP signaling are interacting at the protein level via Smad-1 phosphorylation in prebilaterian animals.

(iii) The function of the major signaling pathways on the level of cell differentiation in prebilaterian animals is largely unknown. We found recently that β-catenin signaling is active in neuronal induction even before BMPs are expressed (Hiroshi Watanabe *et al.*, unpublished). According to this scenario, a primary function of BMP signaling in metazoan evolution might have been the regionalization of the neuronal differentiation, which acts downstream of β-catenin signaling.

It should be also pointed out that also the transcriptional regulation of major signaling pathways is only poorly understood so far. Since these pathways have been repeatedly reused in the context of organ differentiation during embryogenesis of higher animals, the level of *cis*-regulatory elements might provide a clue to understand the increasing complexity of animal body plans on the base of a limited molecular toolkit.

ACKNOWLEDGMENT

Supported by the DFG (FOR 1036).

REFERENCES

Adamska, M., Degnan, S. M., Green, K. M., Adamski, M., Craigie, A., Larroux, C., and Degnan, B. M. (2007). Wnt and TGF-beta expression in the sponge Amphimedon queenslandica and the origin of metazoan embryonic patterning. *PLoS One* **2,** e1031.

Adamska, M., Larroux, C., Adamski, M., Green, K., Lovas, E., Koop, D., Richards, G. S., Zwafink, C., and Degnan, B. M. (2010). Structure and expression of conserved Wnt pathway components in the demosponge Amphimedon queenslandica. *Evol. Dev.* **12,** 494–518.

Adell, T., Salo, E., Boutros, M., and Bartscherer, K. (2009). Smed-Evi/Wntless is required for beta-catenin-dependent and -independent processes during planarian regeneration. *Development* **136,** 905–910.

Adell, T., Cebrià, F., and Saló, E. (2010). Gradients in planarian regeneration and homeostasis. *Cold Spring Harb. Perspect. Biol.* **2,** a000505.

Agathon, A., Thisse, B., and Thisse, C. (2001). Morpholino knock-down of antivin1 and antivin2 upregulates nodal signaling. *Genesis* **30,** 178–182.

Agathon, A., Thisse, C., and Thisse, B. (2003). The molecular nature of the zebrafish tail organizer. *Nature* **424,** 448–452.

Akiyama-Oda, Y., and Oda, H. (2006). Axis specification in the spider embryo: dpp is required for radial-to-axial symmetry transformation and sog for ventral patterning. *Development* **133,** 2347–2357.

Amano, S., and Hori, I. (2001). Metamorphosis of coeloblastula performed by multipotential larval flagellated cells in the calcareous sponge *Leucosolenia laxa. Biol. Bull.* **200,** 20–32.

Arendt, D., and Nubler-Jung, K. (1994). Inversion of dorsoventral axis? *Nature* **371,** 26.

Arendt, D., and Nubler-Jung, K. (1997). Dorsal or ventral: Similarities in fate maps and gastrulation patterns in annelids, arthropods and chordates. *Mech. Dev.* **61,** 7–21.

Arendt, D., Technau, U., and Wittbrodt, J. (2001). Evolution of the bilaterian larval foregut. *Nature* **409,** 81–85.

Augustin, R., Franke, A., Khalturin, K., Kiko, R., Siebert, S., Hemmrich, G., and Bosch, T. C. (2006). Dickkopf related genes are components of the positional value gradient in *Hydra. Dev. Biol.* **296,** 62–70.

Aulehla, A., Wehrle, C., Brand-Saberi, B., Kemler, R., Gossler, A., Kanzler, B., and Herrmann, B. G. (2003). Wnt3a plays a major role in the segmentation clock controlling somitogenesis. *Dev. Cell* **4,** 395–406.

Baguna, J., and Riutort, M. (2004). The dawn of bilaterian animals: The case of acoelomorph flatworms. *Bioessays* **26,** 1046–1057.

Bier, E. (1997). Anti-neural-inhibition: A conserved mechanism for neural induction. *Cell* **89,** 681–684.

Bode, H. R. (2003). Head regeneration in *Hydra. Dev. Dyn.* **226,** 225–236.

Bode, P. M., and Bode, H. R. (1980). Formation of pattern in regenerating tissue pieces of *Hydra attenuata* I. Head-body proportion regulation. *Dev. Biol.* **78,** 484–496.

Bode, M. P., and Bode, H. R. (1984). Formation of pattern in regenerating tissue pieces of *Hydra attenuata* III. The shaping of the body column. *Dev. Biol.* **106,** 315–325.

Borojevic, R. (1967). La ponte el le developpement de Polymastia robusta (Demosponges). *Cah. Biol. Mar.* **8,** 1–6.

Borojevic, R., and Levi, C. (1965). Morphogenese experimentale d'une eponge a partir de cellules de la larve nageante dissociee. *Z. Zellforsch. Mikrosk. Anat.* **68,** 57–69.

Brien, P. (1967). Les Eponges. Leur nature metazoaire[long dash]leur gastrulation[long dash] leur etat colonial. *Ann. Soc. R. Zool. Belg.* **97,** 197–235.
Broun, M., and Bode, H. R. (2002). Characterization of the head organizer in hydra. *Development* **129,** 875–884.
Broun, M., Gee, L., Reinhardt, B., and Bode, H. R. (2005). Formation of the head organizer in hydra involves the canonical Wnt pathway. *Development* **132,** 2907–2916.
Browne, E. N. (1909). The production of new hydranths in hydra by the insertion of small grafts. *J. Exp. Zool.* **7,** 1–37.
Brusca, R. C., and Brusca, G. J. (2002). Invertebrates. 2nd edn. Sinauer, Sunderland, MA.
Bütschli, O. (1884). Bemerkungen zur Gastreatheorie. *Morpholog. Jahrb.* **9,** 415–427.
Chapman, J. A., Kirkness, E. F., Simakov, O., Hampson, S. E., Mitros, T., Weinmaier, T., Rattei, T., Balasubramanian, P. G., Borman, J., Busam, D., Disbennett, K., Pfannkoch, C., *et al.* (2010). The dynamic genome of *Hydra attenuata*. *Nature* **464,** 592–596.
Chea, H. K., Wright, C. V., and Swalla, B. J. (2005). Nodal signaling and the evolution of deuterostome gastrulation. *Dev. Dyn.* **234,** 269–278.
Chen, C. A., Wallace, C. C., and Wolstenholme, J. (2002). Analysis of the mitochondrial 12S rRNA gene supports a two-clade hypothesis of the evolutionary history of scleractinian corals. *Mol. Phylogenet. Evol.* **23,** 137–149.
Ching, W., Hang, H. C., and Nusse, R. (2008). Lipid-independent secretion of a Drosophila Wnt protein. *J. Biol. Chem.* **283,** 17092–17098.
Coates, J. C. (2003). Armadillo repeat proteins: Beyond the animal kingdom. *Trends Cell Biol.* **13,** 463–471.
Collignon, J., Varlet, I., and Robertson, E. J. (1996). Relationship between asymmetric nodal expression and the direction of embryonic turning. *Nature* **381,** 155–158.
Collins, A. G. (1998). Evaluating multiple alternative hypotheses for the origin of Bilateria: An analysis of 18S rRNA molecular evidence. *Proc. Natl. Acad. Sci. USA* **95,** 15458–15463.
Conlon, F. L., Barth, K. S., and Robertson, E. J. (1991). A novel retrovirally induced embryonic lethal mutation in the mouse: Assessment of the developmental fate of embryonic stem cells homozygous for the 413.d proviral integration. *Development* **111,** 969–981.
Conway Morris, S. (1998). The question of metazoan monophyly and the fossil record. *Prog. Mol. Subcell. Biol.* **21,** 1–19.
Conway Morris, S. (2000). The Cambrian "explosion": Slow-fuse or megatonnage? *Proc. Natl. Acad. Sci. USA* **97,** 4426–4429.
Cramer von Laue, C. (2003). Analysis of the dual function of beta-Catenin in Wnt signaling and in Cadherin-mediated cell adhesion of *Hydra attenuata*. Department of Biology, Darmstadt University of Technology, Darmstadtp. 97.
Crawford, K. (2003). Lithium chloride inhibits development along the animal vegetal axis and anterior midline of the squid embryo. *Biol. Bull.* **205,** 181–182.
Darling, J. A., Reitzel, A. R., Burton, P. M., Mazza, M. E., Ryan, J. F., Sullivan, J. C., and Finnerty, J. R. (2005). Rising starlet: The starlet sea anemone, Nematostella vectensis. *Bioessays* **27,** 211–221.
Darras, S., Gerhart, J., Terasaki, M., Kirschner, M., and Lowe, C. J. (2011). beta-Catenin specifies the endomesoderm and defines the posterior organizer of the hemichordate Saccoglossus kowalevskii. *Development* **138,** 959–970.
Davidson, G., Mao, B., del Barco Barrantes, I., and Niehrs, C. (2002). Kremen proteins interact with Dickkopf1 to regulate anteroposterior CNS patterning. *Development* **129,** 5587–5596.
De Robertis, E. M. (2010). Wnt signaling in axial patterning and regeneration: Lessons from planaria. *Sci. Signal.* **3,** pe21.

De Robertis, E. M., and Sasai, Y. (1996). A common plan for dorsoventral patterning in Bilateria. *Nature* **380,** 37–40.

Dellaporta, S. L., Xu, A., Sagasser, S., Jakob, W., Moreno, M. A., Buss, L. W., and Schierwater, B. (2006). Mitochondrial genome of *Trichoplax adhaerens* supports placozoa as the basal lower metazoan phylum. *Proc. Natl. Acad. Sci. USA* **103,** 8751–8756.

Denes, A. S., Jekely, G., Steinmetz, P. R., Raible, F., Snyman, H., Prud'homme, B., Ferrier, D. E., Balavoine, G., and Arendt, D. (2007). Molecular architecture of annelid nerve cord supports common origin of nervous system centralization in bilateria. *Cell* **129,** 277–288.

Dosch, R., Gawantka, V., Delius, H., Blumenstock, C., and Niehrs, C. (1997). Bmp-4 acts as a morphogen in dorsoventral mesoderm patterning in Xenopus. *Development* **124,** 2325–2334.

Duboc, V., and Lepage, T. (2008). A conserved role for the nodal signaling pathway in the establishment of dorso-ventral and left-right axes in deuterostomes. *J. Exp. Zool. B Mol. Dev. Evol.* **310,** 41–53.

Duboc, V., Rottinger, E., Besnardeau, L., and Lepage, T. (2004). Nodal and BMP2/4 signaling organizes the oral-aboral axis of the sea urchin embryo. *Dev. Cell* **6,** 397–410.

Duboc, V., Lapraz, F., Besnardeau, L., and Lepage, T. (2008). Lefty acts as an essential modulator of Nodal activity during sea urchin oral-aboral axis formation. *Dev. Biol.* **320,** 49–59.

Dunn, C. W., Hejnol, A., Matus, D. Q., Pang, K., Browne, W. E., Smith, S. A., Seaver, E., Rouse, G. W., Obst, M., Edgecombe, G. D., Sorensen, M. V., Haddock, S. H., *et al.* (2008). Broad phylogenomic sampling improves resolution of the animal tree of life. *Nature.*

Dunty, W. C., Jr., Biris, K. K., Chalamalasetty, R. B., Taketo, M. M., Lewandoski, M., and Yamaguchi, T. P. (2008). Wnt3a/beta-catenin signaling controls posterior body development by coordinating mesoderm formation and segmentation. *Development* **135,** 85–94.

Erter, C. E., Wilm, T. P., Basler, N., Wright, C. V., and Solnica-Krezel, L. (2001). Wnt8 is required in lateral mesendodermal precursors for neural posteriorization in vivo. *Development* **128,** 3571–3583.

Extavour, C. G. (2005). The fate of isolated blastomeres with respect to germ cell formation in the amphipod crustacean *Parhyale hawaiensis. Dev. Biol.* **277,** 387–402.

Fedders, H., Augustin, R., and Bosch, T. C. (2004). A Dickkopf- 3-related gene is expressed in differentiating nematocytes in the basal metazoan *Hydra attenuata. Dev. Genes Evol.* **214,** 72–80.

Finnerty, J. R., Pang, K., Burton, P., Paulson, D., and Martindale, M. Q. (2004). Origins of bilateral symmetry: Hox and dpp expression in a sea anemone. *Science* **304,** 1335–1337.

Flowers, V. L., Courteau, G. R., Poustka, A. J., Weng, W., and Venuti, J. M. (2004). Nodal/activin signaling establishes oral-aboral polarity in the early sea urchin embryo. *Dev. Dyn.* **231,** 727–740.

Fritzenwanker, J. H., and Technau, U. (2002). Induction of gametogenesis in the basal cnidarian Nematostella vectensis (Anthozoa). *Dev. Genes Evol.* **212,** 99–103.

Fritzenwanker, J. H., Saina, M., and Technau, U. (2004). Analysis of forkhead and snail expression reveals epithelial-mesenchymal transitions during embryonic and larval development of Nematostella vectensis. *Dev. Biol.* **275,** 389–402.

Fritzenwanker, J. H., Genikhovich, G., Kraus, Y., and Technau, U. (2007). Early development and axis specification in the sea anemone Nematostella vectensis. *Dev. Biol.* **310,** 264–279.

Garcia Abreu, J., Coffinier, C., Larrain, J., Oelgeschlager, M., and De Robertis, E. M. (2002). Chordin-like CR domains and the regulation of evolutionarily conserved extracellular signaling systems. *Gene* **287,** 39–47.

Geoffroy St-Hilaire, E. (1822). Considérations générales sur la vertèbre. *Mém. Mus. Hist. Nat.* **9,** 89–119.

Gierer, A., Berking, S., Bode, H., David, C. N., Flick, K., Hansmann, G., Schaller, H., and Trenkner, E. (1972). Regeneration of hydra from reaggregated cells. *Nat. New Biol.* **239,** 98–101.

Grande, C., and Patel, N. H. (2009a). Lophotrochozoa get into the game: The nodal pathway and left/right asymmetry in bilateria. *Cold Spring Harb. Symp. Quant. Biol.* **74,** 281–287.

Grande, C., and Patel, N. H. (2009b). Nodal signalling is involved in left-right asymmetry in snails. *Nature* **457,** 1007–1011.

Grell, K. G., and Benwitz, G. (1971). Die Ultrastruktur von Trichoplax adhaerens F.E. Schulze. *Cytobiologie* **4,** 216–240.

Grimson, M. J., Coates, J. C., Reynolds, J. P., Shipman, M., Blanton, R. L., and Harwood, A. J. (2000). Adherens junctions and beta-catenin-mediated cell signalling in a non-metazoan organism. *Nature* **408,** 727–731.

Grobben, K. (1908). Die systematische Einteilung des Tierreichs. *Verh. Zool. Bot. Ges. Wien.* **58,** 491–511.

Guder, C., Philipp, I., Lengfeld, T., Watanabe, H., Hobmayer, B., and Holstein, T. W. (2006a). The Wnt code: Cnidarians signal the way. *Oncogene* **25,** 7450–7460.

Guder, C., Pinho, S., Nacak, T. G., Schmidt, H. A., Hobmayer, B., Niehrs, C., and Holstein, T. W. (2006b). An ancient Wnt-Dickkopf antagonism in *Hydra. Development* **133,** 901–911.

Gurley, K. A., Rink, J. C., and Sanchez Alvarado, A. (2008). Beta-catenin defines head versus tail identity during planarian regeneration and homeostasis. *Science* **319,** 323–327.

Hampe, W., Urny, J., Franke, I., Hoffmeister-Ullerich, S. A. H., Herrmann, D., Petersen, C. M., Lohmann, J., and Schaller, H. C. (1999). A head-activator binding protein is present in hydra in a soluble and a membrane-anchored form. *Development* **126,** 4077–4086, hampe@uke.uni-hamburg.de.

Hayward, D. C., Samuel, G., Pontynen, P. C., Catmull, J., Saint, R., Miller, D. J., and Ball, E. E. (2002). Localized expression of a dpp/BMP2/4 ortholog in a coral embryo. *Proc. Natl. Acad. Sci. USA* **99,** 8106–8111.

Hayward, D. C., Miller, D. J., and Ball, E. E. (2004). Snail expression during embryonic development of the coral Acropora: Blurring the diploblast/triploblast divide? *Dev. Genes Evol.* **214,** 257–260.

Hejnol, A., and Martindale, M. Q. (2009). Coordinated spatial and temporal expression of Hox genes during embryogenesis in the acoel Convolutriloba longifissura. *BMC Biol.* **7,** 65.

Heldin, C. H., Landstrom, M., and Moustakas, A. (2009). Mechanism of TGF-beta signaling to growth arrest, apoptosis, and epithelial-mesenchymal transition. *Curr. Opin. Cell Biol.* **21,** 166–176.

Henry, J. Q., Perry, K. J., Wever, J., Seaver, E., and Martindale, M. Q. (2008). Beta-catenin is required for the establishment of vegetal embryonic fates in the nemertean, *Cerebratulus lacteus. Dev. Biol.* **317,** 368–379.

Heuberger, J., and Birchmeier, W. (2010). Interplay of cadherin-mediated cell adhesion and canonical Wnt signaling. *Cold Spring Harb. Perspect. Biol.* **2,** a002915.

Hobmayer, E., Hatta, M., Fischer, R., Fujisawa, T., Holstein, T. W., and Sugiyama, T. (1996). Identification of a *Hydra* homologue of the beta-catenin/plakoglobin/armadillo gene family. *Gene* **172,** 155–159.

Hobmayer, B., Rentzsch, F., Kuhn, K., Happel, C. M., Cramer von Laue, C., Snyder, P., Rothbacher, U., and Holstein, T. W. (2000a). WNT signalling molecules act in axis formation in the diploblastic metazoan *Hydra. Nature (London)* **407,** 186–189.

Hobmayer, B., Rentzsch, F., Kuhn, K., Happel, C. M., von Laue, C. C., Snyder, P., Rothbacher, U., and Holstein, T. W. (2000b). WNT signalling molecules act in axis formation in the diploblastic metazoan *Hydra. Nature* **407,** 186–189.

Holland, L. Z. (2002). Heads or tails? Amphioxus and the evolution of anterior-posterior patterning in deuterostomes. *Dev. Biol.* **241,** 209–228.

Holland, P. (2004). Developmental biology. The ups and downs of a sea anemone. *Science* **304,** 1255–1256.

Holley, S. A., Jackson, P. D., Sasai, Y., Lu, B., De Robertis, E. M., Hoffmann, F. M., and Ferguson, E. L. (1995). A conserved system for dorsal-ventral patterning in insects and vertebrates involving sog and chordin. *Nature* **376,** 249–253.

Holstein, T. W., Hobmayer, E., and Technau, U. (2003). Cnidarians: An evolutionarily conserved model system for regeneration? *Dev. Dyn.* **226,** 257–267.

Huelsken, J., Vogel, R., Brinkmann, V., Erdmann, B., Birchmeier, C., and Birchmeier, W. (2000). Requirement for beta-catenin in anterior-posterior axis formation in mice. *J. Cell Biol.* **148,** 567–578.

Hyman, H. L. (1940). The invertebrates. Protozoa Through Ctenophora. Vol. 1. McGraw-Hill, New York, London.

Iglesias, M., Gomez-Skarmeta, J. L., Salo, E., and Adell, T. (2008). Silencing of Smed-betacatenin1 generates radial-like hypercephalized planarians. *Development* **135,** 1215–1221.

Imai, K., Takada, N., Satoh, N., and Satou, Y. (2000). (beta)-Catenin mediates the specification of endoderm cells in ascidian embryos. *Development* **127,** 3009–3020.

Ivanov, A. (1973). Trichoplax adherens—A phygocytella-like animal. *Zool. Zhurnal* **52,** 1117–1131.

Jägersten, G. (1955). On the early phylogeny of the Metazoa: The bilatero-gastraea theory. *Zool. Bidr. Uppsala* **30,** 321–354.

Jones, S. E., and Jomary, C. (2002). Secreted Frizzled-related proteins: Searching for relationships and patterns. *Bioessays* **24,** 811–820.

Jones, C. M., and Smith, J. C. (1998). Establishment of a BMP-4 morphogen gradient by long-range inhibition. *Dev. Biol.* **194,** 12–17.

Joubin, K., and Stern, C. D. (1999). Molecular interactions continuously define the organizer during the cell movements of gastrulation. *Cell* **98,** 559–571.

Kaletta, T., Schnabel, H., and Schnabel, R. (1997). Binary specification of the embryonic lineage in *Caenorhabditis elegans*. *Nature* **390,** 294–298.

Kawano, Y., and Kypta, R. (2003). Secreted antagonists of the Wnt signalling pathway. *J. Cell Sci.* **116,** 2627–2634.

Kelly, C., Chin, A. J., Leatherman, J. L., Kozlowski, D. J., and Weinberg, E. S. (2000). Maternally controlled (beta)-catenin-mediated signaling is required for organizer formation in the zebrafish. *Development* **127,** 3899–3911.

Kiecker, C., and Niehrs, C. (2001). A morphogen gradient of Wnt/beta-catenin signalling regulates anteroposterior neural patterning in Xenopus. *Development* **128,** 4189–4201.

Kim, C. H., Oda, T., Itoh, M., Jiang, D., Artinger, K. B., Chandrasekharappa, S. C., Driever, W., and Chitnis, A. B. (2000). Repressor activity of Headless/Tcf3 is essential for vertebrate head formation. *Nature* **407,** 913–916.

King, N., Hittinger, C. T., and Carroll, S. B. (2003). Evolution of key cell signaling and adhesion protein families predates animal origins. *Science* **301,** 361–363.

King, N., Westbrook, M. J., Young, S. L., Kuo, A., Abedin, M., Chapman, J., Fairclough, S., Hellsten, U., Isogai, Y., Letunic, I., Marr, M., Pincus, D., *et al.* (2008). The genome of the choanoflagellate Monosiga brevicollis and the origin of metazoans. *Nature* **451,** 783–788.

Klingensmith, J., Ang, S. L., Bachiller, D., and Rossant, J. (1999). Neural induction and patterning in the mouse in the absence of the node and its derivatives. *Dev. Biol.* **216,** 535–549.

Knecht, A. K., and Harland, R. M. (1997). Mechanisms of dorsal-ventral patterning in noggin-induced neural tissue. *Development* **124,** 2477–2488.

Kraus, Y., and Technau, U. (2006). Gastrulation in the sea anemone Nematostella vectensis occurs by invagination and immigration: An ultrastructural study. *Dev. Genes Evol.* **216,** 119–132.
Kusserow, A., Pang, K., Sturm, C., Hrouda, M., Lentfer, J., Schmidt, H. A., Technau, U., von Haeseler, A., Hobmayer, B., Martindale, M. Q., and Holstein, T. W. (2005). Unexpected complexity of the Wnt gene family in a sea anemone. *Nature* **433,** 156–160.
Lapebie, P., Gazave, E., Ereskovsky, A., Derelle, R., Bezac, C., Renard, E., Houliston, E., and Borchiellini, C. (2009). WNT/beta-catenin signalling and epithelial patterning in the homoscleromorph sponge Oscarella. *PLoS One* **4,** e5823.
Lee, H. C., Kim, M., and Wands, J. R. (2006a). Wnt/Frizzled signaling in hepatocellular carcinoma. *Front. Biosci.* **11,** 1901–1915.
Lee, P. N., Pang, K., Matus, D. Q., and Martindale, M. Q. (2006b). A WNT of things to come: Evolution of Wnt signaling and polarity in cnidarians. *Semin. Cell Dev. Biol.* **17,** 157–167.
Lee, P. N., Kumburegama, S., Marlow, H. Q., Martindale, M. Q., and Wikramanayake, A. H. (2007). Asymmetric developmental potential along the animal-vegetal axis in the anthozoan cnidarian, *Nematostella vectensis*, is mediated by Dishevelled. *Dev. Biol.* **310,** 169–186.
Lengfeld, T. (2009). Der Wnt/ß-Cateninsignalweg und die axiale Musterbildung im Süßwasserpolyp *Hydra*. *In* Molecular Evolution and Genomics Heidelberg University, Heidelbergp. 200.
Lengfeld, T., Watanabe, H., Simakov, O., Lindgens, D., Gee, L., Law, L., Schmidt, H. A., Ozbek, S., Bode, H., and Holstein, T. W. (2009). Multiple Wnts are involved in *Hydra* organizer formation and regeneration. *Dev. Biol.* **330,** 186–199.
Levin, M. (1998). Left-right asymmetry and the chick embryo. *Semin. Cell Dev. Biol.* **9,** 67–76.
Levin, M., Johnson, R. L., Stern, C. D., Kuehn, M., and Tabin, C. (1995). A molecular pathway determining left-right asymmetry in chick embryogenesis. *Cell* **82,** 803–814.
Leys, S. P. (2004). Gastrulation in sponges. *In* "Gastrulation," (C. D. Stern, ed.), pp. 23–31. Cold Spring Harbor Laboratory Press, Cold Spring Harbor, New York.
Lie, D. C., Colamarino, S. A., Song, H. J., Desire, L., Mira, H., Consiglio, A., Lein, E. S., Jessberger, S., Lansford, H., Dearie, A. R., and Gage, F. H. (2005). Wnt signalling regulates adult hippocampal neurogenesis. *Nature* **437,** 1370–1375.
Lin, R., Hill, R. J., and Priess, J. R. (1998). POP-1 and anterior-posterior fate decisions in *C. elegans* embryos. *Cell* **92,** 229–239.
Liu, P., Wakamiya, M., Shea, M. J., Albrecht, U., Behringer, R. R., and Bradley, A. (1999). Requirement for Wnt3 in vertebrate axis formation. *Nat. Genet.* **22,** 361–365.
Logan, C. Y., and Nusse, R. (2004). The Wnt signaling pathway in development and disease. *Annu. Rev. Cell Dev. Biol.* **20,** 781–810.
Logan, C. Y., Miller, J. R., Ferkowicz, M. J., and McClay, D. R. (1999). Nuclear beta-catenin is required to specify vegetal cell fates in the sea urchin embryo. *Development* **126,** 345–357.
Long, S., Ahmad, N., and Rebagliati, M. (2003). The zebrafish nodal-related gene southpaw is required for visceral and diencephalic left-right asymmetry. *Development* **130,** 2303–2316.
Lowe, C. J., Terasaki, M., Wu, M., Freeman, R. M., Jr., Runft, L., Kwan, K., Haigo, S., Aronowicz, J., Lander, E., Gruber, C., Smith, M., Kirschner, M., *et al.* (2006). Dorsoventral patterning in hemichordates: Insights into early chordate evolution. *PLoS Biol.* **4,** e291.
Lustig, K. D., Kroll, K., Sun, E., Ramos, R., Elmendorf, H., and Kirschner, M. W. (1996). A Xenopus nodal-related gene that acts in synergy with noggin to induce complete secondary axis and notochord formation. *Development* **122,** 3275–3282.
MacWilliams, H. K. (1983a). Hydra transplantation phenomena and the mechanism of hydra head regeneration. I. Properties of the head inhibition. *Dev. Biol.* **96,** 217–238.

MacWilliams, H. K. (1983b). Hydra transplantation phenomena and the mechanism of Hydra head regeneration. II. Properties of the head activation. *Dev. Biol.* **96,** 239–257.
Magie, C. R., Pang, K., and Martindale, M. Q. (2005). Genomic inventory and expression of Sox and Fox genes in the cnidarian *Nematostella vectensis*. *Dev. Genes Evol.* **215,** 618–630.
Manning, G., Young, S. L., Miller, W. T., and Zhai, Y. (2008). The protist, Monosiga brevicollis, has a tyrosine kinase signaling network more elaborate and diverse than found in any known metazoan. *Proc. Natl. Acad. Sci. USA* **105,** 9674–9679.
Manuel, M. (2009). Early evolution of symmetry and polarity in metazoan body plans. *C. R. Biol.* **332,** 184–209.
Mao, B., and Niehrs, C. (2003). Kremen2 modulates Dickkopf2 activity during Wnt/LRP6 signaling. *Gene* **302,** 179–183.
Martin, B. L., and Kimelman, D. (2009). Wnt signaling and the evolution of embryonic posterior development. *Curr. Biol.* **19,** R215–R219.
Martindale, D. (2004). Muscle twitch switch. *Sci. Am.* **291**(22), 24.
Martindale, M. Q., and Hejnol, A. (2009). A developmental perspective: Changes in the position of the blastopore during bilaterian evolution. *Dev. Cell* **17,** 162–174.
Martindale, M. Q., Pang, K., and Finnerty, J. R. (2004). Investigating the origins of triploblasty: 'Mesodermal' gene expression in a diploblastic animal, the sea anemone Nematostella vectensis (phylum, Cnidaria; class, Anthozoa). *Development* **131,** 2463–2474.
Marvin, M. J., Di Rocco, G., Gardiner, A., Bush, S. M., and Lassar, A. B. (2001). Inhibition of Wnt activity induces heart formation from posterior mesoderm. *Genes Dev.* **15,** 316–327.
Massague, J. (1998). TGF-beta signal transduction. *Annu. Rev. Biochem.* **67,** 753–791.
Massague, J., and Wotton, D. (2000). Transcriptional control by the TGF-beta/Smad signaling system. *EMBO J.* **19,** 1745–1754.
Matus, D. Q., Pang, K., Marlow, H., Dunn, C. W., Thomsen, G. H., and Martindale, M. Q. (2006a). Molecular evidence for deep evolutionary roots of bilaterality in animal development. *Proc. Natl. Acad. Sci. USA* **103,** 11195–11200.
Matus, D. Q., Thomsen, G. H., and Martindale, M. Q. (2006b). Dorso/ventral genes are asymmetrically expressed and involved in germ-layer demarcation during cnidarian gastrulation. *Curr. Biol.* **16,** 499–505.
McGregor, A. P., Pechmann, M., Schwager, E. E., Feitosa, N. M., Kruck, S., Aranda, M., and Damen, W. G. (2008). Wnt8 is required for growth-zone establishment and development of opisthosomal segments in a spider. *Curr. Biol.* **18,** 1619–1623.
Meinhardt, H. (2002). The radial-symmetric hydra and the evolution of the bilateral body plan: An old body became a young brain. *Bioessays* **24,** 185–191.
Meinhardt, H. (2004). Models for the generation of the embryonic body axes: Ontogenetic and evolutionary aspects. *Curr. Opin. Genet. Dev.* **14,** 446–454.
Meinhardt, H. (2006). Primary body axes of vertebrates: Generation of a near-Cartesian coordinate system and the role of Spemann-type organizer. *Dev. Dyn.* **235,** 2907–2919.
Miller, D. J., Ball, E. E., and Technau, U. (2005). Cnidarians and ancestral genetic complexity in the animal kingdom. *Trends Genet.* **21,** 536–539.
Minobe, S., Fei, K., Yan, L., Sarras, M., Jr., and Werle, M. (2000). Identification and characterization of the epithelial polarity receptor "Frizzled" in Hydra vulgaris. *Dev. Genes Evol.* **210,** 258–262.
Miyawaki, A. (2003). Fluorescent proteins found in cnidarian animals. *Tanpakushitsu Kakusan Koso* **48,** 1568–1572.
Miyawaki, K., Mito, T., Sarashina, I., Zhang, H., Shinmyo, Y., Ohuchi, H., and Noji, S. (2004). Involvement of Wingless/Armadillo signaling in the posterior sequential segmentation in the cricket, Gryllus bimaculatus (Orthoptera), as revealed by RNAi analysis. *Mech. Dev.* **121,** 119–130.

Mizutani, C. M., Meyer, N., Roelink, H., and Bier, E. (2006). Threshold-dependent BMP-mediated repression: A model for a conserved mechanism that patterns the neuroectoderm. *PLoS Biol.* **4,** e313.

Molina, M. D., Salo, E., and Cebria, F. (2007). The BMP pathway is essential for re-specification and maintenance of the dorsoventral axis in regenerating and intact planarians. *Dev. Biol.* **311,** 79–94.

Momose, T., and Houliston, E. (2007). Two oppositely localised frizzled RNAs as axis determinants in a cnidarian embryo. *PLoS Biol.* **5,** e70.

Momose, T., and Schmid, V. (2006). Animal pole determinants define oral-aboral axis polarity and endodermal cell-fate in hydrozoan jellyfish *Podocoryne carnea. Dev. Biol.* **292,** 371–380.

Momose, T., Derelle, R., and Houliston, E. (2008). A maternally localised Wnt ligand required for axial patterning in the cnidarian Clytia hemisphaerica. *Development* **135,** 2105–2113.

Morokuma, J., Ueno, M., Kawanishi, H., Saiga, H., and Nishida, H. (2002). HrNodal, the ascidian nodal-related gene, is expressed in the left side of the epidermis, and lies upstream of HrPitx. *Dev. Genes Evol.* **212,** 439–446.

Moustakas, A., and Heldin, C. H. (2009). The regulation of TGFbeta signal transduction. *Development* **136,** 3699–3714.

Muller, W., Frank, U., Teo, R., Mokady, O., Guette, C., and Plickert, G. (2007). Wnt signaling in hydroid development: Ectopic heads and giant buds induced by GSK-3beta inhibitors. *Int. J. Dev. Biol.* **51,** 211–220.

Nakamura, Yukio, Tsiairis, Charisios D., Özbek, Suat, and Holstein, Thomas W. (2011). Autoregulatory and repressive inputs localize *Hydra* Wnt3 to the head organizer. *Proc. Natl. Acad. Sci. USA* **108**(22), 9137–9142.

Neave, B., Holder, N., and Patient, R. (1997). A graded response to BMP-4 spatially coordinates patterning of the mesoderm and ectoderm in the zebrafish. *Mech. Dev.* **62,** 183–195.

Nelson, W. J., and Nusse, R. (2004). Convergence of Wnt, beta-catenin, and cadherin pathways. *Science* **303,** 1483–1487.

Nichols, S. A., Dirks, W., Pearse, J. S., and King, N. (2006). Early evolution of animal cell signaling and adhesion genes. *Proc. Natl. Acad. Sci. USA* **103,** 12451–12456.

Niehrs, C. (2001). The Spemann organizer and embryonic head induction. *EMBO J.* **20,** 631–637.

Niehrs, C. (2004). Regionally specific induction by the Spemann-Mangold organizer. *Nat. Rev. Genet.* **5,** 425–434.

Niehrs, C. (2010). On growth and form: A Cartesian coordinate system of Wnt and BMP signaling specifies bilaterian body axes. *Development* **137,** 845–857.

Nielsen, C. (2001). Animal Evolution: Interrelationships of the Living Phyla. 2nd edn. Oxford University Press, Oxford, pp563.

Nielsen, C. (2008). Six major steps in animal evolution: Are we derived sponge larvae? *Evol. Dev.* **10**(2), 241–257.

Nusslein-Volhard, C., and Wieschaus, E. (1980). Mutations affecting segment number and polarity in *Drosophila. Nature* **287,** 795–801.

Onai, T., Sasai, N., Matsui, M., and Sasai, Y. (2004). Xenopus XsalF: Anterior neuroectodermal specification by attenuating cellular responsiveness to Wnt signaling. *Dev. Cell* **7,** 95–106.

Onai, T., Lin, H. C., Schubert, M., Koop, D., Osborne, P. W., Alvarez, S., Alvarez, R., Holland, N. D., and Holland, L. Z. (2009). Retinoic acid and Wnt/beta-catenin have complementary roles in anterior/posterior patterning embryos of the basal chordate amphioxus. *Dev. Biol.* **332,** 223–233.

Pang, K., and Martindale, M. Q. (2008). Comb jellies (ctenophora): A model for Basal metazoan evolution and development. *CSH Protoc. pdb emo106.*

Pang, K., Ryan, J. F., Mullikin, J. C., Baxevanis, A. D., and Martindale, M. Q. (2010). Genomic insights into Wnt signaling in an early diverging metazoan, the ctenophore Mnemiopsis leidyi. *Evodevo* **1,** 10.

Petersen, C. P., and Reddien, P. W. (2008). Smed-betacatenin-1 is required for anteroposterior blastema polarity in planarian regeneration. *Science* **319,** 327–330.

Petersen, C. P., and Reddien, P. W. (2009). Wnt signaling and the polarity of the primary body axis. *Cell* **139,** 1056–1068.

Philipp, I., Holstein, T. W., and Hobmayer, B. (2005). HvJNK, a *Hydra* member of the c-Jun NH2-terminal kinase gene family, is expressed during nematocyte differentiation. *Gene Expr. Patterns* **5,** 397–402.

Philipp, I., Aufschnaiter, R., Ozbek, S., Pontasch, S., Jenewein, M., Watanabe, H., Rentzsch, F., Holstein, T. W., and Hobmayer, B. (2009). Wnt/beta-catenin and noncanonical Wnt signaling interact in tissue evagination in the simple eumetazoan *Hydra*. *Proc. Natl. Acad. Sci. USA* **106,** 4290–4295.

Pick, K. S., Philippe, H., Schreiber, F., Erpenbeck, D., Jackson, D. J., Wrede, P., Wiens, M., Alie, A., Morgenstern, B., Manuel, M., and Worheide, G. (2010). Improved phylogenomic taxon sampling noticeably affects nonbilaterian relationships. *Mol. Biol. Evol.* **27,** 1983–1987.

Plickert, G., Jacoby, V., Frank, U., Muller, W., and Mokady, O. (2006). Wnt signaling in hydroid development: Formation of the primary body axis in embryogenesis and its subsequent patterning. *Dev. Biol.* **298,** 368–378.

Pohl, C., and Bao, Z. (2010). Chiral forces organize left-right patterning in *C. elegans* by uncoupling midline and anteroposterior axis. *Dev. Cell* **19,** 402–412.

Price, M. A. (2006). CKI, there's more than one: Casein kinase I family members in Wnt and Hedgehog signaling. *Genes Dev.* **20,** 399–410.

Putnam, N. H., Srivastava, M., Hellsten, U., Dirks, B., Chapman, J., Salamov, A., Terry, A., Shapiro, H., Lindquist, E., Kapitonov, V. V., Jurka, J., Genikhovich, G., *et al.* (2007). Sea anemone genome reveals ancestral eumetazoan gene repertoire and genomic organization. *Science* **317,** 86–94.

Reddien, P. W., and Sanchez Alvarado, A. (2004). Fundamentals of planarian regeneration. *Annu. Rev. Cell Dev. Biol.* **20,** 725–757.

Reddien, P. W., Bermange, A. L., Kicza, A. M., and Sanchez Alvarado, A. (2007). BMP signaling regulates the dorsal planarian midline and is needed for asymmetric regeneration. *Development* **134,** 4043–4051.

Remane, A. (1950). Die Entstehung der Metamerie der Wirbellosen. *Zool. Anz.* **14,** 18–23.

Renfer, E., Amon-Hassenzahl, A., Steinmetz, P. R., and Technau, U. (2010). A muscle-specific transgenic reporter line of the sea anemone, Nematostella vectensis. *Proc. Natl. Acad. Sci. USA* **107,** 104–108.

Rentzsch, F., Hobmayer, B., and Holstein, T. W. (2005). Glycogen synthase kinase 3 has a proapoptotic function in *Hydra* gametogenesis. *Dev. Biol.* **278,** 1–12.

Rentzsch, F., Anton, R., Saina, M., Hammerschmidt, M., Holstein, T. W., and Technau, U. (2006). Asymmetric expression of the BMP antagonists chordin and gremlin in the sea anemone Nematostella vectensis: Implications for the evolution of axial patterning. *Dev. Biol.* **296,** 375–387.

Rentzsch, F., Guder, C., Vocke, D., Hobmayer, B., and Holstein, T. W. (2007). An ancient chordin-like gene in organizer formation of *Hydra*. *Proc. Natl. Acad. Sci. USA* **104,** 3249–3254.

Rokas, A., Kruger, D., and Carroll, S. B. (2005). Animal evolution and the molecular signature of radiations compressed in time. *Science* **310,** 1933–1938.

Ruiz-Trillo, I., Riutort, M., Littlewood, D. T., Herniou, E. A., and Baguna, J. (1999). Acoel flatworms: Earliest extant bilaterian Metazoans, not members of Platyhelminthes. *Science* **283,** 1919–1923.

Ruiz-Trillo, I., Roger, A. J., Burger, G., Gray, M. W., and Lang, B. F. (2008). A phylogenomic investigation into the origin of metazoa. *Mol. Biol. Evol.* **25,** 664–672.
Schierwater, B., Eitel, M., Jakob, W., Osigus, H. J., Hadrys, H., Dellaporta, S. L., Kolokotronis, S. O., and Desalle, R. (2009). Concatenated analysis sheds light on early metazoan evolution and fuels a modern "urmetazoon" hypothesis. *PLoS Biol.* **7,** e20.
Schneider, S. Q., and Bowerman, B. (2007). beta-Catenin asymmetries after all animal/ vegetal- oriented cell divisions in Platynereis dumerilii embryos mediate binary cell-fate specification. *Dev. Cell* **13,** 73–86.
Scholz, C. B., and Technau, U. (2003). The ancestral role of Brachyury: Expression of NemBra1 in the basal cnidarian Nematostella vectensis (Anthozoa). *Dev. Genes Evol.* **212,** 563–570.
Schubert, M., Holland, L. Z., Stokes, M. D., and Holland, N. D. (2001). Three amphioxus Wnt genes (AmphiWnt3, AmphiWnt5, and AmphiWnt6) associated with the tail bud: The evolution of somitogenesis in chordates. *Dev. Biol.* **240,** 262–273.
Shimizu, T., Bae, Y. K., Muraoka, O., and Hibi, M. (2005). Interaction of Wnt and caudal-related genes in zebrafish posterior body formation. *Dev. Biol.* **279,** 125–141.
Signorovitch, A. Y., Dellaporta, S. L., and Buss, L. W. (2005). Molecular signatures for sex in the Placozoa. *Proc. Natl. Acad. Sci. USA* **102,** 15518–15522.
Signorovitch, A. Y., Buss, L. W., and Dellaporta, S. L. (2007). Comparative genomics of large mitochondria in placozoans. *PLoS Genet.* **3,** e13.
Sikes, J. M., and Bely, A. E. (2010). Making heads from tails: Development of a reversed anterior-posterior axis during budding in an acoel. *Dev. Biol.* **338,** 86–97.
Spitz, F., and Duboule, D. (2005). Developmental biology: Reproduction in clusters. *Nature* **434,** 715–716.
Spitz, F., Gonzalez, F., and Duboule, D. (2003). A global control region defines a chromosomal regulatory landscape containing the HoxD cluster. *Cell* **113,** 405–417.
Spring, J., Yanze, N., Middel, A. M., Stierwald, M., Groger, H., and Schmid, V. (2000). The mesoderm specification factor twist in the life cycle of jellyfish. *Dev. Biol.* **228,** 363–375.
Spring, J., Yanze, N., Josch, C., Middel, A. M., Winninger, B., and Schmid, V. (2002). Conservation of Brachyury, Mef2, and Snail in the myogenic lineage of jellyfish: A connection to the mesoderm of bilateria. *Dev. Biol.* **244,** 372–384.
Srivastava, M., Begovic, E., Chapman, J., Putnam, N. H., Hellsten, U., Kawashima, T., Kuo, A., Mitros, T., Salamov, A., Carpenter, M. L., Signorovitch, A. Y., Moreno, M. A., *et al.* (2008). The Trichoplax genome and the nature of placozoans. *Nature* **454,** 955–960.
Srivastava, M., Larroux, C., Lu, D. R., Mohanty, K., Chapman, J., Degnan, B. M., and Rokhsar, D. S. (2010a). Early evolution of the LIM homeobox gene family. *BMC Biol.* **8,** 4.
Srivastava, M., Simakov, O., Chapman, J., Fahey, B., Gauthier, M. E., Mitros, T., Richards, G. S., Conaco, C., Dacre, M., Hellsten, U., Larroux, C., Putnam, N. H., *et al.* (2010b). The *Amphimedon queenslandica* genome and the evolution of animal complexity. *Nature* **466,** 720–726.
Steele, R. E. (2002). Developmental signaling in *Hydra*: What does it take to build a "simple" animal? *Dev. Biol.* **248,** 199–219.
Steele, R. E. (2006). Trembley's polyps go transgenic. *Proc. Natl. Acad. Sci. USA* **103,** 6415–6416.
Steele, R. E., David, C. N., and Technau, U. (2011). A genomic view of 500 million years of cnidarian evolution. *Trends Genet.* **27,** 7–13.
Streit, A., Lee, K. J., Woo, I., Roberts, C., Jessell, T. M., and Stern, C. D. (1998). Chordin regulates primitive streak development and the stability of induced neural cells, but is not sufficient for neural induction in the chick embryo. *Development* **125,** 507–519.
Strutt, H., Price, M. A., and Strutt, D. (2006). Planar polarity is positively regulated by casein kinase Iepsilon in Drosophila. *Curr. Biol.* **16,** 1329–1336.

Sullivan, J. C., Ryan, J. F., Mullikin, J. C., and Finnerty, J. R. (2007). Conserved and novel Wnt clusters in the basal eumetazoan Nematostella vectensis. *Dev. Genes Evol.* **217,** 235–239.

Sutherland, D. J., Li, M., Liu, X. Q., Stefancsik, R., and Raftery, L. A. (2003). Stepwise formation of a SMAD activity gradient during dorsal-ventral patterning of the Drosophila embryo. *Development* **130,** 5705–5716.

Szathmary, E., and Smith, J. M. (1995). The major evolutionary transitions. *Nature* **374,** 227–232.

Technau, U., and Bode, H. R. (1999). HyBra1, a Brachyury homologue, acts during head formation in *Hydra. Development* **126,** 999–1010.

Technau, U., and Scholz, C. B. (2003). Origin and evolution of endoderm and mesoderm. *Int. J. Dev. Biol.* **47,** 531–539.

Technau, U., and Steele, R. E. (2011). Evolutionary crossroads in developmental biology: Cnidaria. *Development* **138**(8), 1447–1458.

Technau, U., Cramer von Laue, C., Rentzsch, F., Luft, S., Hobmayer, B., Bode, H. R., and Holstein, T. W. (2000). Parameters of self-organization in Hydra aggregates. *Proc. Natl. Acad. Sci. USA* **97,** 12127–12131.

Technau, U., Rudd, S., Maxwell, P., Gordon, P. M., Saina, M., Grasso, L. C., Hayward, D. C., Sensen, C. W., Saint, R., Holstein, T. W., Ball, E. E., and Miller, D. J. (2005). Maintenance of ancestral complexity and non-metazoan genes in two basal cnidarians. *Trends Genet.* **21,** 633–639.

Teo, R., Mohrlen, F., Plickert, G., Muller, W. A., and Frank, U. (2006). An evolutionary conserved role of Wnt signaling in stem cell fate decision. *Dev. Biol.* **289,** 91–99.

Thorpe, C. J., Weidinger, G., and Moon, R. T. (2005). Wnt/beta-catenin regulation of the Sp1-related transcription factor sp5l promotes tail development in zebrafish. *Development* **132,** 1763–1772.

Torras, R., Yanze, N., Schmid, V., and Gonzalez-Crespo, S. (2004). nanos expression at the embryonic posterior pole and the medusa phase in the hydrozoan Podocoryne carnea. *Evol. Dev.* **6,** 362–371.

Tribulo, C., Aybar, M. J., Nguyen, V. H., Mullins, M. C., and Mayor, R. (2003). Regulation of Msx genes by a Bmp gradient is essential for neural crest specification. *Development* **130,** 6441–6452.

von Salvini-Plawen, L. (1978). On the origin and evolution of the lower Metazoa. *Z. Zool. Syst. Evolutionsf.* **16,** 40–88.

Wikramanayake, A. H., Huang, L., and Klein, W. H. (1998). beta-Catenin is essential for patterning the maternally specified animal-vegetal axis in the sea urchin embryo. *Proc. Natl. Acad. Sci. USA* **95,** 9343–9348.

Wikramanayake, A. H., Hong, M., Lee, P. N., Pang, K., Byrum, C. A., Bince, J. M., Xu, R., and Martindale, M. Q. (2003). An ancient role for nuclear beta-catenin in the evolution of axial polarity and germ layer segregation. *Nature* **426,** 446–450.

Wikramanayake, A. H., Peterson, R., Chen, J., Huang, L., Bince, J. M., McClay, D. R., and Klein, W. H. (2004). Nuclear beta-catenin-dependent Wnt8 signaling in vegetal cells of the early sea urchin embryo regulates gastrulation and differentiation of endoderm and mesodermal cell lineages. *Genesis* **39,** 194–205.

Windsor, P. J., and Leys, S. P. (2010). Wnt signaling and induction in the sponge aquiferous system: Evidence for an ancient origin of the organizer. *Evol. Dev.* **12,** 484–493.

Wittlieb, J., Khalturin, K., Lohmann, J. U., Anton-Erxleben, F., and Bosch, T. C. (2006). Transgenic *Hydra* allow in vivo tracking of individual stem cells during morphogenesis. *Proc. Natl. Acad. Sci. USA* **103,** 6208–6211.

Wu, M. Y., and Hill, C. S. (2009). Tgf-beta superfamily signaling in embryonic development and homeostasis. *Dev. Cell* **16,** 329–343.

Xiao, S., and Laflamme, M. (2009). On the eve of animal radiation: Phylogeny, ecology and evolution of the Ediacara biota. *Trends Ecol. Evol.* **24,** 31–40.

Yost, C., Torres, M., Miller, J. R., Huang, E., Kimelman, D., and Moon, R. T. (1996). The axis-inducing activity, stability, and subcellular distribution of beta-catenin is regulated in Xenopus embryos by glycogen synthase kinase 3. *Genes Dev.* **10,** 1443–1454.

Yu, J. K., Holland, L. Z., and Holland, N. D. (2002). An amphioxus nodal gene (AmphiNodal) with early symmetrical expression in the organizer and mesoderm and later asymmetrical expression associated with left-right axis formation. *Evol. Dev.* **4,** 418–425.

Yu, J. K., Satou, Y., Holland, N. D., Shin, I. T., Kohara, Y., Satoh, N., Bronner-Fraser, M., and Holland, L. Z. (2007). Axial patterning in cephalochordates and the evolution of the organizer. *Nature* **445,** 613–617.

Zakin, L., and De Robertis, E. M. (2010). Extracellular regulation of BMP signaling. *Curr. Biol.* **20,** R89–R92.

CHAPTER SEVEN

FGF and ROR2 Receptor Tyrosine Kinase Signaling in Human Skeletal Development

Sigmar Stricker* *and* Stefan Mundlos†

Contents

* Development and Disease Group, Max Planck-Institute for Molecular Genetics, Berlin, Germany
† Institute for Medical Genetics, Charité University Medicine Berlin, Berlin, Germany

Current Topics in Developmental Biology, Volume 97
ISSN 0070-2153, DOI: 10.1016/B978-0-12-385975-4.00013-9

Abstract

Skeletal malformations are among the most frequent developmental disturbances in humans. In the past years, progress has been made in unraveling the molecular mechanisms that govern skeletal development by the use of animal models as well as by the identification of numerous mutations that cause human skeletal syndromes. Receptor tyrosine kinases have critical roles in embryonic development. During formation of the skeletal system, the fibroblast growth factor receptor (FGFR) family plays major roles in the formation of cranial, axial, and appendicular bones. Another player of relevance to skeletal development is the unusual receptor tyrosine kinase ROR2, the function of which is as interesting as it is complex. In this chapter, we review the involvement of FGFR signaling in human skeletal disease and provide an update on the growing knowledge of ROR2.

1. Introduction

1.1. Human skeletal malformations

Skeletal malformations that manifest themselves during human embryonic or fetal development present either as isolated defects or as part of complex syndromes. Features may include isolated malformations of the craniofacial, axial, and appendicular skeleton, or a combination of these. The processes affected include the specification and expansion of skeletal progenitors, the patterning and shaping of the skeletal elements, or their growth during fetal development and childhood. The molecular basis for numerous human genetic skeletal syndromes has been defined in the past 20 years. This has brought a tremendous amount of information on the players and pathways that govern the formation, differentiation, and growth of the skeleton. Mutations in components of the bone morphogenetic protein (BMP), Hedgehog (Hh), fibroblast growth factor (FGF), and WNT signaling pathways have been identified in human skeletal syndromes. For example, it has become clear that the BMP pathway is important for the development of the appendicular skeleton, as mutations in components of the cascade are found in most brachydactyly (i.e., short digit) syndromes (Mundlos, 2009). Digit development is further controlled by a Hh and WNT/ROR2 signaling network (Stricker and Mundlos, 2011; Witte *et al.*, 2010b). The development of the skull is complex and regulated by several pathways, with FGF being the most prominent (Johnson and Wilkie, 2011). In general, there are two distinct mechanisms for the generation of the skeleton called intramembranous and endochondral ossification. These mechanisms will be briefly introduced.

1.2. Skeletal development: Endochondral ossification

Most of the axial and appendicular skeleton develops by endochondral ossification. In this process, the skeletal elements are prefigured by a cartilaginous template that is progressively replaced by bone. Endochondral ossification begins with the condensation of mesenchymal cells. The condensed cells undergo differentiation into chondrocytes, which form the cartilage template of the future bone. Once the cartilage element is established, chondrocytes start to proliferate and undergo a stereotypical series of differentiation events. This process culminates in the apoptosis of the chondrocytes that are then replaced by bone.

Coordinated proliferation and differentiation of chondrocytes take place inside a specialized structure, the cartilage growth plate (Fig. 7.1A), which allows growth of skeletal elements until the end of puberty. In the growth plates, which reside on both ends of the long bones, small round reserve zone chondrocytes generate proliferating chondrocytes that form clonal stacks called columnar chondrocytes. These undergo further differentiation to prehypertrophic chondrocytes that coordinate proliferation, differentiation, and the induction of osteoblasts in the adjacent perichondrium. Prehypertrophic cells finally form hypertrophic chondrocytes that secrete a specialized extracellular matrix. After the apoptotic demise of the hypertrophic chondrocytes, osteoclasts remove cell debris and extracellular matrix and thus make way for the bone marrow cavity. Blood vessels invade the area and deliver hematopoietic cells. Alongside the blood vessels osteoblasts migrate to the cavity, where they form the trabecular bone. In parallel, osteoblasts arise in the perichondrium adjacent to the prehypertrophic chondrocytes and start to deposit cortical bone.

1.3. Skeletal development: Intramembranous ossification

During the process of intramembranous ossification (also referred to as desmal ossification) mesenchymal cells differentiate into osteoblasts, which directly start to deposit bone. This mechanism generates the flat bones of the skull and the lateral clavicles. The first step in intramembranous ossification is the formation of mesenchymal condensations, which differentiate into proliferating preosteoblasts and finally become bone-depositing osteoblasts (Fig. 7.1B). The flat bones of the skull are formed in connective tissue layers called the skeletogenic membrane, which is located between the dura mater (the membrane that ensheathes the brain) and the dermis. These bones constantly grow by new osteogenic differentiation and the deposition of new bone material at their margins. In their growth phase, these bones do not fuse but remain separated by specialized structures, the sutures (Hall and Miyake, 2000). Growth of the skull is required to meet the space requirements of the growing brain and depends on an exchange of signals between

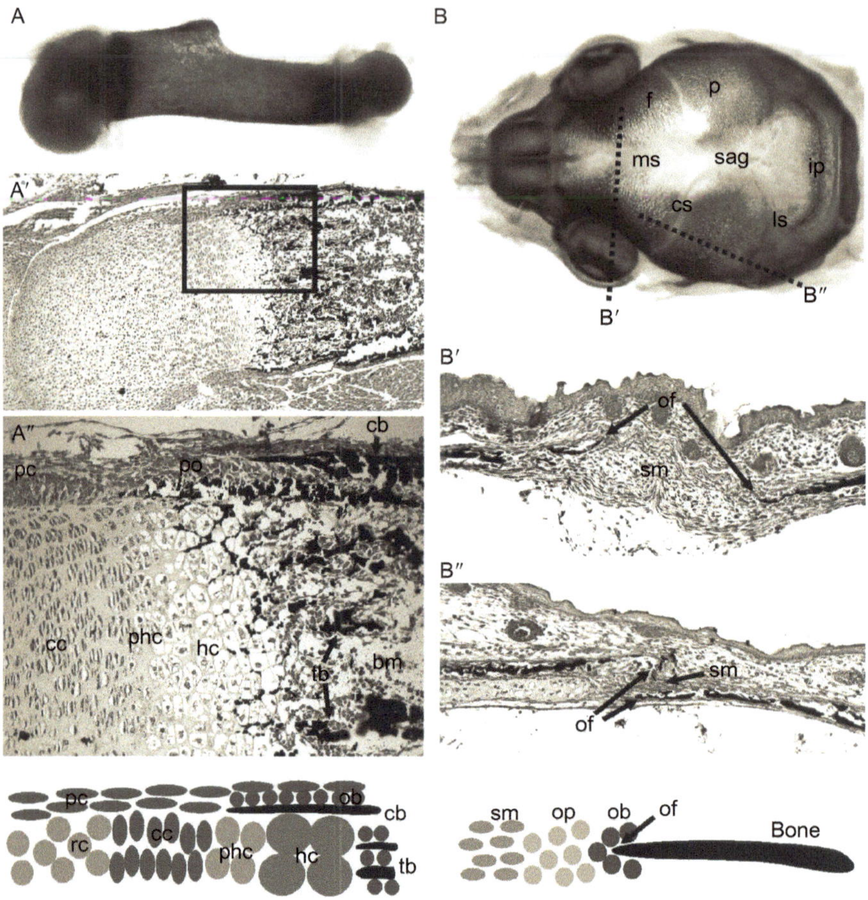

Figure 7.1 Development of the skeleton by (A) endochondral and (B) intramembranous ossification. (A) A skeletal preparation of a mouse humerus at embryonic day 17.5 (E17.5). (A′) Van Kossa-stained section of the proximal humerus at E17.5. (A″) Magnification of the area boxed in (A′). Note that the calcified areas stain black. At the bottom, a schematic display of the differentiating cell types in the growth plate is shown. (B) Mouse skull at E17.5 stained with Alcian blue (cartilage) and Alizarin red (bone); the skull base was removed for clarity. (B′) and (B″) show Van Kossa-stained sections of the metopic suture and a coronal suture, respectively, in which calcified areas appear black. Note the different morphologies of metopic and coronary sutures, that is, the abutting osteogenic fronts in the metopic and the overlapping osteogenic fronts in the coronal suture. At the bottom, a schematic display of the differentiating cell types in the cranial suture is shown. *Abbreviations*: Bm, bone marrow; cb, cortical bone; cc, columnar chondrocytes; cs, coronal suture; f, frontal bone; hc, hypertrophic chondrocytes; ip, interparietal bone; ls, lambdoid suture; ms, metopic suture; ob, osteoblasts; of, osteogenic front; op, osteoprogenitors; p, parietal bone; pc, perichondrium; phc, prehypertrophic chondrocytes; po, periosteum; rc, reserve chondrocytes; sag, sagittal suture; sm, suture mesenchyme; tb, trabecular bone. See Mundlos (2000) for staining methods. (See Color Insert.)

mesenchyme, the osteogenic front and the dura mater (for review, see, e.g., Lenton *et al.*, 2005; Slater *et al.*, 2008). This is an intricately balanced process, and a premature differentiation of mesenchymal cells results in premature fusion of the sutures and craniosynostosis.

2. FGF Signaling: Craniosynostosis, Digit Abnormalities, and Short Stature

FGFs are a family of signaling molecules that comprise 22 members in mammals. FGFs play multiple roles during development and in adult life, and individual members of the family have distinct biological properties. FGFs 11–14 are intracellular proteins that function in a receptor-independent manner, FGFs 15, 19, 21, and 23 are hormone-like factors that circulate in the blood stream, while the FGFs 1–10, 16–18, 20, and 22, the so called "canonical FGFs", act as local growth factors in a paracrine manner (Goldfarb, 2005; Itoh and Ornitz, 2008; Thisse and Thisse, 2005). Four FGF receptor (*FGFR1–4*) genes exist and further diversity arises by alternative splicing (Itoh and Ornitz, 2008). FGF receptors contain an extracellular ligand binding domain with three immunoglobulin (Ig)-like motives, a transmembrane domain, and an intracellular "split" tyrosine kinase (TK) domain (Fig. 7.2A). Particular variability exists in the third

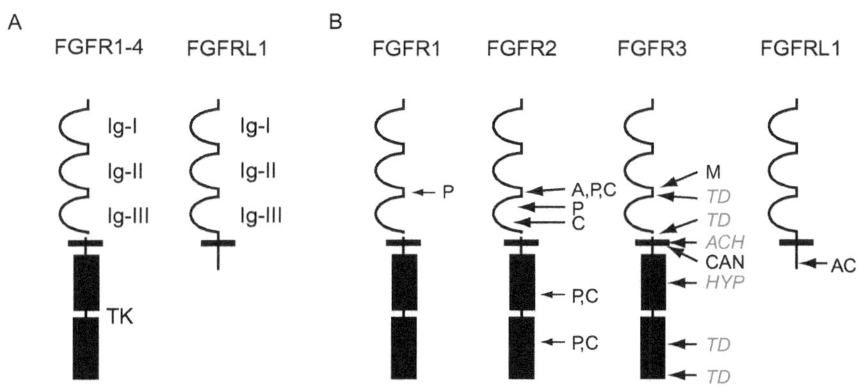

Figure 7.2 FGF receptor structures and mutations observed in skeletal disorders. (A) Structure of FGF receptors displaying the extracellular Ig domains, the transmembrane sequence, and the cytoplasmic domains. IgI–IgIII, immunoglobolin-like domains; TK, tyrosine kinase domain. (B) Mutations in craniosynostosis (black) and chondrodysplasia (gray) syndromes. A, Apert syndrome; AC, asyndromic craniosynostosis; ACH, achondroplasia; C, Crouzon syndrome; CAN, Crouzon syndrome with acanthosis nigricans; HYP, hypochondroplasia; M, Muenke syndrome; P, Pfeiffer syndrome; TD, thanatophoric dysplasia.

Ig-like domain (IgIII) due to alternative splicing, generating isoforms that differ in their affinities to FGFs. For instance, two splice variants of FGFR1–3 exist that contain either the IgIIIb or IgIIIc exon (FGFR1–3b and FGFR1–3c). In addition, one FGF receptor-like 1 (FGFRL1) exists that displays typical extracellular and transmembrane domains but lacks the TK domain.

FGF ligand/FGF receptor signaling control numerous developmental processes. However, mutations identified in human FGFRs cause particular and striking malformations of the craniofacial skeleton, the digits, and the cartilage growth plate of long bones.

2.1. FGF receptor mutations causing craniosynostosis syndromes

The first gene associated with craniosynostosis was *MSX2*, a gene encoding a transcription factor (Jabs *et al.*, 1993), but mutation of *MSX2* accounts for only few clinical cases. Mutations in *FGFR2* are the prevalent cause for three overlapping syndromes, Apert syndrome (MIM #101200), Pfeiffer syndrome (MIM #101600), and Crouzon syndrome (MIM #123500) (Jabs *et al.*, 1994; Muenke and Schell, 1995; Oldridge *et al.*, 1997, 1999; Park *et al.*, 1995; Reardon *et al.*, 1994; Wilkie *et al.*, 1995; recently reviewed by Johnson and Wilkie, 2011). These syndromes are characterized by craniosynostosis and varying limb malformations, that is, midface hypoplasia and (bony) syndactyly in Apert syndrome, midface hypoplayia, broadened thumbs and big toes, and variable cutaneous syndactyly in Pfeiffer syndrome and facial but no limb abnormalities in Crouzon syndrome. Further, mutations in *FGFR2* can cause Jackson–Weiss syndrome (JWS), which shows variable features of Apert, Pfeiffer, and Crouzon syndromes (MIM #123150) (Jabs *et al.*, 1994). All of these syndromes display an autosomal-dominant inheritance pattern. Additionally, mutation in *FGFR2* was identified in a single patient with Saethe–Chotzen syndrome (MIM #101400) characterized by acrocephaly, asymmetry of the skull, soft tissue syndactyly, and variable craniosynostosis (Paznekas *et al.*, 1998). Finally, rare mutations in *FGFR1* have been described in mild forms of Pfeiffer syndrome (Muenke *et al.*, 1994; Rossi *et al.*, 2003).

Mutations in *FGFR3* cause Muenke syndrome (MIM #602849) characterized by coronal synostosis combined with variable and low penetrance limb malformations (Bellus *et al.*, 1996; Muenke *et al.*, 1997), as well as Crouzon syndrome with acanthosis nigricans (MIM #612247; Meyers *et al.*, 1995). The discovery of *FGFRL1* frameshift mutations in a single patient with asyndromic craniosynostosis also implicated FGFRL1 in craniofacial development (Rieckmann *et al.*, 2009). Interestingly, *Fgfrl1* null mice show features of Wolf–Hirschhorn syndrome (WHS; MIM #194190) characterized by several defects including craniofacial malformations (Catela *et al.*,

2009). WHS is caused by hemizygous rearrangements on Chromosome 4p (Battaglia *et al.*, 2001; Bergemann *et al.*, 2005), with two minimal critical regions identified to date on 4p16.3. The human WHS-critical regions (WHSCR1 and WHSCR2) do not contain *FGFRL1*, and mice lacking WHSCR1 or WHSCR2 do not develop the full WHS phenotypic spectrum (Naf *et al.*, 2001). This might indicate that the human WHS deletions affect regulatory sequences in the *FGFRL1* gene whose positions are not conserved in evolution. *Fgfrl1* is expressed in craniofacial cartilage (Trueb and Taeschler, 2006; Trueb *et al.*, 2003), making it an excellent candidate to cause at least the craniofacial features of WHS. This is further supported by the recent finding of a patient with craniofacial characteristics of WHS that carried a deletion of *FGFRL1* but not of WHSCR1 (Engbers *et al.*, 2009). Human and mouse FGFRL1 bind FGFs and heparin but due to the lack of a kinase domain were proposed to function as decoy receptors and to fine-tune signaling of the other FGFRs (Sleeman *et al.*, 2001; Wiedemann and Trueb, 2000). Mutations in FGF ligands were as yet not associated with human craniosynostosis. However, mutations in *FGF8* and *FGFR1*, *FGFR2*, and *FGFR3* can cause nonsyndromic cleft palate (Riley *et al.*, 2007).

2.2. Limb malformations associated with FGFR mutations

Apert, Crouzon, and Pfeiffer syndromes share a spectrum of craniofacial malformations but display distinct limb phenotypes (see above). Several further syndromes affecting the limb but not the craniofacial skeleton are associated with mutations in the FGF signaling pathway. Lacrimo-auriculo-dento-digital syndrome (LADD; MIM #149739) is caused by mutations in *FGFR2*, *FGFR3* (Rohmann *et al.*, 2006), and *FGF10* (Milunsky *et al.*, 2006), a ligand for FGFR2 (Ornitz *et al.*, 1996). LADD syndrome includes clinodactyly of the fifth finger, bifid thumb, triphalangeal thumb, and syndactyly and thus partially overlaps with the spectrum of malformations seen in Apert and Pfeiffer syndromes. *FGFR2* mutations associated with Apert syndrome cause bony fusions (synostoses), and multiple synostoses syndrome 3 (SYNS3; MIM #612961) is caused by mutations in *FGF9* (Wu *et al.*, 2009), a ligand for FGFR2 (Bellus *et al.*, 1995). In the mouse, mutations in *Fgf9* cause radiohumeral and tibiofemoral synostoses and also craniosynostosis (Harada *et al.*, 2009).

2.3. FGF signaling and enchondral ossification: Chondrodysplasia syndromes

Skeletal growth occurs until the end of puberty and is accomplished by the cartilage growth plates of the long bones. Growth is achieved by a coordinated proliferation and differentiation of chondrocytes that is controlled on multiple levels by endocrine and paracrine signals (see, e.g., Karsenty and Wagner, 2002; Siebler *et al.*, 2001). FGF signaling is one of the major regulators of these processes, and particularly, chondrocyte proliferation in the growth plate is tightly controlled by FGFR3 as evidenced by the seminal discovery of mutations in *FGFR3* in the most common human short-stature syndrome, achondroplasia (ACH; MIM #100800) (Rousseau *et al.*, 1994; Shiang *et al.*, 1994). Most patients with ACH carry a G380R mutation in FGFR3, which in 80% of the cases arises *de novo* (Horton *et al.*, 2007). The less severe syndrome, hypochondroplasia (HCH; MIM #146000), and the lethal condition thanatophoric dysplasia (TD; MIM #187600, #187601) are also caused by mutations in *FGFR3* (Bellus *et al.*, 1995; Rousseau *et al.*, 1995; Tavormina *et al.*, 1995). Further, mutations in *FGFR3* can result in "severe achondroplasia with developmental delay and acanthosis nigricans" (SADDAN; listed in OMIM as a subgroup of TD) (Tavormina *et al.*, 1999).

The only FGF ligand known to be mutated in a human syndrome associated with global skeletal alterations is FGF23 (see Krejci *et al.*, 2009). Autosomal dominant hypophosphatemic rickets (ADHR; MIM #193100), also known as vitamin D-resistant rickets, characterized by hypophosphatemia, defective cartilage, and bone mineralization, and short stature (Econs and McEnery, 1997) is caused by missense mutations in *FGF23* (ADHR Consortium, 2000). FGF23 belongs to the hormone-like FGFs and controls phosphate homeostasis. The mutations in ADHR are thought to be gain-of-function mutations that inhibit the proteolytic cleavage and inactivation of FGF23 (White *et al.*, 2001a). In accordance, FGF23 is a decisive factor in tumor-induced osteomalacia (Shimada *et al.*, 2001; White *et al.*, 2001b). Conversely, loss-of-function mutations in *FGF23* are thought to cause familial hyperhosphatemic tumoral calcinosis (MIM #211900) associated with aberrant deposition of calcium phosphate in periarticular spaces (Chefetz *et al.*, 2005; Larsson *et al.*, 2004).

2.4. Molecular consequences of mutations in FGFRs

The mutational spectrum of the FGFRs in Apert, Pfeiffer, and Crouzon syndromes is complex; however, most mutations cluster at specific points in the protein (reviewed in Johnson and Wilkie, 2011; Morriss-Kay and Wilkie, 2005; Wilkie *et al.*, 2002). An overview of mutations identified in craniosynostosis syndromes and in chondrodysplasia syndromes is provided in Fig. 7.2B. Most mutations change the extracellular domain of the FGFR proteins. The most common Apert syndrome mutations in FGFR2 (S252W and P253R) are located in the linker region between the IgII and IgIII domains. These mutations have been extensively analyzed and result in a broadened spectrum of ligand binding and increased affinity for FGF ligands (Ibrahimi *et al.*, 2005). Most mutations causing either Pfeiffer or Crouzon syndrome are found within the IgIIIa or IgIIIc domain of FGFR2 and either remove or add a cysteine residue, leading to the formation of disulfide bridges between two receptor molecules. This constitutive receptor dimerization results in constitutive signaling. In addition, mutations at the splice site of the IIIc exon, which result in misexpression of the IIIb isoform, cause Pfeiffer and Apert syndrome (Wilkie *et al.*, 2002). Further, putative activating mutations in the TK domain of FGFR2 were identified in Crouzon and Pfeiffer patients (Kan *et al.*, 2002). Similarly, mutations in FGFR3 causing Muenke syndrome map to the IgII–IgIII linker sequence. This P250R substitution is the equivalent of the Apert mutation P253R (Ibrahimi *et al.*, 2005). A mutation of the corresponding position in FGFR1 was also found in Pfeiffer syndrome (P252R).

In the chondrodysplasia syndromes, a partially overlapping spectrum of mutations was identified, that is, mutations generating odd numbers of extracellular cysteine residues, the recurrent G380R mutation in the transmembrane domain, and mutations in the TK domain. However, mutations in the Ig domains of the protein are not associated with chondrodysplasia syndromes.

Altogether, mutations in FGFRs lead to a gain of function by different mechanisms in both craniosynostosis and chondrodysplasia syndromes: covalent cross-linking mutations (i.e., mutations in cystein residues), mutations in the TK domains, and mutation of the transmembrane domain thought to stabilize ligand-induced FGFR3 dimers (see Horton *et al.*, 2007 for a recent review). Specific for the craniosynostosis syndromes are mutations in the receptors that confer a higher affinity for FGF ligands, mutations that allow binding of illegitimate FGF ligands, and mutations leading to the expression of illegitimate splice variants.

2.5. FGF signaling in cranial suture development

Mutations in several FGF receptors interfere with the maintenance of the cranial sutures, demonstrating that FGF signaling is a key regulator of this process. Specifically, craniosynostosis (the premature fusion of sutures) is thought to be caused by increased or premature differentiation of mesenchymal cells to bone-forming osteoblasts (De Pollack *et al.*, 1996; Lomri *et al.*, 1998). Several FGFs were reported to be expressed in the suture mesenchyme, the dura, or the calvarial bones. FGFR2 is mainly expressed in differentiating osteoprogenitors, FGFR1 in mesenchyme and osteoblasts, and FGFR3 at low levels in the osteogenic front (Morriss-Kay and Wilkie, 2005; Ornitz and Marie, 2002). Several lines of evidence suggest that FGFR signaling promotes osteoblast differentiation in cranial osteogenesis (Eswarakumar *et al.*, 2004; Holmes *et al.*, 2009; Iseki *et al.*, 1999; Yang *et al.*, 2008). This correlates with activation of the intracellular MEK-ERK1/2 pathway in a mouse model for Apert syndrome (FGFR2 p. S252W) (Shukla *et al.*, 2007).

RUNX2 is a transcription factor pivotal for osteoblast differentiation. Mice with homozygous loss-of-function mutations in Runx2 lack bone (Komori *et al.*, 1997; Otto *et al.*, 1997). RUNX2 function is particularly critical for intramembranous ossification, and heterozygous mutations in *RUNX2* result in cleidocranial dysplasia (MIM #119600) characterized by open fontanelles and a delay in suture closure (Mundlos *et al.*, 1997). There is substantial evidence that FGFR signaling controls *RUNX2* transcription and protein stability/activity (Baroni *et al.*, 2005; Kim *et al.*, 2006; Miraoui *et al.*, 2009; Park *et al.*, 2010). In this context, it is noteworthy that TWIST1, mutated in Saethe–Chotzen syndrome, binds to and regulates the activity of RUNX2 (Glass *et al.*, 2005). TWIST1 appears to regulate the expression of FGFR2 in a negative manner (Guenou *et al.*, 2005; Miraoui *et al.*, 2010). Conversely, FGF2 induces *TWIST1* expression in the cranial suture (Rice *et al.*, 2000, 2003), forming a negative feedback loop.

2.6. Crosstalk of TGFβ/BMP and FGF signaling in cranial suture development

TGFβ signaling is an important regulator of skeletal development. This is underscored by *TGFBR2* mutations causing Loeys–Dietz syndrome type 1, an aortic aneurism syndrome that is accompanied by craniosynostosis (Loeys *et al.*, 2005, 2006). TGFβ2 and TGFβ3 are ligands for TGFBRs in the cranial suture (Rawlins and Opperman, 2008), and TGFβ2 signaling in cranial sutures impinges on ERK1/2 (Opperman *et al.*, 2006). TGFβ2 induced the expression of *ERK1/2*, thus facilitating FGFR signaling, and also enhanced ERK1/2 phosphorylation, thus cooperating with FGFR signaling. Further, the FGFR pathway intersects with signaling of BMPs. BMPs are critical regulators of chondrogenesis and osteogenesis and are expressed in sutures in the osteogenic front and in the dura mater (Kim *et al.*, 1998). BMP signaling is regulated at the extracellular level by secreted antagonists such as NOGGIN (NOG). In the mouse, *Nog* expression in patent, but not fusing cranial sutures, was suppressed by FGF signaling, and recombinant NOG protein administration was sufficient to prevent suture closure *in vivo* (Hillier *et al.*, 2004). Importantly, FGFRs carrying Apert or Crouzon mutations showed ligand-independent suppression of *NOG*, suggesting that the failure of *NOG* expression causes premature suture closure in these syndromes (Hillier *et al.*, 2004).

2.7. FGF signaling in limb development

The distal outgrowth of the limb is under control of FGF signaling emanating from a specialized ectodermal structure, the apical ectodermal ridge (AER) that forms the distal dorso-ventral border of the limb bud (AER; Niswander *et al.*, 1993; Sun *et al.*, 2002). The AER expresses several FGFs, most prominently FGF8 and FGF4, and at lower levels FGF9 and FGF17. These FGFs act in a partially redundant manner, and FGF8 is the only ligand that is essential (Sun *et al.*, 2000). AER–FGFs signal to the underlying mesenchyme to maintain a progenitor pool, that is, stimulate proliferation and inhibit differentiation of mesenchymal progenitors (Ornitz and Marie, 2002). A positive feedback loop between AER-expressed FGF8 and mesenchymally produced FGF10 ensures a continuous outgrowth of the limb (Niswander *et al.*, 1994). As mentioned above, mutations in FGF receptors in Apert and Pfeiffer syndromes affect mainly autopod patterning and differentiation. Continuous condensation of mesenchymal cells proceeds from the proximal to the distal limb during formation of the limb skeleton. Later, the cartilage rod is subdivided into the individual skeletal elements by the insertion of synovial joints (Shubin and Alberch, 1986). During autopod development in animal models, FGF signaling from the AER plays a major role in elongation of the digital rays (Casanova and Sanz-Ezquerro, 2007; Sanz-Ezquerro and Tickle, 2003) and also represses interdigital apoptosis (Pajni-Underwood *et al.*, 2007).

FGF receptors are expressed in specific patterns in the developing limb bud. The ectoderm expresses FGFR2b receiving the mesenchymal FGF10 signal, while the mesenchyme expresses FGFR1c receiving the ectodermal FGF8 signal. Further, during cartilage condensation, FGFR2c is expressed in the condensing limb mesenchyme (Ornitz and Marie, 2002) and at later developmental stages is restricted to the interdigital mesenchyme (Ota *et al.*, 2007).

It appears that Apert and Pfeiffer mutations in FGFR2 affect the ectoderm (AER) and the mesoderm, and thus alter distinct developmental events. The length of the AER is associated with digit number, and anterior expansion of the AER results in polydactyly and triphalangeal thumbs in mice (Ovchinnikov *et al.*, 2006). Increased FGFR signaling in the ectoderm may lead to an anterior expansion of the AER. In late autopod development, the AER regresses in interdigital domains, which correlates with the onset of interdigital apoptosis (Merino *et al.*, 1998; Montero and Hurle, 2010). Hence increased or sustained FGF signaling in the AER may prevent normal AER regression and interfere with formation of the interdigital web, thus causing cunateous syndactyly. Enhanced FGFR2 signaling might enhance the size of mesenchymal condensations causing bony syndactyly and synostoses.

2.8. FGF signaling in the cartilage growth plate

Once the mesenchymal condensations are formed and committed chondrocytes express markers such as collagen type 2a1, the cartilage starts to express FGFR3. At late developmental stages, FGFR3 is also strongly expressed in differentiating and proliferating chondrocytes of the growth plate (Peters *et al.*, 1993). Loss- and gain-of-function mutations in the mouse show that FGFR3 is a major negative regulator of chondrocyte proliferation (Colvin *et al.*, 1996; Naski *et al.*, 1998). Further genetic studies suggest that murine FGF18, expressed in the perichondrium, might be the intrinsic activator for FGFR3 (Liu *et al.*, 2002; Ohbayashi *et al.*, 2002). The antiproliferative effect of FGFR3 is mediated by the activation of STAT1 (Li *et al.*, 1999; Sahni *et al.*, 1999) and the MAPK pathway (Krejci *et al.*, 2008; Murakami *et al.*, 2004; Zhang *et al.*, 2006).

In the growth plate, cross talk between FGF and several other pathways controls the development of the limb skeleton. Chondrocyte proliferation and differentiation are under control of Indian hedgehog (IHH). IHH is expressed in the prehypertrophic chondrocytes and diffuses across the growth plate toward the joints. IHH stimulates chondrocyte proliferation directly (Karp *et al.*, 2000) and regulates chondrocyte hypertrophic differentiation indirectly via the induction of parathyroid hormone-like hormone (PTHLH; also called parathyroid hormone-related peptide PTHRP) in the periarticular cartilage. PTHLH signals to its receptor, PTHR1 (parathyroid hormone receptor 1), which is expressed in proliferating chondrocytes preceding the onset of IHH expression. PTHR1 signaling prevents chondrocytes from undergoing differentiation to prehypertrophic cells and thus depletes the pool of IHH-expressing cells, resulting in reduced IHH expression. This IHH/PTHLH feedback is a major regulator of hypertrophic cartilage differentiation (St-Jacques *et al.*, 1999; Vortkamp *et al.*, 1996). *Runx2* is a target of IHH, and RUNX2 controls FGF18 expression in the perichondrium and thus FGFR3 activation and chondrocyte proliferation in the growth plate (Horton and Degnin, 2009 and references therein). FGFs antagonize BMP signaling in the developing growth plate (Minina *et al.*, 2002). In addition, cross talk between FGFR3 and C-type natriuretic peptide (CNP) signaling (the latter is mediated by the natriuretic peptide receptor B (NPR-B)) occurs. CNP/NPR-B signaling blocks ERK/RAF-1 and thus inhibits FGFR3 signaling (Krejci *et al.*, 2005). Interestingly, mutations in NRP-B cause acromesomelic chondrodysplasia type Maroteaux (MIM #602875), characterized by a shortening of skeletal elements (Bartels *et al.*, 2004).

3. The Atypical Receptor TK ROR2 and WNT Signaling

Wnt signals mediated by Frizzled receptors activate the "canonical" pathway employing β-catenin as well as several "noncanonical" pathways (for a detailed review, see other contributions on WNT signaling in this volume). WNTs are important regulators in embryonic development, and mutations in murine WNTs or WNT pathway molecules result in skeletal phenotypes (Grigoryan *et al.*, 2008; Hartmann, 2007). In human skeletal syndromes, few mutations in WNT signaling components have been identified. Mutations in *WNT3* and *WNT7A* cause severe phocomelia syndromes (Niemann *et al.*, 2004; Woods *et al.*, 2006), namely autosomal recessive tera-amelia (WNT3; MIM #273395) showing absence of limbs, and Fuhrmann syndrome or Al-Awadi/Raas-Rothschild/Schinzel phocomelia syndrome (WNT7A; MIM #228930 and #276820) showing variable hypoplasia or aplasia of limb skeletal elements. WNT3 is required for AER induction and maintenance (Barrow *et al.*, 2003), and WNT7a is essential for the maintenance of Sonic hedgehog (SHH) in the limb (Yang and Niswander, 1995). Thus, WNT3 and WNT7A do not affect skeletal development as such but regulate the overall outgrowth of the limb.

At this point a new player enters, the receptor TK ROR2. Mutations in ROR2 cause inheritable human skeletal malformations, and ROR2 is now thought to function as coreceptor for WNT signaling.

3.1. Mutations in ROR2: Robinow syndrome and Brachydactyly type B1

ROR2 and its paralog ROR1 were initially described as a putative receptor TK (Masiakowski and Carroll, 1992). The extracellular part of ROR receptors contains an Ig domain, a cysteine-rich domain that resembles sequences present in WNT receptors of the Frizzled family and a Kringle domain. The intracellular domains of ROR1 and ROR2 possess a TK domain and a peculiar C-terminal part that is unique to RORs. This C-terminal part contains two serine–threonine-rich domains (STD1 and STD2), separated by a proline-rich sequence (PRD; see Fig. 7.3).

The importance of ROR2 in human skeletal development was highlighted by the discovery that ROR2 mutations cause autosomal-dominant brachydactyly type B1 (BDB1; MIM #113000) (Oldridge *et al.*, 2000; Schwabe *et al.*, 2000) and autosomal recessive Robinow syndrome (RRS; MIM #268310) (Afzal *et al.*, 2000; van Bokhoven *et al.*, 2000). RRS and BDB1 are characterized by distinctive skeletal features. In RRS, much of the skeleton is affected and the patients exhibit short stature, acromesomelic limb shortening (shortening of the medial and distal skeletal elements of the limbs), craniofacial malformations, and nonskeletal features like heart defects and small external genitalia. BDB1 belongs to the brachydactyly family of digit malformations and is characterized by absent or hypoplastic distal and medial phalanges, which often show distal symphalangism (fusion of phalanges). ROR2 mutations show a clear genotype–phenotype correlation (Fig. 7.3). Nonsense or missense mutations in the extracellular domain or in the intracellular TK domain are found in RRS. Nonsense or frameshift BDB1 mutations affect only linker sequences between the

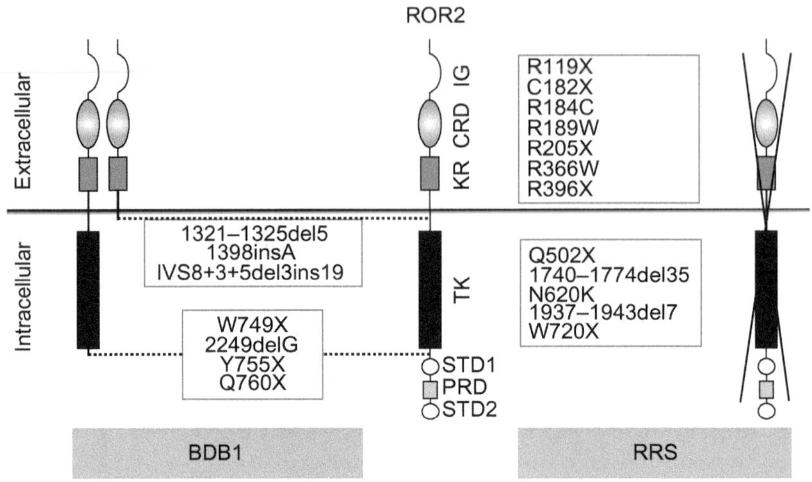

Figure 7.3 Structure of ROR2 and mutations leading to BDB1 or RRS. Left: Frameshift as well as nonsense mutations in the intracellular domain leading to the expression of truncated but membrane-anchored proteins cause BDB1. Right: Various mutations in the extra- and intracellular domains of the protein lead to a loss of function and cause RRS. CRD, cysteine-rich domain; IG, immunoglobulin-like domain; KR, Kringle domain; PRD, proline-rich domain; STD1, STD2, serine–threonine-rich domains; TK, tyrosine kinase domain.

transmembrane domain and the TK domain or between the TK domain and the C-terminal domain. The RRS mutations in extracellular coding sequences have been suggested to be loss-of-function mutations (Chen *et al.*, 2005). We recently showed that the RRS mutations in intracellular sequences are also loss-of-function mutations and lead to ROR2 retention in the endoplasmic reticulum and degradation. In contrast, BDB1 mutations give rise to stable proteins that are transferred to the cell membrane, which might act in a dominant-negative manner (Schwarzer *et al.*, 2009). Since the loss of ROR2 and the BDB1 mutations cause distinct phenotypes, BDB1 mutant variants do not act as classical dominant-negatives interfering with ROR2 function but are thought to impede other pathways.

3.2. ROR2 as a WNT (co)receptor

The ROR2 null mutant mouse recapitulates several features of RRS including short limbs (DeChiara *et al.*, 2000; Schwabe *et al.*, 2004; Takeuchi *et al.*, 2000). Interestingly, *WNT5A* mutations display an overlapping phenotype. Based on this, Oishi *et al.* (2003) used biochemistry to show that WNT5A is a ROR2 ligand and that WNT5A/ROR2 signals result in activation of c-Jun N-terminal kinase (JNK). Several intracellular interaction partners of ROR2 have been identified, and besides JNK, other downstream signaling mechanisms have been discussed (reviewed in Green *et al.*, 2008; Minami *et al.*, 2010). WNT5A appears to be the main ligand for ROR2, and ROR2 might function as a sole receptor for WNT5A or may act in combination with other receptors (Fig. 7.4).

It was noted early that ROR2 possesses kinase activity although it does not share several highly conserved amino acids with other TKs (Masiakowski and Carroll, 1992). However, studies analyzing the *Caenorhabditis elegans* ortholog, CAM-1, indicated that RORs possess kinase-dependent and -independent functions (Forrester *et al.*, 1999). ROR2 can function as WNT5A receptor in a classical RTK-like fashion, that is, upon ligand binding dimerization and activation of the TK are observed (Akbarzadeh *et al.*, 2008; Liu *et al.*, 2007, 2008; Mikels *et al.*, 2009). It has

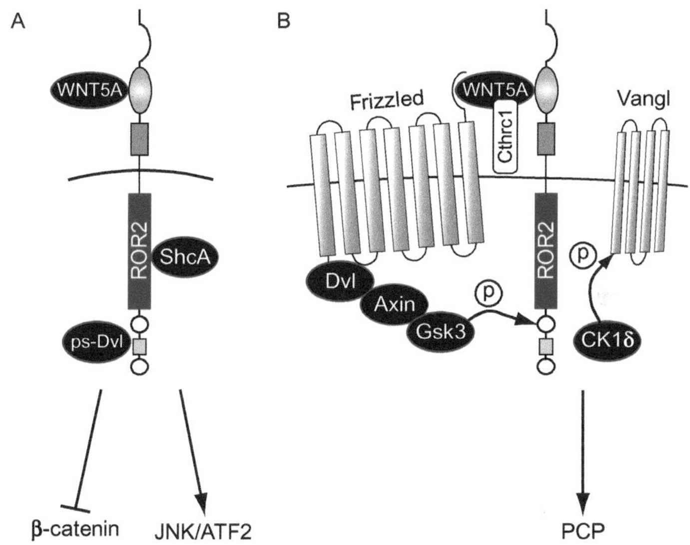

Figure 7.4 ROR2 signaling. (A) WNT5A signals via ROR2 in a kinase-dependent fashion to inhibit canonical WNT/β-catenin signaling. Caseine kinase 1 epsilon (CK1ε) phosphorylated Dishevelled (ps-Dvl) binds the C-terminus of ROR2 and promotes the suppression of canonical WNT signaling. Further, analyses in Xenopus indicate that WNT5A/ROR2 use the SH2 domain protein ShcA to activate the JNK/ATF2 signaling cascade necessary for noncanoncial WNT signaling during CE movements. It is unknown if these functions involve Ror2 association with Frizzled receptors. (B) WNT5A/ROR2 in PCP signaling. ROR2 is an alternative coreceptor for "noncanonical" WNTs. Association of WNT/ROR/Frizzled complexes lead to the recruitment of Dvl, Axin, and Gsk3 to the membrane with subsequent phosphorylation of ROR2 by Gsk3. The WNT/ROR/Frizzled complex is stabilized by Cthrc1. During PCP signaling, ROR2 associates with Vangl2 and promotes the phosphorylation of Vangl2 by caseine kinase 1 delta (CK1δ).

remained controversial whether activation of the TK occurs under physiological conditions (Green *et al.*, 2008). Signaling of WNT5A via ROR2 can inhibit the "canonical" WNT/β-catenin pathway and activate "noncanonical" WNT pathways (Fig. 7.4A).

Canonical WNT signaling is initiated by binding of WNTs to Frizzled receptors, which leads to the recruitment of the signaling mediator Dishevelled to the membrane and the stabilization and nuclear translocation of β-catenin. Nuclear β-catenin binds transcription factors of the LEF/TCF family to initiate transcription of target genes. WNT5A/ROR2 signaling inhibits β-catenin-mediated transcriptional responses (Mikels and Nusse, 2006). This was reported to require the TK activity of ROR2 (Mikels *et al.*, 2009). We recently proposed that ROR2 can affect Dishevelled activity (Dvl). During WNT/Frizzled signaling, Dvl is phosphorylated by CK1ε, which is accompanied by a shift in gel migration. Phosphorylated Dvl binds ROR2 and triggers the inhibition of β-catenin signaling (Witte *et al.*, 2010a). Although the inhibitory effect of ROR2 on the canonical WNT pathway is well documented, ROR2 was also reported to promote canonical signaling and it is thus possible that the effects of ROR2 might depend on the cellular context (Billiard *et al.*, 2005; Winkel *et al.*, 2008).

ROR2 and its effects on noncanonical signaling have been extensively analyzed. In Xenopus, ROR2 participates in convergent extension (CE), a process known to depend on noncanonical WNT signaling (Hikasa *et al.*, 2002). CE leads to the elongation and narrowing of the body axis and requires coordinated cell adhesion and polarization (Wallingford *et al.*, 2002). WNT5A binding to ROR2 activates a TK activity-dependent cascade that employs ShcA, PI3-kinase, Akt, and JNK, which culminates in the phosphorylation of the transcription factors c-Jun and Atf2 and the increased transcription of paraxial protocadherin, a protein important for CE (Feike *et al.*, 2010; Schambony and Wedlich, 2007). Further, ROR activity impinges on the planar cell polarity (PCP) pathway (Fig. 7.4B) which controls planar polarity of inner ear hair cells and cell orientation in cartilagionous condensations of the limb (Gao *et al.*, 2011; Wang *et al.*, 2011; Yamamoto *et al.*, 2008). During polarization of cells in cartilagionous condensations, WNT5A-depenent activation of ROR2 allows the formation of a complex between ROR2 and Vangl2; Vangl2 is a well-characterized component of the PCP pathway. This results in the phosphorylation and activation of Vangl via CK1δ and depends on the C-terminal domain of ROR2 (Gao *et al.*, 2011).

A recent study proposes that canonical and noncanonical WNT signaling relies on the use of distinct Frizzled coreceptors, that is, LRP5/6 in canonical and ROR1/2 in noncanonical signaling. The two modes of signaling can be analyzed using two WNT ligands, WNT3A and WNT5A, that activate canonical and noncanonical signaling, respectively. In both signaling modes, a common set of components, that is, Dvl, Axin, and glycogen synthase kinase 3 (GSK3) are recruited to the receptor complexes, and GSK3 then phosphorylates LRP5/6 in canonical and ROR1/2 in noncanonical signaling. Canonical and noncanonical WNTs inhibit the reciprocal pathways by competition for Frizzled binding. Thus, the specific coupling and phosphorylation of distinct coreceptors might be responsible for the activation of the canonical and noncanonical WNT signaling cascades (Grumolato *et al.*, 2010). Further, the secreted glycoprotein CTHRC1 was recently implicated in the selection of the downstream pathway. CTHRC1 stabilizes the WNT/ROR2/Frizzled complex and activated the PCP pathway but suppressed the canonical pathway (Yamamoto *et al.*, 2008).

3.3. BDB1: ROR2 in digit development

Elongation of cartilaginous condensations determines the length and number of phalanges in fingers and toes. The human brachydactylies are characterized by variable degrees of digit shortening, and the mutations causing brachydactylies provide clues to the molecular mechanisms that govern digit elongation. Several pathways are implicated in this process, most importantly BMP, WNT, and Hh signaling (Stricker and Mundlos, 2011; Witte *et al.*, 2010b).

A knock-in mouse mutant expressing a truncated form of ROR2 (Ror2W749X mutation), which causes BDB1 in humans, revealed the role of ROR2 in digit formation. This mutation in the mouse leads to the absence of middle phalanges. Elongation of the digital rays relies on a signaling center distal to the growing condensation which is called the phalanx-forming region or digit crescent (Montero *et al.*, 2008; Suzuki *et al.*, 2008; Witte *et al.*, 2010b). These cells are exposed to high levels of BMP signaling as evidenced by phosphorylation of SMAD1/5/8. In Ror2W749X homozygous mutant mice, reduced SMAD1/5/8 phosphorylation in the phalanx-forming region is observable. This resulted in a strongly impaired elongation of the digit condensations due to reduced chondrogenic commitment of undifferentiated distal mesenchymal cells. Interestingly, heterozygous Ror2W749X mutations and a heterozygous IHH mutation interact genetically, indicating that ROR2 and IHH cooperate to maintain the phalanx-forming region. The truncated ROR2 produced from the Ror2W749X allele might act as a scavenger for WNT5A and therefore may interfere with ROR2-dependent and -independent WNT5A signaling. WNT5A is known to inhibit canonical WNT signaling not only via ROR2 (Mikels and Nusse, 2006) but also via a Frizzled-dependent pathway involving Ca^{2+}/CamKII (Calmodulin-dependent protein kinase II) (Ishitani *et al.*, 2003). Consistent with this Ror2W749X mice display elevated β-catenin signaling in distal limb mesenchyme (Witte *et al.*, 2010b). The elevation of the canonical WNT/β-catenin signaling pathway might contribute to the digit phenotype, since canonical WNT signaling is known to inhibit chondrogenesis *in vitro* and *in vivo* (Rudnicki and Brown, 1997; ten Berge *et al.*, 2008).

3.4. RRS: ROR2 in the cartilage growth plate

RRS patients show mesomelic limb shortening, that is, a reduction in the middle segment of the limb and short stature. This phenotype is recapitulated in the ROR2 null mouse (Schwabe *et al.*, 2004). The growth plate in these mice shows two prominent phenotypes: (i) a delay in chondrocyte hypertrophic differentiation and (ii) a failure in the formation of columnar chondrocytes in the proliferation zone. However, the proliferation rate of chondrocytes was unchanged (Schwabe *et al.*, 2004). Chondrocytes undergo a series of peculiar changes after cell division that lead to the emergence of chondrocyte stacks. Cells divide perpendicular to the growth axis, but then slip on top of each other, forming a two-pair stack of columnar chondrocytes parallel to longitudinal axis of growth plate. This depends on the adhesive capacities of the chondrocytes, and deletion of β1-integrin in growth plate chondrocytes interferes with the formation of columnar chondrocytes (Aszodi *et al.*, 2003). Primary β1-integrin-mutant chondrocytes display impaired matrix interaction, a change in the reorganization of the actin cytoskeleton, and reduced cell motility. Interestingly, ROR2 also affects cell motility. WNT5A induces cellular motility in several experimental systems and induces the formation of cellular protrusions. Nishita *et al.* (2006) showed that in mouse embryonic fibroblasts (MEFs), filopodia formation and cell migration into a scratch wound depend on the presence of ROR2. The mobilization of the cells is in part mediated by the interaction of ROR2 with Filamin A, an actin-binding protein. Further, JNK is activated in a polarized fashion facing the wounded edge in the scratch assay (Nishita *et al.*, 2006; Nomachi *et al.*, 2008). Thus, the change in formation of columnar chondrocytes in ROR2 mutant mice might be linked to a defect in cell polarization and motility.

REFERENCES

ADHR Consortium (2000). Autosomal dominant hypophosphataemic rickets is associated with mutations in FGF23. *Nat. Genet.* **26,** 345–348.

Afzal, A. R., Rajab, A., Fenske, C. D., Oldridge, M., Elanko, N., Ternes-Pereira, E., Tuysuz, B., Murday, V. A., Patton, M. A., Wilkie, A. O., and Jeffery, S. (2000). Recessive Robinow syndrome, allelic to dominant brachydactyly type B, is caused by mutation of ROR2. *Nat. Genet.* **25,** 419–422.

Akbarzadeh, S., Wheldon, L. M., Sweet, S. M., Talma, S., Mardakheh, F. K., and Heath, J. K. (2008). The deleted in brachydactyly B domain of ROR2 is required for receptor activation by recruitment of Src. *PLoS One* **3,** e1873.

Aszodi, A., Hunziker, E. B., Brakebusch, C., and Fassler, R. (2003). Beta1 integrins regulate chondrocyte rotation, G1 progression, and cytokinesis. *Genes Dev.* **17,** 2465–2479.

Baroni, T., Carinci, P., Lilli, C., Bellucci, C., Aisa, M. C., Scapoli, L., Volinia, S., Carinci, F., Pezzetti, F., Calvitti, M., Farina, A., Conte, C., *et al.* (2005). P253R fibroblast growth factor receptor-2 mutation induces RUNX2 transcript variants and calvarial osteoblast differentiation. *J. Cell. Physiol.* **202,** 524–535.

Barrow, J. R., Thomas, K. R., Boussadia-Zahui, O., Moore, R., Kemler, R., Capecchi, M. R., and McMahon, A. P. (2003). Ectodermal Wnt3/beta-catenin signaling is required for the establishment and maintenance of the apical ectodermal ridge. *Genes Dev.* **17,** 394–409.

Bartels, C. F., Bukulmez, H., Padayatti, P., Rhee, D. K., van Ravenswaaij-Arts, C., Pauli, R. M., Mundlos, S., Chitayat, D., Shih, L. Y., Al-Gazali, L. I., Kant, S., Cole, T., *et al.* (2004). Mutations in the transmembrane natriuretic peptide receptor NPR-B impair skeletal growth and cause acromesomelic dysplasia, type Maroteaux. *Am. J. Hum. Genet.* **75,** 27–34.

Battaglia, A., Carey, J. C., and Wright, T. J. (2001). Wolf-Hirschhorn (4p-) syndrome. *Adv. Pediatr.* **48,** 75–113.

Bellus, G. A., McIntosh, I., Smith, E. A., Aylsworth, A. S., Kaitila, I., Horton, W. A., Greenhaw, G. A., Hecht, J. T., and Francomano, C. A. (1995). A recurrent mutation in the tyrosine kinase domain of fibroblast growth factor receptor 3 causes hypochondroplasia. *Nat. Genet.* **10,** 357–359.

Bellus, G. A., Gaudenz, K., Zackai, E. H., Clarke, L. A., Szabo, J., Francomano, C. A., and Muenke, M. (1996). Identical mutations in three different fibroblast growth factor receptor genes in autosomal dominant craniosynostosis syndromes. *Nat. Genet.* **14,** 174–176.

Bergemann, A. D., Cole, F., and Hirschhorn, K. (2005). The etiology of Wolf-Hirschhorn syndrome. *Trends Genet.* **21,** 188–195.

Billiard, J., Way, D. S., Seestaller-Wehr, L. M., Moran, R. A., Mangine, A., and Bodine, P. V. (2005). The orphan receptor tyrosine kinase Ror2 modulates canonical Wnt signaling in osteoblastic cells. *Mol. Endocrinol.* **19,** 90–101.

Casanova, J. C., and Sanz-Ezquerro, J. J. (2007). Digit morphogenesis: Is the tip different? *Dev. Growth Differ.* **49,** 479–491.

Catela, C., Bilbao-Cortes, D., Slonimsky, E., Kratsios, P., Rosenthal, N., and Te Welscher, P. (2009). Multiple congenital malformations of Wolf-Hirschhorn syndrome are recapitulated in Fgfrl1 null mice. *Dis. Model. Mech.* **2,** 283–294.

Chefetz, I., Heller, R., Galli-Tsinopoulou, A., Richard, G., Wollnik, B., Indelman, M., Koerber, F., Topaz, O., Bergman, R., Sprecher, E., and Schoenau, E. (2005). A novel homozygous missense mutation in FGF23 causes Familial Tumoral Calcinosis associated with disseminated visceral calcification. *Hum. Genet.* **118,** 261–266.

Chen, Y., Bellamy, W. P., Seabra, M. C., Field, M. C., and Ali, B. R. (2005). ER-associated protein degradation is a common mechanism underpinning numerous monogenic diseases including Robinow syndrome. *Hum. Mol. Genet.* **14,** 2559–2569.

Colvin, J. S., Bohne, B. A., Harding, G. W., McEwen, D. G., and Ornitz, D. M. (1996). Skeletal overgrowth and deafness in mice lacking fibroblast growth factor receptor 3. *Nat. Genet.* **12,** 390–397.

De Pollack, C., Renier, D., Hott, M., and Marie, P. J. (1996). Increased bone formation and osteoblastic cell phenotype in premature cranial suture ossification (craniosynostosis). *J. Bone Miner. Res.* **11,** 401–407.

DeChiara, T. M., Kimble, R. B., Poueymirou, W. T., Rojas, J., Masiakowski, P., Valenzuela, D. M., and Yancopoulos, G. D. (2000). Ror2, encoding a receptor-like tyrosine kinase, is required for cartilage and growth plate development. *Nat. Genet.* **24,** 271–274.

Econs, M. J., and McEnery, P. T. (1997). Autosomal dominant hypophosphatemic rickets/osteomalacia: Clinical characterization of a novel renal phosphate-wasting disorder. *J. Clin. Endocrinol. Metab.* **82,** 674–681.

Engbers, H., van der Smagt, J. J., van 't Slot, R., Vermeesch, J. R., Hochstenbach, R., and Poot, M. (2009). Wolf-Hirschhorn syndrome facial dysmorphic features in a patient with a terminal 4p16.3 deletion telomeric to the WHSCR and WHSCR 2 regions. *Eur. J. Hum. Genet.* **17,** 129–132.

Eswarakumar, V. P., Horowitz, M. C., Locklin, R., Morriss-Kay, G. M., and Lonai, P. (2004). A gain-of-function mutation of Fgfr2c demonstrates the roles of this receptor variant in osteogenesis. *Proc. Natl. Acad. Sci. USA* **101,** 12555–12560.

Feike, A. C., Rachor, K., Gentzel, M., and Schambony, A. (2010). Wnt5a/Ror2-induced upregulation of xPAPC requires xShcA. *Biochem. Biophys. Res. Commun.* **400,** 500–506.

Forrester, W. C., Dell, M., Perens, E., and Garriga, G. (1999). A *C. elegans* Ror receptor tyrosine kinase regulates cell motility and asymmetric cell division. *Nature* **400,** 881–885.

Gao, B., Song, H., Bishop, K., Elliot, G., Garrett, L., English, M. A., Andre, P., Robinson, J., Sood, R., Minami, Y., Economides, A. N., and Yang, Y. (2011). Wnt signaling gradients establish planar cell polarity by inducing Vangl2 phosphorylation through Ror2. *Dev. Cell* **20,** 163–176.

Glass, D. A., 2nd, Bialek, P., Ahn, J. D., Starbuck, M., Patel, M. S., Clevers, H., Taketo, M. M., Long, F., McMahon, A. P., Lang, R. A., and Karsenty, G. (2005). Canonical Wnt signaling in differentiated osteoblasts controls osteoclast differentiation. *Dev. Cell* **8,** 751–764.

Goldfarb, M. (2005). Fibroblast growth factor homologous factors: Evolution, structure, and function. *Cytokine Growth Factor Rev.* **16,** 215–220.

Green, J. L., Kuntz, S. G., and Sternberg, P. W. (2008). Ror receptor tyrosine kinases: Orphans no more. *Trends Cell Biol.* **18,** 536–544.

Grigoryan, T., Wend, P., Klaus, A., and Birchmeier, W. (2008). Deciphering the function of canonical Wnt signals in development and disease: Conditional loss- and gain-of-function mutations of beta-catenin in mice. *Genes Dev.* **22,** 2308–2341.

Grumolato, L., Liu, G., Mong, P., Mudbhary, R., Biswas, R., Arroyave, R., Vijayakumar, S., Economides, A. N., and Aaronson, S. A. (2010). Canonical and noncanonical Wnts use a common mechanism to activate completely unrelated coreceptors. *Genes Dev.* **24,** 2517–2530.

Guenou, H., Kaabeche, K., Mee, S. L., and Marie, P. J. (2005). A role for fibroblast growth factor receptor-2 in the altered osteoblast phenotype induced by twist haploinsufficiency in the Saethre-Chotzen syndrome. *Hum. Mol. Genet.* **14,** 1429–1439.

Hall, B. K., and Miyake, T. (2000). All for one and one for all: Condensations and the initiation of skeletal development. *Bioessays* **22,** 138–147.

Harada, M., Murakami, H., Okawa, A., Okimoto, N., Hiraoka, S., Nakahara, T., Akasaka, R., Shiraishi, Y., Futatsugi, N., Mizutani-Koseki, Y., Kuroiwa, A., Shirouzu, M., *et al.* (2009). FGF9 monomer-dimer equilibrium regulates extracellular matrix affinity and tissue diffusion. *Nat. Genet.* **41,** 289–298.

Hartmann, C. (2007). Skeletal development: Wnts are in control. *Mol. Cells* **24,** 177–184.

Hikasa, H., Shibata, M., Hiratani, I., and Taira, M. (2002). The Xenopus receptor tyrosine kinase Xror2 modulates morphogenetic movements of the axial mesoderm and neuroectoderm via Wnt signaling. *Development* **129,** 5227–5239.

Hillier, L. W., Miller, W., Birney, E., Warren, W., Hardison, R. C., Ponting, C. P., Bork, P., Burt, D. W., Groenen, M. A., Delany, M. E., Dodgson, J. B., Chinwalla, A. T., *et al.* (2004). Sequence and comparative analysis of the chicken genome provide unique perspectives on vertebrate evolution. *Nature* **432,** 695–716.

Holmes, G., Rothschild, G., Roy, U. B., Deng, C. X., Mansukhani, A., and Basilico, C. (2009). Early onset of craniosynostosis in an Apert mouse model reveals critical features of this pathology. *Dev. Biol.* **328,** 273–284.

Horton, W. A., and Degnin, C. R. (2009). FGFs in endochondral skeletal development. *Trends Endocrinol. Metab.* **20,** 341–348.

Horton, W. A., Hall, J. G., and Hecht, J. T. (2007). Achondroplasia. *Lancet* **370,** 162–172.

Ibrahimi, O. A., Chiu, E. S., McCarthy, J. G., and Mohammadi, M. (2005). Understanding the molecular basis of Apert syndrome. *Plast. Reconstr. Surg.* **115,** 264–270.

Iseki, S., Wilkie, A. O., and Morriss-Kay, G. M. (1999). Fgfr1 and Fgfr2 have distinct differentiation- and proliferation-related roles in the developing mouse skull vault. *Development* **126,** 5611–5620.

Ishitani, T., Kishida, S., Hyodo-Miura, J., Ueno, N., Yasuda, J., Waterman, M., Shibuya, H., Moon, R. T., Ninomiya-Tsuji, J., and Matsumoto, K. (2003). The TAK1-NLK mitogen-activated protein kinase cascade functions in the Wnt-5a/Ca (2+) pathway to antagonize Wnt/beta-catenin signaling. *Mol. Cell. Biol.* **23,** 131–139.

Itoh, N., and Ornitz, D. M. (2008). Functional evolutionary history of the mouse Fgf gene family. *Dev. Dyn.* **237,** 18–27.

Jabs, E. W., Muller, U., Li, X., Ma, L., Luo, W., Haworth, I. S., Klisak, I., Sparkes, R., Warman, M. L., Mulliken, J. B., *et al.* (1993). A mutation in the homeodomain of the human MSX2 gene in a family affected with autosomal dominant craniosynostosis. *Cell* **75,** 443–450.

Jabs, E. W., Li, X., Scott, A. F., Meyers, G., Chen, W., Eccles, M., Mao, J. I., Charnas, L. R., Jackson, C. E., and Jaye, M. (1994). Jackson-Weiss and Crouzon syndromes are allelic with mutations in fibroblast growth factor receptor 2. *Nat. Genet.* **8,** 275–279.

Johnson, D., and Wilkie, A. O. (2011). Craniosynostosis. *Eur. J. Hum. Genet.* **19**(4), 369–376.

Kan, S. H., Elanko, N., Johnson, D., Cornejo-Roldan, L., Cook, J., Reich, E. W., Tomkins, S., Verloes, A., Twigg, S. R., Rannan-Eliya, S., McDonald-McGinn, D. M., Zackai, E. H., *et al.* (2002). Genomic screening of fibroblast growth-factor receptor 2 reveals a wide spectrum of mutations in patients with syndromic craniosynostosis. *Am. J. Hum. Genet.* **70,** 472–486.

Karp, S. J., Schipani, E., St-Jacques, B., Hunzelman, J., Kronenberg, H., and McMahon, A. P. (2000). Indian hedgehog coordinates endochondral bone growth and morphogenesis via parathyroid hormone related-protein-dependent and -independent pathways. *Development* **127,** 543–548.

Karsenty, G., and Wagner, E. F. (2002). Reaching a genetic and molecular understanding of skeletal development. *Dev. Cell* **2,** 389–406.

Kim, S. H., Yamamoto, A., Bouwmeester, T., Agius, E., and Robertis, E. M. (1998). The role of paraxial protocadherin in selective adhesion and cell movements of the mesoderm during Xenopus gastrulation. *Development* **125,** 4681–4690.

Kim, B. G., Kim, H. J., Park, H. J., Kim, Y. J., Yoon, W. J., Lee, S. J., Ryoo, H. M., and Cho, J. Y. (2006). Runx2 phosphorylation induced by fibroblast growth factor-2/protein kinase C pathways. *Proteomics* **6,** 1166–1174.

Komori, T., Yagi, H., Nomura, S., Yamaguchi, A., Sasaki, K., Deguchi, K., Shimizu, Y., Bronson, R. T., Gao, Y. H., Inada, M., Sato, M., Okamoto, R., *et al.* (1997). Targeted disruption of Cbfa1 results in a complete lack of bone formation owing to maturational arrest of osteoblasts. *Cell* **89,** 755–764.

Krejci, P., Masri, B., Fontaine, V., Mekikian, P. B., Weis, M., Prats, H., and Wilcox, W. R. (2005). Interaction of fibroblast growth factor and C-natriuretic peptide signaling in regulation of chondrocyte proliferation and extracellular matrix homeostasis. *J. Cell Sci.* **118,** 5089–5100.

Krejci, P., Salazar, L., Goodridge, H. S., Kashiwada, T. A., Schibler, M. J., Jelinkova, P., Thompson, L. M., and Wilcox, W. R. (2008). STAT1 and STAT3 do not participate in FGF-mediated growth arrest in chondrocytes. *J. Cell Sci.* **121,** 272–281.

Krejci, P., Prochazkova, J., Bryja, V., Kozubik, A., and Wilcox, W. R. (2009). Molecular pathology of the fibroblast growth factor family. *Hum. Mutat.* **30,** 1245–1255.

Larsson, T., Marsell, R., Schipani, E., Ohlsson, C., Ljunggren, O., Tenenhouse, H. S., Juppner, H., and Jonsson, K. B. (2004). Transgenic mice expressing fibroblast growth factor 23 under the control of the alpha1(I) collagen promoter exhibit growth retardation, osteomalacia, and disturbed phosphate homeostasis. *Endocrinology* **145,** 3087–3094.

Lenton, K. A., Nacamuli, R. P., Wan, D. C., Helms, J. A., and Longaker, M. T. (2005). Cranial suture biology. *Curr. Top. Dev. Biol.* **66,** 287–328.

Li, C., Chen, L., Iwata, T., Kitagawa, M., Fu, X. Y., and Deng, C. X. (1999). A Lys644Glu substitution in fibroblast growth factor receptor 3 (FGFR3) causes dwarfism in mice by activation of STATs and ink4 cell cycle inhibitors. *Hum. Mol. Genet.* **8,** 35–44.

Liu, Z., Xu, J., Colvin, J. S., and Ornitz, D. M. (2002). Coordination of chondrogenesis and osteogenesis by fibroblast growth factor 18. *Genes Dev.* **16,** 859–869.

Liu, Y., Ross, J. F., Bodine, P. V., and Billiard, J. (2007). Homodimerization of Ror2 tyrosine kinase receptor induces 14-3-3(beta) phosphorylation and promotes osteoblast differentiation and bone formation. *Mol. Endocrinol.* **21,** 3050–3061.

Liu, Y., Rubin, B., Bodine, P. V., and Billiard, J. (2008). Wnt5a induces homodimerization and activation of Ror2 receptor tyrosine kinase. *J. Cell. Biochem.* **105,** 497–502.

Loeys, B. L., Chen, J., Neptune, E. R., Judge, D. P., Podowski, M., Holm, T., Meyers, J., Leitch, C. C., Katsanis, N., Sharifi, N., Xu, F. L., Myers, L. A., *et al.* (2005). A syndrome of altered cardiovascular, craniofacial, neurocognitive and skeletal development caused by mutations in TGFBR1 or TGFBR2. *Nat. Genet.* **37,** 275–281.

Loeys, B. L., Schwarze, U., Holm, T., Callewaert, B. L., Thomas, G. H., Pannu, H., De Backer, J. F., Oswald, G. L., Symoens, S., Manouvrier, S., Roberts, A. E., Faravelli, F., *et al.* (2006). Aneurysm syndromes caused by mutations in the TGF-beta receptor. *N. Engl. J. Med.* **355,** 788–798.

Lomri, A., Lemonnier, J., Hott, M., de Parseval, N., Lajeunie, E., Munnich, A., Renier, D., and Marie, P. J. (1998). Increased calvaria cell differentiation and bone matrix formation induced by fibroblast growth factor receptor 2 mutations in Apert syndrome. *J. Clin. Invest.* **101,** 1310–1317.

Masiakowski, P., and Carroll, R. D. (1992). A novel family of cell surface receptors with tyrosine kinase-like domain. *J. Biol. Chem.* **267,** 26181–26190.

Merino, R., Ganan, Y., Macias, D., Economides, A. N., Sampath, K. T., and Hurle, J. M. (1998). Morphogenesis of digits in the avian limb is controlled by FGFs, TGFbetas, and noggin through BMP signaling. *Dev. Biol.* **200,** 35–45.

Meyers, G. A., Orlow, S. J., Munro, I. R., Przylepa, K. A., and Jabs, E. W. (1995). Fibroblast growth factor receptor 3 (FGFR3) transmembrane mutation in Crouzon syndrome with acanthosis nigricans. *Nat. Genet.* **11,** 462–464.

Mikels, A. J., and Nusse, R. (2006). Purified Wnt5a protein activates or inhibits beta-catenin-TCF signaling depending on receptor context. *PLoS Biol.* **4,** e115.

Mikels, A., Minami, Y., and Nusse, R. (2009). Ror2 receptor requires tyrosine kinase activity to mediate Wnt5A signaling. *J. Biol. Chem.* **284,** 30167–30176.

Milunsky, J. M., Zhao, G., Maher, T. A., Colby, R., and Everman, D. B. (2006). LADD syndrome is caused by FGF10 mutations. *Clin. Genet.* **69,** 349–354.

Minami, Y., Oishi, I., Endo, M., and Nishita, M. (2010). Ror-family receptor tyrosine kinases in noncanonical Wnt signaling: Their implications in developmental morphogenesis and human diseases. *Dev. Dyn.* **239,** 1–15.

Minina, E., Kreschel, C., Naski, M. C., Ornitz, D. M., and Vortkamp, A. (2002). Interaction of FGF, Ihh/Pthlh, and BMP signaling integrates chondrocyte proliferation and hypertrophic differentiation. *Dev. Cell* **3,** 439–449.

Miraoui, H., Oudina, K., Petite, H., Tanimoto, Y., Moriyama, K., and Marie, P. J. (2009). Fibroblast growth factor receptor 2 promotes osteogenic differentiation in mesenchymal cells via ERK1/2 and protein kinase C signaling. *J. Biol. Chem.* **284,** 4897–4904.

Miraoui, H., Severe, N., Vaudin, P., Pages, J. C., and Marie, P. J. (2010). Molecular silencing of Twist1 enhances osteogenic differentiation of murine mesenchymal stem cells: Implication of FGFR2 signaling. *J. Cell. Biochem.* **110,** 1147–1154.

Montero, J. A., and Hurle, J. M. (2010). Sculpturing digit shape by cell death. *Apoptosis* **15,** 365–375.

Montero, J. A., Lorda-Diez, C. I., Ganan, Y., Macias, D., and Hurle, J. M. (2008). Activin/TGFbeta and BMP crosstalk determines digit chondrogenesis. *Dev. Biol.* **321,** 343–356.

Morriss-Kay, G. M., and Wilkie, A. O. (2005). Growth of the normal skull vault and its alteration in craniosynostosis: Insights from human genetics and experimental studies. *J. Anat.* **207,** 637–653.

Muenke, M., and Schell, U. (1995). Fibroblast-growth-factor receptor mutations in human skeletal disorders. *Trends Genet.* **11,** 308–313.

Muenke, M., Schell, U., Hehr, A., Robin, N. H., Losken, H. W., Schinzel, A., Pulleyn, L. J., Rutland, P., Reardon, W., Malcolm, S., *et al.* (1994). A common mutation in the fibroblast growth factor receptor 1 gene in Pfeiffer syndrome. *Nat. Genet.* **8,** 269–274.

Muenke, M., Gripp, K. W., McDonald-McGinn, D. M., Gaudenz, K., Whitaker, L. A., Bartlett, S. P., Markowitz, R. I., Robin, N. H., Nwokoro, N., Mulvihill, J. J., Losken, H. W., Mulliken, J. B., *et al.* (1997). A unique point mutation in the fibroblast growth factor receptor 3 gene (FGFR3) defines a new craniosynostosis syndrome. *Am. J. Hum. Genet.* **60,** 555–564.

Mundlos, S. (2000). Skeletal morphogenesis. *Methods Mol. Biol.* **136,** 61–70.

Mundlos, S. (2009). The brachydactylies: A molecular disease family. *Clin. Genet.* **76,** 123–136.

Mundlos, S., Otto, F., Mundlos, C., Mulliken, J. B., Aylsworth, A. S., Albright, S., Lindhout, D., Cole, W. G., Henn, W., Knoll, J. H., Owen, M. J., Mertelsmann, R., *et al.* (1997). Mutations involving the transcription factor CBFA1 cause cleidocranial dysplasia. *Cell* **89,** 773–779.

Murakami, S., Balmes, G., McKinney, S., Zhang, Z., Givol, D., and de Crombrugghe, B. (2004). Constitutive activation of MEK1 in chondrocytes causes Stat1-independent achondroplasia-like dwarfism and rescues the Fgfr3-deficient mouse phenotype. *Genes Dev.* **18,** 290–305.

Naf, D., Wilson, L. A., Bergstrom, R. A., Smith, R. S., Goodwin, N. C., Verkerk, A., van Ommen, G. J., Ackerman, S. L., Frankel, W. N., and Schimenti, J. C. (2001). Mouse models for the Wolf-Hirschhorn deletion syndrome. *Hum. Mol. Genet.* **10,** 91–98.

Naski, M. C., Colvin, J. S., Coffin, J. D., and Ornitz, D. M. (1998). Repression of hedgehog signaling and BMP4 expression in growth plate cartilage by fibroblast growth factor receptor 3. *Development* **125,** 4977–4988.

Niemann, S., Zhao, C., Pascu, F., Stahl, U., Aulepp, U., Niswander, L., Weber, J. L., and Muller, U. (2004). Homozygous WNT3 mutation causes tetra-amelia in a large consanguineous family. *Am. J. Hum. Genet.* **74,** 558–563.

Nishita, M., Yoo, S. K., Nomachi, A., Kani, S., Sougawa, N., Ohta, Y., Takada, S., Kikuchi, A., and Minami, Y. (2006). Filopodia formation mediated by receptor tyrosine kinase Ror2 is required for Wnt5a-induced cell migration. *J. Cell Biol.* **175,** 555–562.

Niswander, L., Tickle, C., Vogel, A., Booth, I., and Martin, G. R. (1993). FGF-4 replaces the apical ectodermal ridge and directs outgrowth and patterning of the limb. *Cell* **75,** 579–587.

Niswander, L., Jeffrey, S., Martin, G. R., and Tickle, C. (1994). A positive feedback loop coordinates growth and patterning in the vertebrate limb. *Nature* **371,** 609–612.

Nomachi, A., Nishita, M., Inaba, D., Enomoto, M., Hamasaki, M., and Minami, Y. (2008). Receptor tyrosine kinase Ror2 mediates Wnt5a-induced polarized cell migration by activating c-Jun N-terminal kinase via actin-binding protein filamin A. *J. Biol. Chem.* **283,** 27973–27981.

Ohbayashi, N., Shibayama, M., Kurotaki, Y., Imanishi, M., Fujimori, T., Itoh, N., and Takada, S. (2002). FGF18 is required for normal cell proliferation and differentiation during osteogenesis and chondrogenesis. *Genes Dev.* **16,** 870–879.

Oishi, I., Suzuki, H., Onishi, N., Takada, R., Kani, S., Ohkawara, B., Koshida, I., Suzuki, K., Yamada, G., Schwabe, G. C., Mundlos, S., Shibuya, H., *et al.* (2003). The receptor tyrosine kinase Ror2 is involved in non-canonical Wnt5a/JNK signalling pathway. *Genes Cells* **8,** 645–654.

Oldridge, M., Lunt, P. W., Zackai, E. H., McDonald-McGinn, D. M., Muenke, M., Moloney, D. M., Twigg, S. R., Heath, J. K., Howard, T. D., Hoganson, G., Gagnon, D. M., Jabs, E. W., *et al.* (1997). Genotype-phenotype correlation for nucleotide substitutions in the IgII-IgIII linker of FGFR2. *Hum. Mol. Genet.* **6,** 137–143.

Oldridge, M., Zackai, E. H., McDonald-McGinn, D. M., Iseki, S., Morriss-Kay, G. M., Twigg, S. R., Johnson, D., Wall, S. A., Jiang, W., Theda, C., Jabs, E. W., and Wilkie, A. O. (1999). De novo alu-element insertions in FGFR2 identify a distinct pathological basis for Apert syndrome. *Am. J. Hum. Genet.* **64,** 446–461.

Oldridge, M., Fortuna, A. M., Maringa, M., Propping, P., Mansour, S., Pollitt, C., DeChiara, T. M., Kimble, R. B., Valenzuela, D. M., Yancopoulos, G. D., and Wilkie, A. O. (2000). Dominant mutations in ROR2, encoding an orphan receptor tyrosine kinase, cause brachydactyly type B. *Nat. Genet.* **24,** 275–278.

Opperman, L. A., Fernandez, C. R., So, S., and Rawlins, J. T. (2006). Erk1/2 signaling is required for Tgf-beta 2-induced suture closure. *Dev. Dyn.* **235,** 1292–1299.

Ornitz, D. M., and Marie, P. J. (2002). FGF signaling pathways in endochondral and intramembranous bone development and human genetic disease. *Genes Dev.* **16,** 1446–1465.

Ornitz, D. M., Xu, J., Colvin, J. S., McEwen, D. G., MacArthur, C. A., Coulier, F., Gao, G., and Goldfarb, M. (1996). Receptor specificity of the fibroblast growth factor family. *J. Biol. Chem.* **271,** 15292–15297.

Ota, S., Zhou, Z. Q., Keene, D. R., Knoepfler, P., and Hurlin, P. J. (2007). Activities of N-Myc in the developing limb link control of skeletal size with digit separation. *Development* **134,** 1583–1592.

Otto, F., Thornell, A. P., Crompton, T., Denzel, A., Gilmour, K. C., Rosewell, I. R., Stamp, G. W., Beddington, R. S., Mundlos, S., Olsen, B. R., Selby, P. B., and Owen, M. J. (1997). Cbfa1, a candidate gene for cleidocranial dysplasia syndrome, is essential for osteoblast differentiation and bone development. *Cell* **89,** 765–771.

Ovchinnikov, D. A., Selever, J., Wang, Y., Chen, Y. T., Mishina, Y., Martin, J. F., and Behringer, R. R. (2006). BMP receptor type IA in limb bud mesenchyme regulates distal outgrowth and patterning. *Dev. Biol.* **295,** 103–115.

Pajni-Underwood, S., Wilson, C. P., Elder, C., Mishina, Y., and Lewandoski, M. (2007). BMP signals control limb bud interdigital programmed cell death by regulating FGF signaling. *Development* **134,** 2359–2368.

Park, W. J., Meyers, G. A., Li, X., Theda, C., Day, D., Orlow, S. J., Jones, M. C., and Jabs, E. W. (1995). Novel FGFR2 mutations in Crouzon and Jackson-Weiss syndromes show allelic heterogeneity and phenotypic variability. *Hum. Mol. Genet.* **4,** 1229–1233.

Park, O. J., Kim, H. J., Woo, K. M., Baek, J. H., and Ryoo, H. M. (2010). FGF2-activated ERK mitogen-activated protein kinase enhances Runx2 acetylation and stabilization. *J. Biol. Chem.* **285,** 3568–3574.

Paznekas, W. A., Cunningham, M. L., Howard, T. D., Korf, B. R., Lipson, M. H., Grix, A. W., Feingold, M., Goldberg, R., Borochowitz, Z., Aleck, K., Mulliken, J., Yin, M., *et al.* (1998). Genetic heterogeneity of Saethre-Chotzen syndrome, due to TWIST and FGFR mutations. *Am. J. Hum. Genet.* **62,** 1370–1380.

Peters, K., Ornitz, D., Werner, S., and Williams, L. (1993). Unique expression pattern of the FGF receptor 3 gene during mouse organogenesis. *Dev. Biol.* **155,** 423–430.

Rawlins, J. T., and Opperman, L. A. (2008). Tgf-beta regulation of suture morphogenesis and growth. *Front. Oral Biol.* **12,** 178–196.

Reardon, W., Winter, R. M., Rutland, P., Pulleyn, L. J., Jones, B. M., and Malcolm, S. (1994). Mutations in the fibroblast growth factor receptor 2 gene cause Crouzon syndrome. *Nat. Genet.* **8,** 98–103.

Rice, D. P., Aberg, T., Chan, Y., Tang, Z., Kettunen, P. J., Pakarinen, L., Maxson, R. E., and Thesleff, I. (2000). Integration of FGF and TWIST in calvarial bone and suture development. *Development* **127,** 1845–1855.

Rice, D. P., Rice, R., and Thesleff, I. (2003). Molecular mechanisms in calvarial bone and suture development, and their relation to craniosynostosis. *Eur. J. Orthod.* **25,** 139–148.

Rieckmann, T., Zhuang, L., Fluck, C. E., and Trueb, B. (2009). Characterization of the first FGFRL1 mutation identified in a craniosynostosis patient. *Biochim. Biophys. Acta* **1792,** 112–121.

Riley, B. M., Mansilla, M. A., Ma, J., Daack-Hirsch, S., Maher, B. S., Raffensperger, L. M., Russo, E. T., Vieira, A. R., Dode, C., Mohammadi, M., Marazita, M. L., and Murray, J. C. (2007). Impaired FGF signaling contributes to cleft lip and palate. *Proc. Natl. Acad. Sci. USA* **104,** 4512–4517.

Rohmann, E., Brunner, H. G., Kayserili, H., Uyguner, O., Nurnberg, G., Lew, E. D., Dobbie, A., Eswarakumar, V. P., Uzumcu, A., Ulubil-Emeroglu, M., Leroy, J. G., Li, Y., *et al.* (2006). Mutations in different components of FGF signaling in LADD syndrome. *Nat. Genet.* **38,** 414–417.

Rossi, M., Jones, R. L., Norbury, G., Bloch-Zupan, A., and Winter, R. M. (2003). The appearance of the feet in Pfeiffer syndrome caused by FGFR1 P252R mutation. *Clin. Dysmorphol.* **12,** 269–274.

Rousseau, F., Bonaventure, J., Legeai-Mallet, L., Pelet, A., Rozet, J. M., Maroteaux, P., Le Merrer, M., and Munnich, A. (1994). Mutations in the gene encoding fibroblast growth factor receptor-3 in achondroplasia. *Nature* **371,** 252–254.

Rousseau, F., Saugier, P., Le Merrer, M., Munnich, A., Delezoide, A. L., Maroteaux, P., Bonaventure, J., Narcy, F., and Sanak, M. (1995). Stop codon FGFR3 mutations in thanatophoric dwarfism type 1. *Nat. Genet.* **10,** 11–12.

Rudnicki, J. A., and Brown, A. M. (1997). Inhibition of chondrogenesis by Wnt gene expression in vivo and in vitro. *Dev. Biol.* **185,** 104–118.

Sahni, M., Ambrosetti, D. C., Mansukhani, A., Gertner, R., Levy, D., and Basilico, C. (1999). FGF signaling inhibits chondrocyte proliferation and regulates bone development through the STAT-1 pathway. *Genes Dev.* **13,** 1361–1366.

Sanz-Ezquerro, J. J., and Tickle, C. (2003). Fgf signaling controls the number of phalanges and tip formation in developing digits. *Curr. Biol.* **13,** 1830–1836.

Schambony, A., and Wedlich, D. (2007). Wnt-5A/Ror2 regulate expression of XPAPC through an alternative noncanonical signaling pathway. *Dev. Cell* **12,** 779–792.

Schwabe, G. C., Tinschert, S., Buschow, C., Meinecke, P., Wolff, G., Gillessen-Kaesbach, G., Oldridge, M., Wilkie, A. O., Komec, R., and Mundlos, S. (2000). Distinct mutations in the receptor tyrosine kinase gene ROR2 cause brachydactyly type B. *Am. J. Hum. Genet.* **67,** 822–831.

Schwabe, G. C., Trepczik, B., Suring, K., Brieske, N., Tucker, A. S., Sharpe, P. T., Minami, Y., and Mundlos, S. (2004). Ror2 knockout mouse as a model for the developmental pathology of autosomal recessive Robinow syndrome. *Dev. Dyn.* **229,** 400–410.

Schwarzer, W., Witte, F., Rajab, A., Mundlos, S., and Stricker, S. (2009). A gradient of ROR2 protein stability and membrane localization confers brachydactyly type B or Robinow syndrome phenotypes. *Hum. Mol. Genet.* **18,** 4013–4021.

Shiang, R., Thompson, L. M., Zhu, Y. Z., Church, D. M., Fielder, T. J., Bocian, M., Winokur, S. T., and Wasmuth, J. J. (1994). Mutations in the transmembrane domain of FGFR3 cause the most common genetic form of dwarfism, achondroplasia. *Cell* **78,** 335–342.

Shimada, T., Mizutani, S., Muto, T., Yoneya, T., Hino, R., Takeda, S., Takeuchi, Y., Fujita, T., Fukumoto, S., and Yamashita, T. (2001). Cloning and characterization of FGF23 as a causative factor of tumor-induced osteomalacia. *Proc. Natl. Acad. Sci. USA* **98,** 6500–6505.

Shubin, N. H., and Alberch, P. (1986). A morphogenetic approach to the origin and basic organization of the tetrapod limb. *In* "Evolutionary Biology," (M. K. Hecht, B. Wallace, and G. T. Prance, eds.), pp. 319–387. Plenum Press, New York.

Shukla, V., Coumoul, X., Wang, R. H., Kim, H. S., and Deng, C. X. (2007). RNA interference and inhibition of MEK-ERK signaling prevent abnormal skeletal phenotypes in a mouse model of craniosynostosis. *Nat. Genet.* **39,** 1145–1150.

Siebler, T., Robson, H., Shalet, S. M., and Williams, G. R. (2001). Glucocorticoids, thyroid hormone and growth hormone interactions: Implications for the growth plate. *Horm. Res.* **56**(Suppl 1), 7–12.

Slater, B. J., Lenton, K. A., Kwan, M. D., Gupta, D. M., Wan, D. C., and Longaker, M. T. (2008). Cranial sutures: A brief review. *Plast. Reconstr. Surg.* **121,** 170e–178e.

Sleeman, M., Fraser, J., McDonald, M., Yuan, S., White, D., Grandison, P., Kumble, K., Watson, J. D., and Murison, J. G. (2001). Identification of a new fibroblast growth factor receptor, FGFR5. *Gene* **271,** 171–182.

St-Jacques, B., Hammerschmidt, M., and McMahon, A. P. (1999). Indian hedgehog signaling regulates proliferation and differentiation of chondrocytes and is essential for bone formation. *Genes Dev.* **13,** 2072–2086.

Stricker, S., and Mundlos, S. (2011). Mechanisms of digit formation: Human malformation syndromes tell the story. *Dev. Dyn.* **240,** 990–1004.

Sun, X., Lewandoski, M., Meyers, E. N., Liu, Y. H., Maxson, R. E., Jr., and Martin, G. R. (2000). Conditional inactivation of Fgf4 reveals complexity of signalling during limb bud development. *Nat. Genet.* **25,** 83–86.

Sun, X., Mariani, F. V., and Martin, G. R. (2002). Functions of FGF signalling from the apical ectodermal ridge in limb development. *Nature* **418,** 501–508.

Suzuki, T., Hasso, S. M., and Fallon, J. F. (2008). Unique SMAD1/5/8 activity at the phalanx-forming region determines digit identity. *Proc. Natl. Acad. Sci. USA* **105,** 4185–4190.

Takeuchi, S., Takeda, K., Oishi, I., Nomi, M., Ikeya, M., Itoh, K., Tamura, S., Ueda, T., Hatta, T., Otani, H., Terashima, T., Takada, S., *et al.* (2000). Mouse Ror2 receptor tyrosine kinase is required for the heart development and limb formation. *Genes Cells* **5,** 71–78.

Tavormina, P. L., Shiang, R., Thompson, L. M., Zhu, Y. Z., Wilkin, D. J., Lachman, R. S., Wilcox, W. R., Rimoin, D. L., Cohn, D. H., and Wasmuth, J. J. (1995). Thanatophoric dysplasia (types I and II) caused by distinct mutations in fibroblast growth factor receptor 3. *Nat. Genet.* **9,** 321–328.

Tavormina, P. L., Bellus, G. A., Webster, M. K., Bamshad, M. J., Fraley, A. E., McIntosh, I., Szabo, J., Jiang, W., Jabs, E. W., Wilcox, W. R., Wasmuth, J. J., Donoghue, D. J., *et al.* (1999). A novel skeletal dysplasia with developmental delay and acanthosis nigricans is caused by a Lys650Met mutation in the fibroblast growth factor receptor 3 gene. *Am. J. Hum. Genet.* **64,** 722–731.

ten Berge, D., Brugmann, S. A., Helms, J. A., and Nusse, R. (2008). Wnt and FGF signals interact to coordinate growth with cell fate specification during limb development. *Development* **135,** 3247–3257.

Thisse, B., and Thisse, C. (2005). Functions and regulations of fibroblast growth factor signaling during embryonic development. *Dev. Biol.* **287,** 390–402.

Trueb, B., and Taeschler, S. (2006). Expression of FGFRL1, a novel fibroblast growth factor receptor, during embryonic development. *Int. J. Mol. Med.* **17,** 617–620.

Trueb, B., Zhuang, L., Taeschler, S., and Wiedemann, M. (2003). Characterization of FGFRL1, a novel fibroblast growth factor (FGF) receptor preferentially expressed in skeletal tissues. *J. Biol. Chem.* **278,** 33857–33865.

van Bokhoven, H., Celli, J., Kayserili, H., van Beusekom, E., Balci, S., Brussel, W., Skovby, F., Kerr, B., Percin, E. F., Akarsu, N., and Brunner, H. G. (2000). Mutation of the gene encoding the ROR2 tyrosine kinase causes autosomal recessive Robinow syndrome. *Nat. Genet.* **25,** 423–426.

Vortkamp, A., Lee, K., Lanske, B., Segre, G. V., Kronenberg, H. M., and Tabin, C. J. (1996). Regulation of rate of cartilage differentiation by Indian hedgehog and PTH-related protein. *Science* **273,** 613–622.

Wallingford, J. B., Fraser, S. E., and Harland, R. M. (2002). Convergent extension: The molecular control of polarized cell movement during embryonic development. *Dev. Cell* **2,** 695–706.

Wang, B., Sinha, T., Jiao, K., Serra, R., and Wang, J. (2011). Disruption of PCP signaling causes limb morphogenesis and skeletal defects and may underlie Robinow syndrome and brachydactyly type B. *Hum. Mol. Genet.* **20,** 271–285.

White, K. E., Carn, G., Lorenz-Depiereux, B., Benet-Pages, A., Strom, T. M., and Econs, M. J. (2001a). Autosomal-dominant hypophosphatemic rickets (ADHR) mutations stabilize FGF-23. *Kidney Int.* **60,** 2079–2086.

White, K. E., Jonsson, K. B., Carn, G., Hampson, G., Spector, T. D., Mannstadt, M., Lorenz-Depiereux, B., Miyauchi, A., Yang, I. M., Ljunggren, O., Meitinger, T., Strom, T. M., *et al.* (2001b). The autosomal dominant hypophosphatemic rickets (ADHR) gene is a secreted polypeptide overexpressed by tumors that cause phosphate wasting. *J. Clin. Endocrinol. Metab.* **86,** 497–500.

Wiedemann, M., and Trueb, B. (2000). Characterization of a novel protein (FGFRL1) from human cartilage related to FGF receptors. *Genomics* **69,** 275–279.

Wilkie, A. O., Slaney, S. F., Oldridge, M., Poole, M. D., Ashworth, G. J., Hockley, A. D., Hayward, R. D., David, D. J., Pulleyn, L. J., Rutland, P., *et al.* (1995). Apert syndrome results from localized mutations of FGFR2 and is allelic with Crouzon syndrome. *Nat. Genet.* **9,** 165–172.

Wilkie, A. O., Patey, S. J., Kan, S. H., van den Ouweland, A. M., and Hamel, B. C. (2002). FGFs, their receptors, and human limb malformations: Clinical and molecular correlations. *Am. J. Med. Genet.* **112,** 266–278.

Winkel, A., Stricker, S., Tylzanowski, P., Seiffart, V., Mundlos, S., Gross, G., and Hoffmann, A. (2008). Wnt-ligand-dependent interaction of TAK1 (TGF-beta-activated kinase-1) with the receptor tyrosine kinase Ror2 modulates canonical Wnt-signalling. *Cell. Signal.* **20,** 2134–2144.

Witte, F., Bernatik, O., Kirchner, K., Masek, J., Mahl, A., Krejci, P., Mundlos, S., Schambony, A., Bryja, V., and Stricker, S. (2010a). Negative regulation of Wnt signaling mediated by CK1-phosphorylated Dishevelled via Ror2. *FASEB J.* **24,** 2417–2426.

Witte, F., Chan, D., Economides, A. N., Mundlos, S., and Stricker, S. (2010b). Receptor tyrosine kinase-like orphan receptor 2 (ROR2) and Indian hedgehog regulate digit outgrowth mediated by the phalanx-forming region. *Proc. Natl. Acad. Sci. USA* **107,** 14211–14216.

Woods, C. G., Stricker, S., Seemann, P., Stern, R., Cox, J., Sherridan, E., Roberts, E., Springell, K., Scott, S., Karbani, G., Sharif, S. M., Toomes, C., *et al.* (2006). Mutations in WNT7A cause a range of limb malformations, including Fuhrmann syndrome and Al-Awadi/Raas-Rothschild/Schinzel Phocomelia syndrome. *Am. J. Hum. Genet.* **79,** 402–408.

Wu, X. L., Gu, M. M., Huang, L., Liu, X. S., Zhang, H. X., Ding, X. Y., Xu, J. Q., Cui, B., Wang, L., Lu, S. Y., Chen, X. Y., Zhang, H. G., *et al.* (2009). Multiple synostoses syndrome is due to a missense mutation in exon 2 of FGF9 gene. *Am. J. Hum. Genet.* **85,** 53–63.

Yamamoto, S., Nishimura, O., Misaki, K., Nishita, M., Minami, Y., Yonemura, S., Tarui, H., and Sasaki, H. (2008). Cthrc1 selectively activates the planar cell polarity pathway of Wnt signaling by stabilizing the Wnt-receptor complex. *Dev. Cell* **15,** 23–36.

Yang, Y., and Niswander, L. (1995). Interaction between the signaling molecules WNT7a and SHH during vertebrate limb development: Dorsal signals regulate anteroposterior patterning. *Cell* **80,** 939–947.

Yang, F., Wang, Y., Zhang, Z., Hsu, B., Jabs, E. W., and Elisseeff, J. H. (2008). The study of abnormal bone development in the Apert syndrome Fgfr2+/S252W mouse using a 3D hydrogel culture model. *Bone* **43,** 55–63.

Zhang, R., Murakami, S., Coustry, F., Wang, Y., and de Crombrugghe, B. (2006). Constitutive activation of MKK6 in chondrocytes of transgenic mice inhibits proliferation and delays endochondral bone formation. *Proc. Natl. Acad. Sci. USA* **103,** 365–370.

Index

Contents of Previous Volumes

Volume 47

Volume 48

Volume 50

Volume 51

Volume 52

Volume 53

Volume 54

Volume 57

Volume 60

Volume 61

Volume 62

Volume 65

Volume 66

Volume 67

Volume 68

Volume 69

Volume 70

Volume 71

Volume 72

Volume 73

Volume 74

Volume 75

Volume 76

Volume 77

Volume 78

Volume 79

Volume 81

Volume 82

Volume 83

Volume 84

Volume 85

Volume 86

Volume 87

Volume 88

Volume 89

Volume 90

Volume 91

Volume 92

Volume 93

Volume 94

Volume 95

Volume 96

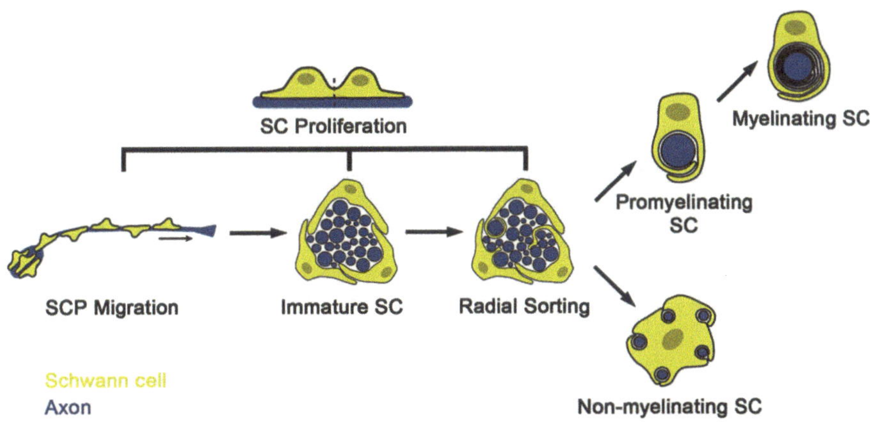

Alya R. Raphael and William S. Talbot, Figure 1.1 Multiple roles for Nrg1/ErbB signaling during Schwann cell development. Schwann cells (yellow) migrate along axons (blue) as Schwann cell precursors (SCP; Jessen and Mirsky, 2005). Once migration is complete, SCPs differentiate into immature Schwann cells (SC) and begin the process of radial sorting by inserting their processes into the axon bundle (Jessen and Mirsky, 2005; Webster *et al.*, 1973). Immature Schwann cells can become either promyelinating and then myelinating Schwann cells, if they are associated with one axon, or non-myelinating Schwann cells, which associate with multiple small axons (Nave and Salzer, 2006). Schwann cell proliferation occurs in SCPs and in immature Schwann cells and during radial sorting; however, once immature Schwann cells differentiate, they exit the cell cycle (Jessen and Mirsky, 2005; Martin and Webster, 1973; Webster *et al.*, 1973). Schwann cells divide parallel to the axons (Martin and Webster, 1973). Steps that require Nrg1/ErbB signaling are underlined and include Schwann cell migration, Schwann cell proliferation, radial sorting, myelination, and the formation of Remak bundles—nonmyelinating Schwann cells ensheathing small caliber axons (Dong *et al.*, 1995; Garratt *et al.*, 2000a,b; Lyons *et al.*, 2005; Michailov *et al.*, 2004; Morrissey *et al.*, 1995; Perlin *et al.*, 2011; Raphael *et al.*, 2011; Riethmacher *et al.*, 1997; Taveggia *et al.*, 2005; Woldeyesus *et al.*, 1999).

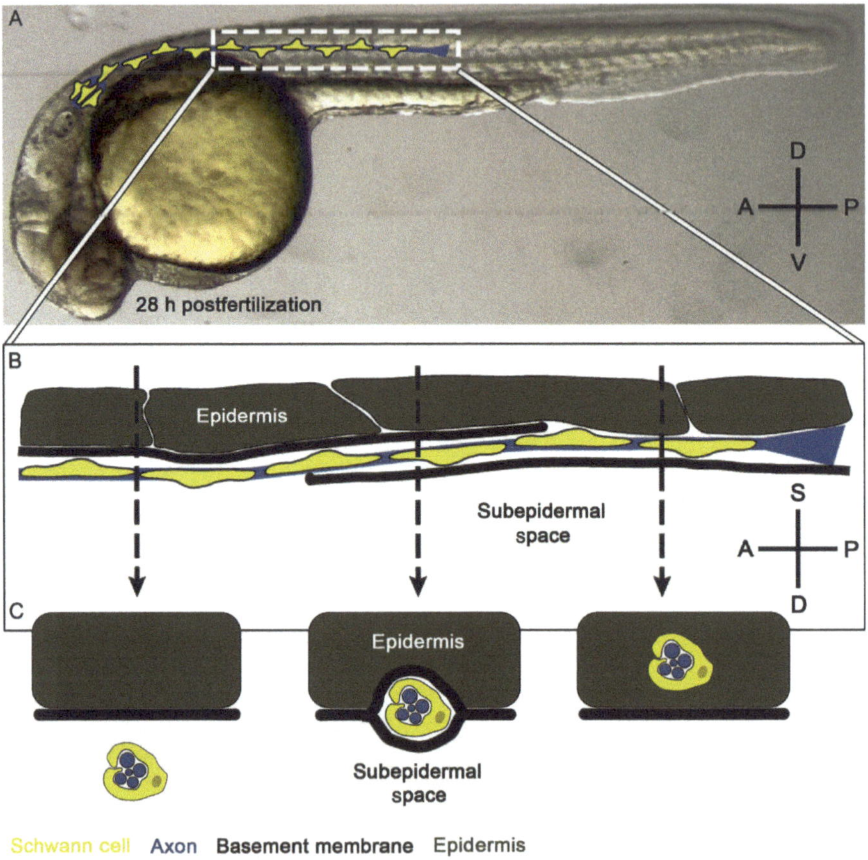

Alya R. Raphael and William S. Talbot, Figure 1.2 Schwann cells are required for repositioning a peripheral nerve. (A) The posterior lateral line nerve in cartoon superimposed on a 28-h postfertilization zebrafish embryo. Schwann cells (yellow) comigrate with axons (blue; Gilmour *et al.*, 2002). Anterior (A) is to the left, posterior (P) to the right, dorsal (D) up, ventral down (v). (B) Zoom of dashed region in (A), showing the transition of the posterior lateral line nerve across the epidermal basement membrane (dark gray) from the epidermis (light gray) into the subepidermal space (unlabeled), with anterior-most portions of the nerve transitioning prior to posterior portions closer to the outgrowing axonal growth cones (Raphael *et al.*, 2010). Dashed lines indicate cross sections shown in (C). Superficial is up (S), deep (D) down, anterior (A) left, posterior (P) right. (C) Cross sections through the posterior lateral line nerve showing the position of the nerve with respect to the basement membrane (Raphael *et al.*, 2010). Anterior (far left), the nerve has transitioned across the basement membrane and is embedded in the subepidermal space. Middle, the nerve is transitioning, with basement membrane both superficial and medial to the nerve. Posterior (far right), the nerve is still within the epidermis, superficial to the basement membrane. (A, B) adapted from Raphael *et al.* (2010).

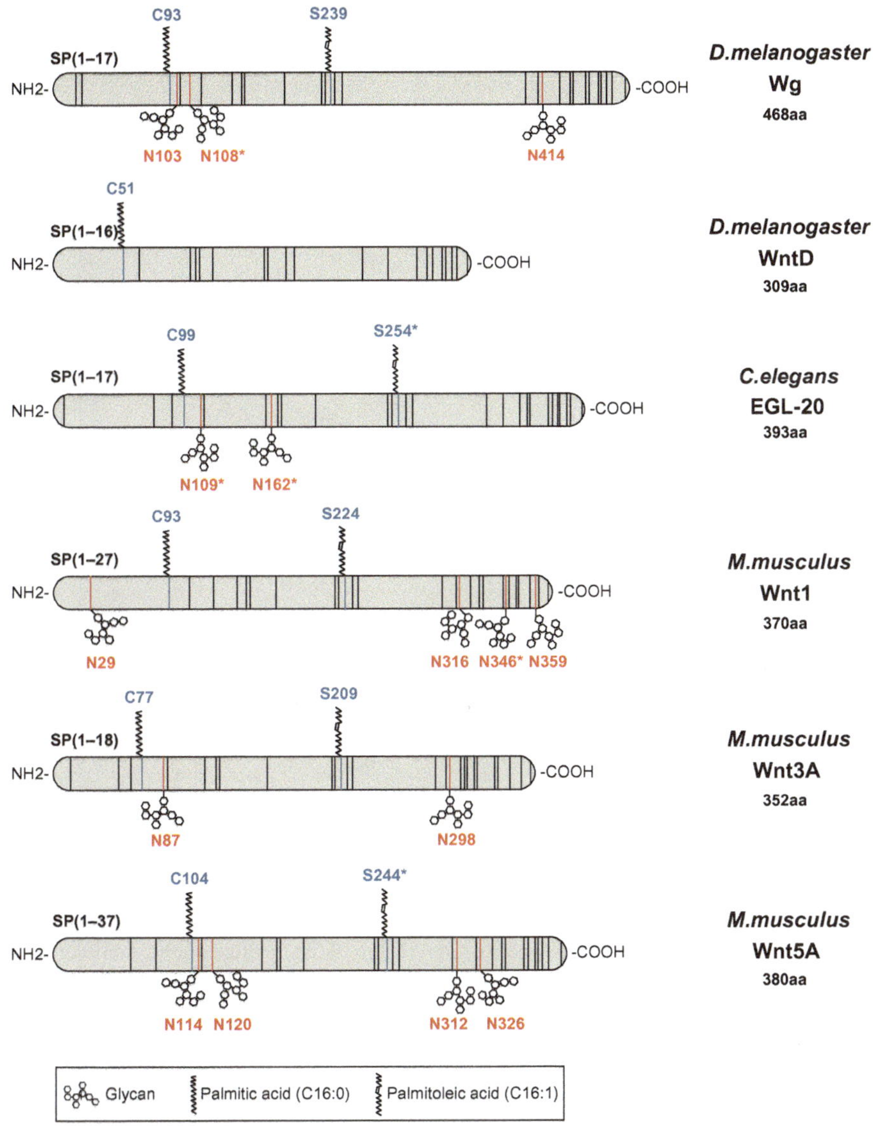

Tina Buechling and Michael Boutros, Figure 2.3 Structure and modifications of Wnt proteins. The canonical Wnt proteins Wg, mWnt1, and mWnt3A as well as the noncanonical mWnt5A are secreted glycoproteins with an N-terminal signal peptide (SP). All Wnt proteins share a signature of highly conserved cysteines (black lines), one of which is palmitoylated (C16:0, in blue). In addition to *S*-palmitoylation, Wg, mWnt1, and mWnt3A contain an unsaturated palmitoleate residue at a conserved serine (C16:1, in blue). According serine residues have been predicted for other Wnt proteins due to the highly conserved amino acid motif surrounding the palmitoleated serine residue. *Drosophila* WntD/Wnt8 is thought to be the only nonlipid modified Wnt protein as it lacks the conserved palmitoleated serine residue, but has been shown to contain a palmitoylated cysteine (C51). *N*-Linked glycosylation sites are indicated in red. Predicted glycosylation or acylation sites are indicated by asterisks. *N*-glycosylation sites were predicted using NetNGlyc1.0, signal peptides were determined using SignalP 3.0. Note that for Wnt5A, N312, and N326 correspond to N311 and N325 determined by Kurayoshi *et al.*, 2007, respectively.

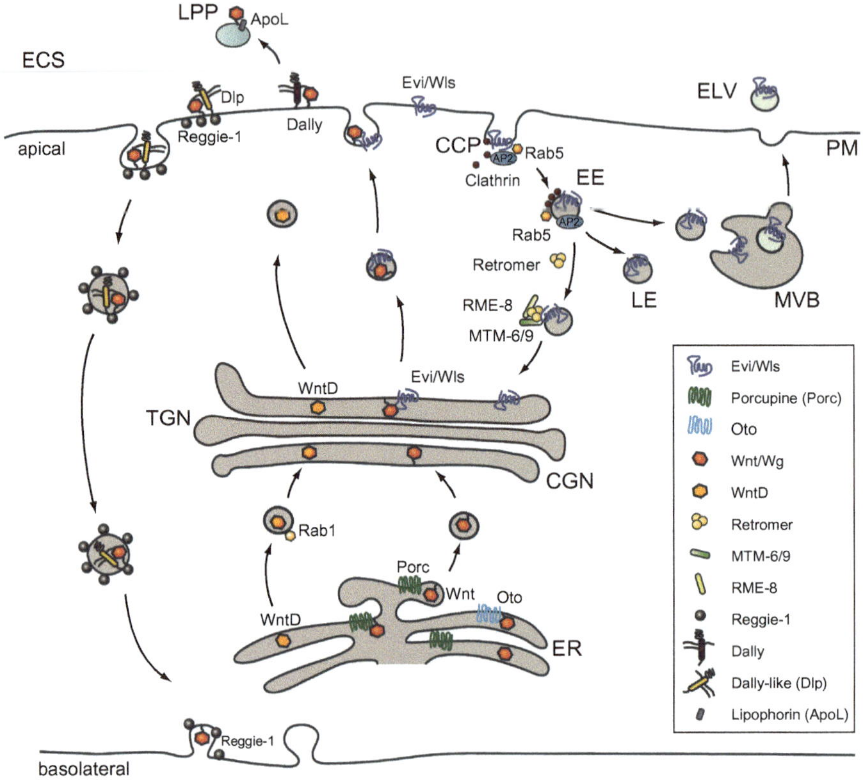

Tina Buechling and Michael Boutros, Figure 2.4 Current model of Wnt protein secretion. In the endoplasmatic reticulum (ER), Wnt proteins are palmitoylated through the *N*-acyltransferase Porcupine (Porc) and transported to the cis-Golgi network (CGN). The ER-resident (GPI)-inositol-deacylase Oto/PGAP-1 has been implicated in GPI-modifications of Wnt proteins. The multipass transmembrane protein Evi/Wls functions as a carrier protein for transport of Wnt proteins to the plasma membrane (PM). Evi/Wls itself is internalized to early endosomes (EE) through clathrin-coated pits (CCP), a process which depends on the μ subunit of AP-2 and the GTPase Rab5, and is recycled back to the trans-Golgi network (TGN) through the Retromer complex. RME-8 and MTM-6/9 have been found to bind to Retromer for Evi/Wls recycling. In the absence of Retromer, RME-8 or MTMs, Evi/Wls is targeted to late endosomes (LE) for lysosomal degradation. Wnt proteins are apically secreted from producing cells and associate with lipophorin (ApoL) on lipoprotein particles (LPPs) for short-range signaling and trafficking through the extracellular space (ECS). The glypican-type heparan sulfate proteoglycan (HSPG) Dally functions as a Wnt coreceptor promoting short-range signaling. In contrast, Wnts associate with the noncaveolin microdomain component Reggie-1 and the glypican Dally-like (Dlp) for transcytosis to the basolateral membrane to mediate long-range signaling. Evi/Wls itself has been shown to be secreted across the synaptic cleft in neuromuscular junctions through exosome-like vesicles (ELV) generated via inward budding of multivesicular bodies (MVB). To date, WntD is the only Wnt protein secreted in a Porc- and Evi/Wls-independent manner. However, WntD secretion maintains a requirement for the small GTPase Rab1 for ER to cis-Golgi (CGN) trafficking suggesting an alternative route of nonlipid modified Wnts through the secretory pathway.

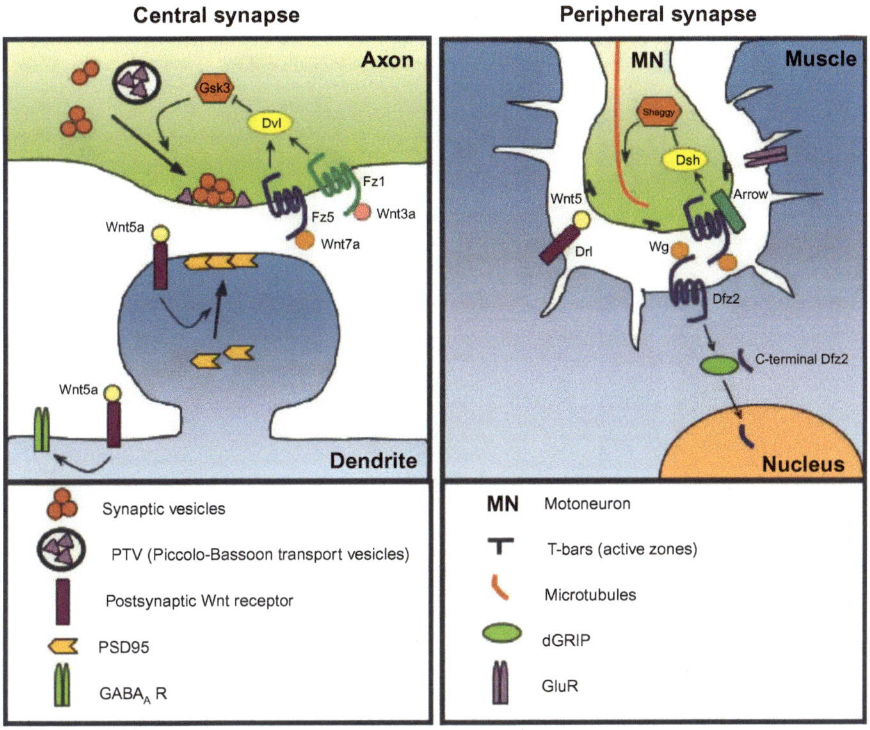

Macarena Sahores and Patricia C. Salinas, Figure 5.1 Regulation of synaptic development by Wnt signaling. At central synapses, the binding of Wnt7a to presynaptic Fz5 leads to Dvl activation resulting in the inhibition of GSK3β. This inhibition leads to the recruitment of synaptic vesicles and cytomatrix proteins such as Bassoon to the zone of nascent synapses. Wnt3a has been shown to induce vesicle clustering through Fz1. Postsynaptically, Wnt5a triggers the clustering of PSD-95 and regulates GABA$_A$R clustering through an unidentified receptor. At peripheral synapses, Wg binds to Dfz2, which is located at pre- and postsynaptic sites. Presynaptically, Dsh inhibits Shaggy by a divergent canonical pathway that requires the coreceptor Arrow and leads to changes in microtubule dynamics. Postsynaptically, Wg triggers the endocytosis and cleavage of the C-terminal of Dfz2, which is transported to the nucleus by dGRIP. In addition to Wg, Wnt5 also regulates presynaptic development. Wnt5 binds to its postsynaptic Drl receptor to regulate bouton development.

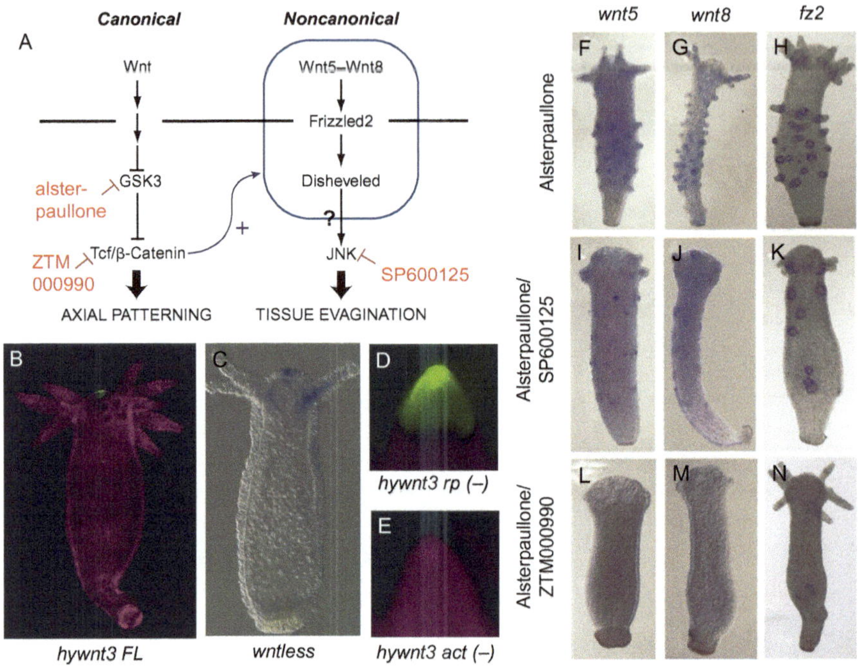

Thomas W. Holstein *et al.*, Figure 6.2 Wnt signaling in the cnidarian model *Hydra*. (A) Model of the interaction between canonical and noncanonical Wnt signaling in *Hydra* showing that noncanonical Wnt signaling acts downstream of canonical Wnt signaling in the formation of signaling centers. (B) *Hydra vulgaris* carrying the *hyactin-RFP::hywnt3-EGFP* construct where *hyactin-RFP* is expressed ubiquitously (pink), while *hywnt3FL-EGFP* (green) is localized to the hypostome, which is derived from the blastopore of the gastrula stage and represents the polyp's major signaling center, that is, the "head organizer." (C) *wntless* expression in the hypostome and at the sites of tentacle formation. (D and E) Autoregulatory and repressive elements restrict *wnt3* expression to the organizer. (F–N) Polyps treated with 5 M alsterpaullone show multiple ectopic *hywnt5, hywnt8, hyfz2* expressing cell clusters and ectopic tentacle formation. (E–G) Cotreatment with 5 M alsterpaullone/25 M SP600125 results in formation of multiple ectopic cell clusters expressing noncanonical *wnt* genes, but not in ectopic tentacle formation. (L–N) Cotreatment with 5 M alsterpaullone/25 M ZTM000990 blocks ectopic activation of noncanonical *wnt* genes and tentacle formation.

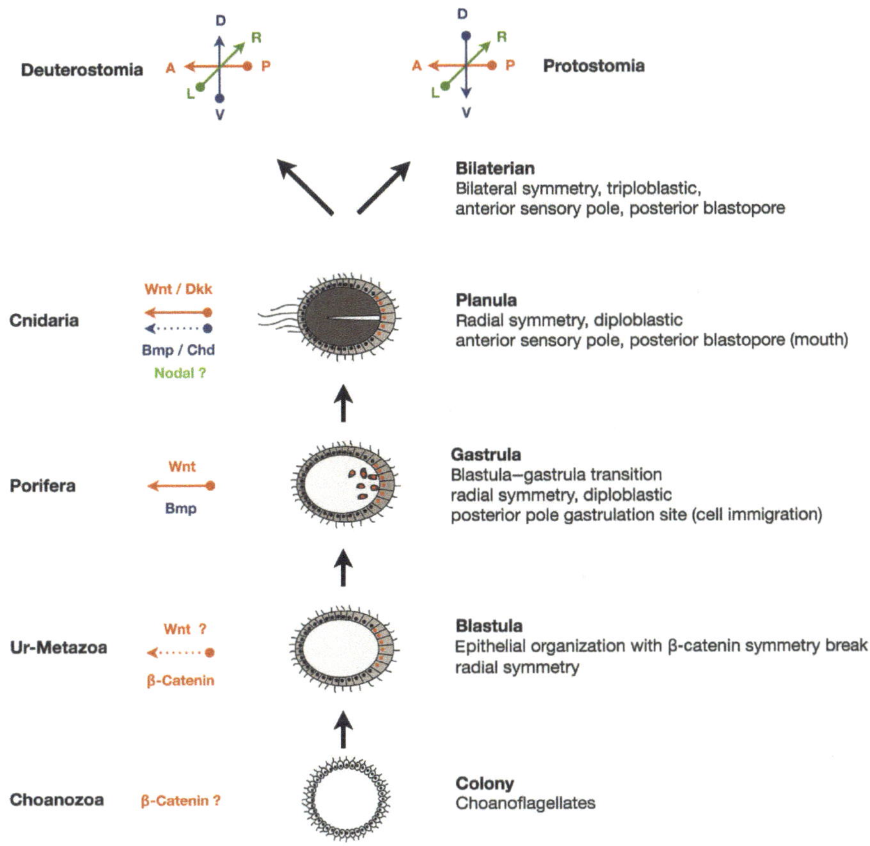

Thomas W. Holstein *et al.*, Figure 6.3 Signaling pathways and axis formation in the basal metazoans. Major evolutionary transitions show the function of β-catenin and Wnt signaling in the formation of the primary body axis (red). In cnidarians, a functional BMP pathway (blue) is acting in parallel to the Wnt pathway and gives rise to the second (DV) body axis in bilaterians. In bilaterians, the nodal pathway (green) determines the third spatial coordinate, the left–right axis. It is yet unclear whether this pathway is already acting in cnidarians.

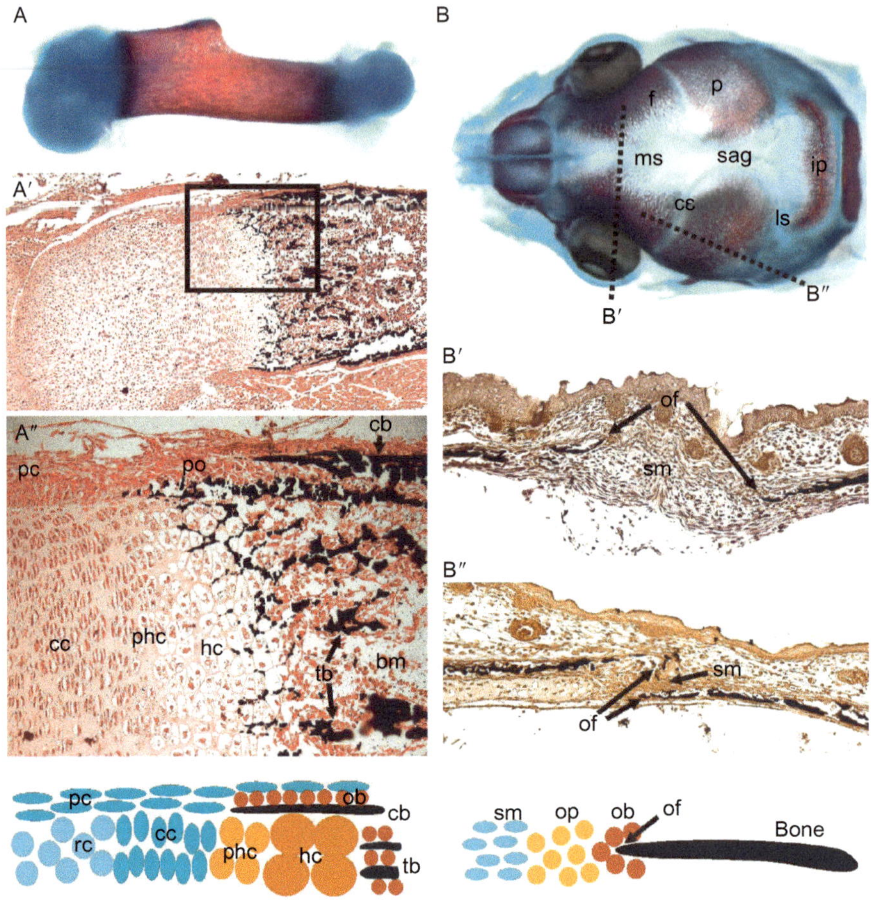

Sigmar Stricker and Stefan Mundlos, Figure 7.1 Development of the skeleton by (A) endochondral and (B) intramembranous ossification. (A) A skeletal preparation of a mouse humerus at embryonic day 17.5 (E17.5). (A′) Van Kossa-stained section of the proximal humerus at E17.5. (A″) Magnification of the area boxed in (A′). Note that the calcified areas stain black. At the bottom, a schematic display of the differentiating cell types in the growth plate is shown. (B) Mouse skull at E17.5 stained with Alcian blue (cartilage) and Alizarin red (bone); the skull base was removed for clarity. (B′) and (B″) show Van Kossa-stained sections of the metopic suture and a coronal suture, respectively, in which calcified areas appear black. Note the different morphologies of metopic and coronary sutures, that is, the abutting osteogenic fronts in the metopic and the overlapping osteogenic fronts in the coronal suture. At the bottom, a schematic display of the differentiating cell types in the cranial suture is shown. *Abbreviations*: Bm, bone marrow; cb, cortical bone; cc, columnar chondrocytes; cs, coronal suture; f, frontal bone; hc, hypertrophic chondrocytes; ip, interparietal bone; ls, lambdoid suture; ms, metopic suture; ob, osteoblasts; of, osteogenic front; op, osteoprogenitors; p, parietal bone; pc, perichondrium; phc, prehypertrophic chondrocytes; po, periosteum; rc, reserve chondrocytes; sag, sagittal suture; sm, suture mesenchyme; tb, trabecular bone. See Mundlos (2000) for staining methods.

FSC
www.fsc.org
MIX
Papier | Fördert
gute Waldnutzung
FSC® C083411

Kolibri 360 GmbH
Ferdinand-Jühlke-Straße 7
99095 Erfurt, Deutschland
produktsicherheit@kolibri360.de